MEMOIRS
of the
American Mathematical Society

ISSN 0065-9266 (print)
ISSN 1947-6221 (online)

Number 1407

Infinite Time Blow-Up Solutions to the Energy Critical Wave Maps Equation

Mohandas Pillai

April 2023 • Volume 284 • Number 1407 (third of 6 numbers)

AMS AMERICAN MATHEMATICAL SOCIETY

Library of Congress Cataloging-in-Publication Data

Cataloging-in-Publication Data has been applied for by the AMS.
See http://www.loc.gov/publish/cip/.
DOI: https://doi.org/10.1090/memo/1407

Memoirs of the American Mathematical Society

This journal is devoted entirely to research in pure and applied mathematics.

Subscription information. Beginning in 2023, Memoirs will be published monthly through 2026. Memoirs is also accessible from www.ams.org/journals. The 2023 subscription begins with volume 281 and consists of twelve mailings, each containing one or more numbers. Individual subscription prices for 2023 are as follows. For electronic only: US$984. For paper delivery: US$1,129. Add US$22 for delivery within the United States; US$85 for surface delivery outside the United States. Upon request, subscribers to paper delivery of this journal are also entitled to receive electronic delivery. For information on institutional pricing, please visit `https://www.ams.org/publications/journals/subscriberinfo`. Subscription renewals are subject to late fees. See `www.ams.org/journal-faq` for more journal subscription information. Each number may be ordered separately; please specify number when ordering an individual number.

Back number information. For back issues see `www.ams.org/backvols`.

Subscriptions and orders should be addressed to the American Mathematical Society, P. O. Box 845904, Boston, MA 02284-5904 USA. *All orders must be accompanied by payment.* Other correspondence should be addressed to 201 Charles Street, Providence, RI 02904-2213 USA.

Memoirs of the American Mathematical Society (ISSN 0065-9266 (print); 1947-6221 (online)) is published bimonthly (each volume consisting usually of more than one number) by the American Mathematical Society at 201 Charles Street, Providence, RI 02904-2213 USA. Periodicals postage paid at Providence, RI. Postmaster: Send address changes to Memoirs, American Mathematical Society, 201 Charles Street, Providence, RI 02904-2213 USA.

This publication is indexed in *Mathematical Reviews®*, *Zentralblatt MATH*, *Science Citation Index®*, *Science Citation Index™-Expanded*, *ISI Alerting Services℠*, *SciSearch®*, *Research Alert®*, *CompuMath Citation Index®*, *Current Contents®/Physical, Chemical & Earth Sciences*.
This publication is archived in *Portico* and *CLOCKSS*.
Printed in the United States of America.

∞ The paper used in this book is acid-free and falls within the guidelines established to ensure permanence and durability.
Visit the AMS home page at `https://www.ams.org/`

10 9 8 7 6 5 4 3 2 1 28 27 26 25 24 23 23

Contents

Abstract

We consider the wave maps problem with domain $\mathbb{R}^{2+1}$ and target $\mathbb{S}^2$ in the 1-equivariant, topological degree one setting. In this setting, we recall that the soliton is a harmonic map from $\mathbb{R}^2$ to $\mathbb{S}^2$, with polar angle equal to $Q_1(r) = 2\arctan(r)$. By applying the scaling symmetry of the equation, $Q_\lambda(r) = Q_1(r\lambda)$ is also a harmonic map, and the family of all such Q_λ are the unique minimizers of the harmonic map energy among finite energy, 1-equivariant, topological degree one maps. In this work, we construct infinite time blowup solutions along the Q_λ family. More precisely, for $b > 0$, and for all $\lambda_{0,0,b} \in C^\infty([100,\infty))$ satisfying, for some $C_l, C_{m,k} > 0$,

$$\frac{C_l}{\log^b(t)} \leq \lambda_{0,0,b}(t) \leq \frac{C_m}{\log^b(t)}, \quad |\lambda_{0,0,b}^{(k)}(t)| \leq \frac{C_{m,k}}{t^k \log^{b+1}(t)}, k \geq 1 \quad t \geq 100$$

there exists a wave map with the following properties. If u_b denotes the polar angle of the wave map into $\mathbb{S}^2$, we have

$$u_b(t,r) = Q_{\frac{1}{\lambda_b(t)}}(r) + v_2(t,r) + v_e(t,r), \quad t \geq T_0$$

where

$$-\partial_{tt}v_2 + \partial_{rr}v_2 + \frac{1}{r}\partial_r v_2 - \frac{v_2}{r^2} = 0$$

$$||\partial_t(Q_{\frac{1}{\lambda_b(t)}} + v_e)||^2_{L^2(rdr)} + ||\frac{v_e}{r}||^2_{L^2(rdr)} + ||\partial_r v_e||^2_{L^2(rdr)} \leq \frac{C}{t^2\log^{2b}(t)}, \quad t \geq T_0$$

and

$$\lambda_b(t) = \lambda_{0,0,b}(t) + O\left(\frac{1}{\log^b(t)\sqrt{\log(\log(t))}}\right)$$

Received by the editor June 28, 2019, and, in revised form, December 9, 2019.

Article electronically published on March 21, 2023.

DOI: `https://doi.org/10.1090/memo/1407`

2020 *Mathematics Subject Classification.* Primary 35L05, 35Q75.

The author is affiliated with the Department of Mathematics, University of California, Berkeley. Email: mpillai@berkeley.edu.

This material is based upon work partially supported by the National Science Foundation under Grant No. DMS-1800294.

CHAPTER 1

Introduction

This monograph considers the wave maps equation, with domain $\mathbb{R}^{2+1}$ and target $\mathbb{S}^2$. This equation is the Euler-Lagrange equation associated to the functional

$$\mathcal{L}(\Phi) = \int_{\mathbb{R}^{2+1}} \langle \partial^{\alpha}\Phi(t,x), \partial_{\alpha}\Phi(t,x)\rangle_{g(\Phi(t,x))} dtdx$$

where g denotes the round metric on $\mathbb{S}^2$, and $\Phi : \mathbb{R}^{2+1} \to \mathbb{S}^2$. We will only work with 1-equivariant maps Φ, which we describe by first regarding Φ as a map into $\mathbb{R}^3$ with unit norm, and then writing

$$\Phi_u(t,r,\phi) = (\cos(\phi)\sin(u(t,r)), \sin(\phi)\sin(u(t,r)), \cos(u(t,r)))$$

where (r,ϕ) are polar coordinates on $\mathbb{R}^2$. Then, the wave maps equation becomes

$$-\partial_{tt}u + \partial_{rr}u + \frac{1}{r}\partial_r u = \frac{\sin(2u)}{2r^2} \tag{1.1}$$

[**19**] studied (a more general problem which includes) the Cauchy problem associated to (1.1), with data (u_0, u_1) such that

$$(x_1,x_2) \mapsto (\frac{x_1 u_0(r)}{r}, \frac{x_2 u_0(r)}{r}) \in H^1_{\text{loc}}(\mathbb{R}^2)$$
$$(x_1,x_2) \mapsto (\frac{x_1 u_1(r)}{r}, \frac{x_2 u_1(r)}{r}) \in L^2_{\text{loc}}(\mathbb{R}^2)$$

We will say that u is a finite energy solution to (1.1) if u is a distributional solution, with $\Phi_u \in C^0_t \dot{H}^1(\mathbb{R}^2)$ and $\partial_t \Phi_u \in C^0_t L^2(\mathbb{R}^2)$. Define the energy by

$$E_{\text{WM}}(u,v) = \pi \int_0^{\infty} \left(v^2 + \frac{\sin^2(u)}{r^2} + (\partial_r u)^2\right) rdr \tag{1.2}$$

Then, if u solves (1.1), $E_{\text{WM}}(u, \partial_t u)$ is formally independent of time. We recall the ground state soliton, $Q_1(r) = 2\arctan(r)$ is a solution to (1.1), with the property that the family of all $Q_\lambda(r) = Q_1(r\lambda)$ are the unique minimizers of $E_{\text{WM}}(u,0)$ among finite energy u with Φ_u having topological degree one. In order to state our main theorem, it will also be useful to consider the following non-degenerate energy which will be used to measure perturbations of Q_λ.

$$E(u,v) = \pi \int_0^{\infty} \left(v^2 + (\partial_r u)^2 + \frac{u^2}{r^2}\right) rdr$$

The quantity $E(u, \partial_t u)$ is formally conserved for solutions to the wave equation

$$-\partial_{tt}u + \partial_{rr}u + \frac{1}{r}\partial_r u - \frac{u}{r^2} = 0 \tag{1.3}$$

We consider the problem of constructing 1-equivariant, topological degree one, global, non-scattering solutions to (1.1), which have energy strictly greater than $E_{\text{WM}}((Q_1, 0))$. The authors of [**3**] classified 1-equivariant topological degree one solutions to the wave maps equation with energy strictly between $E_{\text{WM}}((Q_1,0))$ and $3E_{\text{WM}}((Q_1,0))$. The part of their result which is relevant for this paper is the statement that any such solutions which are global in time admit a generic decomposition into the form

$$u = Q_{\frac{1}{\lambda(t)}} + \phi_L + \epsilon \tag{1.4}$$

where ϕ_L solves 1.3, $\lambda(t) = o(t), \quad t \to \infty$, and $\epsilon \to 0$ (in an appropriate sense) as $t \to \infty$. There are many possible asymptotic behaviors of λ allowed by the above result, and, according to [**3**], there were no known constructions of solutions of the above form, with $\lambda(t) \to 0$ or $\lambda(t) \to \infty$.

This paper constructs a family of finite energy solutions to (1.1), say $\{u_b\}_{b>0}$, where each u_b can be decomposed as in (1.4), with

$$\lambda_b(t) = \lambda_{0,0,b}(t) + O\left(\frac{1}{\log^b(t)\sqrt{\log(\log(t))}}\right)$$

(see the main theorem below for the sense in which the ϵ term for our solution, which is called v_e, vanishes as t goes to infinity). Here, $\lambda_{0,0,b} \in C^\infty([100,\infty))$ satisfies, for some $b, C_l, C_{m,k} > 0$,

$$\frac{C_l}{\log^b(t)} \leq \lambda_{0,0,b}(t) \leq \frac{C_m}{\log^b(t)}, \quad |\lambda_{0,0,b}^{(k)}(t)| \leq \frac{C_{m,k}}{t^k \log^{b+1}(t)}, k \geq 1 \quad t \geq 100$$

To the author's knowledge, these are the first examples of such solutions of (1.1). First, we will prove the following theorem, which corresponds to the special case of $\lambda_{0,0,b}(t) = \frac{1}{\log^b(t)}$. Solutions with more general $\lambda_{0,0,b}$ can be constructed with a slight modification of the proof of this special case, outlined in the appendix.

THEOREM 1.1. *For each $b > 0$, there exists $T_0 > 0$ and a (real-valued) finite energy solution u_b to (1.1) for $t \geq T_0$, of the form*

$$u_b(t,r) = Q_{\frac{1}{\lambda_b(t)}}(r) + v_2(t,r) + v_e(t,r)$$

where $\lambda_b \in C^4([T_0,\infty))$,

$$E_{\text{WM}}(u_b, \partial_t u_b) < \infty$$

$$-\partial_{tt}v_2 + \partial_{rr}v_2 + \frac{1}{r}\partial_r v_2 - \frac{v_2}{r^2} = 0, \quad E(v_2, \partial_t v_2) < \infty, \quad v_2 \in C^\infty([T_0,\infty) \times [0,\infty))$$

$$E(v_e, \partial_t\left(Q_{\frac{1}{\lambda_b}} + v_e\right)) \leq \frac{C}{t^2 \log^{2b}(t)}, \quad t \geq T_0$$

and

$$\lambda_b(t) = \frac{1}{\log^b(t)} + e(t), \quad |e(t)| \leq \frac{C}{\log^b(t)\sqrt{\log(\log(t))}}$$

$$|e^{(j)}(t)| \leq \begin{cases} \frac{C}{t^j \log^{b+1}(t)\sqrt{\log(\log(t))}}, & j = 1,2 \\ \frac{C}{t^j \log^{b+1}(t)}, & j = 3,4 \end{cases}$$

Regarding more general $\lambda_{0,0,b}$, we have the following. For $b > 0$, let Λ_b be the set of $f \in C^\infty([100,\infty))$ such that there exist $C_l, C_m, C_{m,k} > 0$ with

$$\frac{C_l}{\log^b(t)} \leq f(t) \leq \frac{C_m}{\log^b(t)}, \quad |f^{(k)}(t)| \leq \frac{C_{m,k}}{t^k \log^{b+1}(t)}, k \geq 1, \quad t \geq 100$$

THEOREM 1.2. *Let $b > 0$. For all $\lambda_{0,0,b} \in \Lambda_b$, there exists $T_0 > 100$ and a (real-valued) finite energy solution u_b to (1.1) for $t \geq T_0$, of the form*

$$u_b(t,r) = Q_{\frac{1}{\lambda_b(t)}}(r) + v_2(t,r) + v_e(t,r)$$

where $\lambda_b \in C^4([T_0,\infty))$,

$$E_{\text{WM}}(u_b, \partial_t u_b) < \infty$$

$$-\partial_{tt}v_2 + \partial_{rr}v_2 + \frac{1}{r}\partial_r v_2 - \frac{v_2}{r^2} = 0, \quad E(v_2, \partial_t v_2) < \infty, \quad v_2 \in C^\infty([T_0,\infty) \times [0,\infty))$$

$$E\left(v_e, \partial_t \left(Q_{\frac{1}{\lambda_b}} + v_e\right)\right) \leq \frac{C}{t^2 \log^{2b}(t)}, \quad t \geq T_0$$

and

$$\lambda_b(t) = \lambda_{0,0,b}(t) + e(t), \quad |e(t)| \leq \frac{C}{\log^b(t)\sqrt{\log(\log(t))}}$$

$$|e^{(j)}(t)| \leq \begin{cases} \frac{C}{t^j \log^{b+1}(t)\sqrt{\log(\log(t))}}, & j = 1,2 \\ \frac{C}{t^j \log^{b+1}(t)}, & j = 3,4 \end{cases}$$

REMARK 1. Our proof yields more information about the regularity of v_e appearing in the main theorem. In particular, we have

$$v_e = v_{e,0} + v_6$$

where the function $v_{e,0}$ is fairly explicit (but complicated), and v_6 is constructed with a fixed point argument, but has more Sobolev regularity than what follows from u having finite energy, namely: if (r,θ) denote polar coordinates on $\mathbb{R}^2$, then

$$(t,r,\theta) \mapsto e^{i\theta} v_6(t,r) \in C^0_t([T_0,\infty), H^2(\mathbb{R}^2))$$
$$(t,r,\theta) \mapsto e^{i\theta} \partial_t v_6(t,r) \in C^0_t([T_0,\infty), H^1(\mathbb{R}^2))$$

REMARK 2. Theorem 1.2 includes solutions with mildly oscillating $\lambda_{0,0,b}$, such as

$$\lambda_{0,0,b} = \frac{2 + \sin(\log(\log(t)))}{\log^b(t)}, \quad t \geq 100$$

REMARK 3. It is expected that the method used in this paper can be extended to allow for the construction of such solutions with $\lambda(t) \to \infty, \quad t \to \infty$.

Finally, this method should also be applicable to higher equivariance classes, the energy critical Yang-Mills problem in 4 dimensions with gauge group $SO(4)$, as well as the quintic, focusing semilinear wave equation in $\mathbb{R}^{1+3}$. All of these extensions are work in progress by the author.

REMARK 4. The leading behavior of $\lambda(t)$ is partially motivated by the fact that, for $k = 1,2$,

$$\frac{|\lambda^{(k)}(t)|}{\lambda(t)} \leq \frac{C}{t^k \log(t)} = o(t^{-k}), \quad t \to \infty$$

This fact is very important so that certain error terms appearing in this work which involve the "transferrence operator" defined in [**12**] can be treated perturbatively. In particular, the method of this work can not be directly applied to construct solutions with $\lambda(t)$ being all powers of t.

In order to understand this work in a larger context, we review previous work regarding dynamical behavior of solutions to the critical wave maps problem with $\mathbb{S}^2$ target. Firstly, the works [**20**] and [**21**] (which apply to much more general targets than $\mathbb{S}^2$) show that for data with energy strictly less than that of the lowest energy non-trivial harmonic map, one has global well-posedness and scattering. In the 1-equivariant setting, [**2**] showed that global existence and scattering holds for smooth, topological degree zero data with energy less than twice that of Q_1 (which is the appropriate threshold in this setting). An analogous result, but without the equivariance assumption, is implied by [**15**]. Then, [**8**] constructed, for all $k \geq 2$, k equivariant, topological degree zero, two-bubble solutions with energy exactly equal to 2 times the energy of $Q_k(r) := Q_1(r^k)$. In [**9**], all k-equivariant solutions (with $k \geq 2$) of topological degree zero, and energy exactly equal to twice the energy of $Q_k(r)$ were classified. A classification of 1-equivariant, topological degree 0 solutions with energy equal to twice $E(Q_1)$ was obtained in [**18**], and shows that the dynamical behavior of such solutions can be quite different than in the higher equivariance case. [**18**] also constructs a finite-time blow-up solution in this setting. The methods used in this work differ significantly from these previous works.

1-equivariant, topological degree one, finite time blow-up solutions have been constructed in [**12**], and the work [**5**] extended the range of possible blow-up rates of these solutions. The method of construction of the ansatz of this paper differs significantly from that used in [**12**]. However, we do eventually use some technical information, most importantly, the distorted Fourier transform, from [**12**] as part of the process to complete our ansatz to an exact solution. Analogs of the solutions of [**12**] for the 4+1-dimensional Yang-Mills equation with gauge group $SO(4)$ and the quintic, focusing nonlinear wave equation in $\mathbb{R}^{1+3}$ were also constructed in [**13**] and [**14**]. The work of [**11**] studies the stability of the solutions of [**12**] and [**5**] under certain equivariant perturbations. In addition, [**17**] constructs finite-time blow-up, $k-$equivariant solutions with $k \geq 4$, and [**16**] constructs finite time blow-up solutions for the $(4+1)$-dimensional Yang-Mills problem with gauge group $SO(4)$, as well as in all equivariance classes for energy critical wave maps. The method of this work is quite different from the methods used in [**16**] and [**17**].

The work of [**1**] constructs modulated soliton solutions, where λ is bounded away from zero and infinity for all time. Some facts from [**1**] about the wave maps equation linearized around Q_1 will be utilized in this paper, but the ansatz construction is again quite different. Finally, infinite time blow-up and infinite time relaxation solutions to the quintic, focusing, nonlinear wave equation in $\mathbb{R}^{1+3}$ have been constructed in [**4**], but the method of this paper is again quite different.

Notation

We will occasionally use the Hankel transform of order 1, and it will be denoted as

$$\widehat{f}(\xi) = \int_0^\infty f(r)J_1(r\xi)rdr$$

The main Fourier transform we will use is the distorted Fourier transform of [**12**], which we denote by

$$\mathcal{F}(f)(\xi) = \int_0^\infty \phi(r,\xi)f(r)dr$$

Briefly, we will make some use of the distorted Fourier transforms of [**1**], which are denoted by $\mathcal{F}_H$ and $\mathcal{F}_{\tilde{H}}$. Finally, we use the same notation as [**1**] for the following norm

$$||f||^2_{\dot{H}^1_e} = ||\partial_r f||^2_{L^2(rdr)} + ||\frac{f}{r}||^2_{L^2(rdr)}$$

We denote by ϕ_0, the zero resonance of the elliptic part of the wave equation linearized around Q_1:

$$\phi_0(r) = \frac{d}{d\lambda}\Big|_{\lambda=1} Q_\lambda(r) = \frac{2r}{1+r^2}$$

(This notation for ϕ_0 is the same as that used in [**1**], but is different from that used in [**12**]).

Acknowledgments

The author thanks his adviser, Daniel Tataru, for suggesting the problem, and many useful discussions.

CHAPTER 2

Overview of the proof

We remind the reader that we will prove theorem 1.1, with the extra arguments needed to establish theorem 1.2 summarized in the appendix. The argument used in this monograph proceeds in two parts. First, we construct an approximate solution, u, to (1.1). Second, we construct an exact solution to (1.1) which is close to our approximate one.

Part 1: Constructing the approximate solution

One way to understand the intuition behind our approximate solution is as follows. One could start by looking for an approximate solution to (1.1) which consists of a dynamically rescaled soliton and a radiation field, along with an appropriate compatibility condition between the soliton length scale, and the radiation field. This would correspond to an approximate solution of the form $u_a = Q_{\frac{1}{\lambda(t)}} + v_2$, where v_2 is some solution to (1.3), $\lambda(t)$ is not yet chosen, and one can look for an appropriate relation between v_2 and $\lambda(t)$ so as to make the approximate solution accurate. Two difficulties immediately arise with this procedure. One is that $\partial_t Q_{\frac{1}{\lambda(t)}}(r) \notin L^2(rdr)$, which means that $E_{\text{WM}}(u_a, \partial_t u_a)$ is not finite. Another key difficulty is that the elliptic part of the wave maps equation linearized around $Q_{\frac{1}{\lambda(t)}}$ has a zero resonance, $\phi_0(\frac{r}{\lambda(t)})$. So, one would like the principal part of the error term of u_a to be orthogonal to $\phi_0(\frac{r}{\lambda(t)})$. On the other hand, the soliton error term

$$\partial_t^2 Q_{\frac{1}{\lambda(t)}}(r) = \frac{-2r\lambda''(t)}{r^2+\lambda(t)^2} + \frac{4r\lambda(t)\lambda'(t)^2}{(r^2+\lambda(t)^2)^2}$$

does not decay fast enough in r for its inner product with $\phi_0(\frac{r}{\lambda(t)})$ to even be defined in the first place.

Hence, we start by introducing an additional correction, v_1, which is independent of v_2, and whose purpose is both to make the ansatz have finite energy and to eliminate an appropriate principal part of the soliton error term, $\partial_t^2 Q_{\frac{1}{\lambda(t)}}$, for large r. More precisely the starting point for our ansatz is

$$u_{a,1} = Q_{\frac{1}{\lambda(t)}}(r) + v_1(t,r) + v_2(t,r)$$

Here, $\lambda(t)$ is not yet chosen, and v_1 solves the nondegenerate wave equation

$$-\partial_{tt}v_1 + \partial_{rr}v_1 + \frac{1}{r}\partial_r v_1 - \frac{v_1}{r^2} = -2\lambda''(t)\frac{r}{1+r^2}$$

with 0 Cauchy data at infinity. As explained in more detail later on, we do not have the entire soliton error term on the right-hand side of the v_1 equation, because doing so would be more difficult and also would hide the leading, linear part in the modulation equation for λ.

On the other hand, v_2 solves (1.3) with Cauchy data

$$v_2(0,r)=0, \quad \partial_t v_2(0,r)=v_{2,0}(r)$$

where $v_{2,0}$ will be a prescribed function depending on a fixed parameter $b>0$. The choices for $v_{2,0}$ and $\lambda(t)$ are closely related. Later, we will choose $\lambda(t)$ to solve a modulation equation involving $v_{2,0}$, thereby correlating the length scale of the dynamically rescaled soliton with the radiation profile.

At this stage, one could choose $\lambda(t)$ by requiring that the principal part of the error term associated to $u_{a,1}$, say $e_{a,1}$, is orthogonal to $\phi_0(\frac{r}{\lambda(t)})$. This is not exactly what is done in our argument, since we will need to add additional corrections to $u_{a,1}$ before imposing the orthogonality condition on the principal part of the error term of our final ansatz. Nevertheless, computing $\langle e_{a,1}(t,R\lambda(t)),\phi_0(R)\rangle_{L^2(RdR)}$ allows one to see, in a simpler context than our final equation for $\lambda(t)$, a relation between the leading order behavior of $\lambda''(t)$ and v_2. For the purposes of this discussion, the principal part of the $u_{a,1}$ error term is

$$\begin{aligned}e_{a,1}(t,r)=\partial_{tt}Q_{\frac{1}{\lambda(t)}}(r)+\frac{2\lambda''(t)r}{1+r^2}+\left(\frac{\cos(2Q_{\frac{1}{\lambda(t)}}(r))-1}{r^2}\right)v_1(t,r)\\+\left(\frac{\cos(2Q_{\frac{1}{\lambda(t)}}(r))-1}{r^2}\right)v_2(t,r)\end{aligned}$$

Using the Hankel transform of order 1 to express v_2, the contribution to

$$\langle e_{a,1}(t,R\lambda(t)),\phi_0(R)\rangle_{L^2(RdR)}$$

from the v_2-related term above is given by

$$\begin{aligned}&\langle\left(\frac{\cos(2Q_1(R))-1}{R^2\lambda(t)^2}\right)v_2(t,R\lambda(t)),\phi_0(R)\rangle_{L^2(RdR)}\\&=-2\int_0^\infty \sin(t\xi)\widehat{v_{2,0}}(\xi)\xi^2K_1(\xi\lambda(t))d\xi\end{aligned}$$

where K_1 denotes the modified Bessel function of the second kind. We thus get

$$\begin{aligned}&\langle e_{a,1}(t,R\lambda(t)),\phi_0(R)\rangle_{L^2(RdR)}\\&=-\frac{4}{\lambda(t)}\int_t^\infty\frac{\lambda''(s)}{1+s-t}ds-2\int_0^\infty\sin(t\xi)\widehat{v_{2,0}}(\xi)\xi^2K_1(\xi\lambda(t))d\xi+\frac{4\lambda''(t)\log(\lambda(t))}{\lambda(t)}\\&+f_1(\lambda(t),\lambda'(t),\lambda''(t))\end{aligned}$$

We will choose $v_{2,0}$ and the principal part of λ, denoted by $\lambda_{0,0}$, in order to have a leading order cancellation in the above equation. The term $f_1(\lambda(t),\lambda'(t),\lambda''(t))$ turns out to be subleading for all λ close (in a C^2 sense) to the choice of $\lambda_{0,0}$ which we make. Our choice of $v_{2,0}$ gives the following equation

$$\begin{aligned}\langle e_{a,1}(t,R\lambda(t)),\phi_0(R)\rangle_{L^2(RdR)}=-\frac{4}{\lambda(t)}\int_t^\infty\frac{\lambda''(s)}{1+s-t}ds+\frac{4b}{\lambda(t)t^2\log^b(t)}\\+\frac{4\lambda''(t)\log(\lambda(t))}{\lambda(t)}+f_2(\lambda(t),\lambda'(t),\lambda''(t))\end{aligned}$$

Again, the term $f_2(\lambda(t),\lambda'(t),\lambda''(t))$ is subleading, for all $\lambda(t)$ close (in a C^2 sense) to $\lambda_{0,0}(t)=\frac{1}{\log^b(t)}$, which is a leading order solution to

$$-\frac{4}{\lambda(t)}\int_t^\infty \frac{\lambda''(s)}{1+s-t}ds+\frac{4b}{\lambda(t)t^2\log^b(t)}+\frac{4\lambda''(t)\log(\lambda(t))}{\lambda(t)}=0$$

The term

$$\frac{4b}{\lambda(t)t^2\log^b(t)}$$

occurs in the above computation, as a consequence of our particular choice of Cauchy data for v_2. Despite the fact that our final equation for $\lambda(t)$ is not exactly the one given above, the leading behavior of our $\lambda(t)$ is indeed $\lambda_{0,0}(t)$, due to the same cancellation seen above.

Even if the principal part of the error term of $u_{a,1}$ is chosen to be orthogonal to $\phi_0(\frac{\cdot}{\lambda(t)})$, it does not have enough decay to be treated perturbatively via our methods. So, we need to add three more corrections to $u_{a,1}$, denoted by v_3,v_4, and v_5, in order to achieve an acceptable error term.

When the error term resulting from $Q_{\frac{1}{\lambda(t)}}$ and v_1 is computed in the renormalized spatial coordinate R, defined by $r=R\lambda(t)$, it has insufficient decay for large R (because factors of $\frac{1}{\lambda(t)}$ will turn out to correspond to logarithmic growth in t, once we choose $\lambda(t)$). On the other hand, if this error term is completely eliminated, the resulting modulation equation for λ becomes much more difficult to study. The third correction, v_3, solves an inhomogeneous version of (1.3) with 0 Cauchy data at infinity in order to correct the problem of this error term for large R while also not complicating the final modulation equation for λ. In the process of doing this, we introduce a small, positive parameter α.

The soliton, along with these three corrections can be regarded as the principal components of the ansatz. However, we introduce two more corrections, v_4 and v_5, both of which solve inhomogeneous versions of (1.3), with 0 Cauchy data at infinity, in order to improve the error terms resulting from the previous terms in the ansatz. More precisely, v_4 eliminates a large r portion of linear error terms associated to v_k, $\quad k=1,2,3$, as well as an error term arising from the combination of the right-hand sides of the v_3 and v_1 equations and $\partial_t^2 Q_{\frac{1}{\lambda(t)}}$. v_5 eliminates error terms associated to the nonlinear interaction between v_k, $\quad k=1,2,3,4$.

Our final ansatz

$$u_{ansatz}=Q_{\frac{1}{\lambda(t)}}(r)+\sum_{k=1}^5 v_k(t,r)$$

satisfies

$$E_{\text{WM}}(u_{ansatz},\partial_t u_{ansatz})<\infty$$

$$-\partial_{tt}u_{ansatz}+\partial_{rr}u_{ansatz}+\frac{1}{r}\partial_r u_{ansatz}-\frac{\sin(2u_{ansatz})}{2r^2}=-\left(F_4+F_5+F_6\right)$$

where F_5+F_6 has sufficiently fast time decay in sufficiently many norms, and is perturbative:

$$\frac{||F_5(t,r)+F_6(t,r)||_{L^2(rdr)}}{\lambda(t)^2}\leq\frac{C}{t^4\log^{3b+2N-1}(t)}$$

$$\frac{||F_5+F_6||_{\dot{H}^1_e}}{\lambda(t)}\leq\frac{C\log^{6+b}(t)}{t^{35/8}}$$

and F_4, the principal part of the error term, does not have as fast decay as F_5+F_6, but satisfies

$$\langle F_4(t,\cdot), \phi_0(\frac{\cdot}{\lambda(t)})\rangle = 0$$

and the following estimate (in a symbol-type fashion, see theorem 3.1 for the precise statement):

$$|F_4(t,r)| \leq \frac{C\mathbb{1}_{\{r\leq \log^N(t)\}}r}{t^2\log^{3b+1-2\alpha b}(t)(r^2+\lambda(t)^2)^2} + \frac{C\mathbb{1}_{\{r\leq \frac{t}{2}\}}r}{t^2\log^{5b+2N-2}(t)(r^2+\lambda(t)^2)^2}$$

where α and N are parameters associated to v_3 and v_4, respectively. α is small relative to b and 1, while N and is large relative to b and 1.

The modulation equation for λ,

$$\langle F_4(t,\cdot), \phi_0(\frac{\cdot}{\lambda(t)})\rangle = 0$$

has a principal part which is of the form of a Volterra equation of the second kind in the variable λ'' (with kernel and coefficients weakly depending on λ). In particular, this equation is of the form

$$\begin{aligned}
&-4\int_t^\infty \frac{\lambda''(s)}{1+s-t}ds + \frac{4b}{t^2\log^b(t)} + 4\alpha\log(\lambda(t))\lambda''(t)\\
&-4\int_t^\infty \frac{\lambda''(s)}{(\lambda(t)^{1-\alpha}+s-t)(1+s-t)^3}ds\\
&= f_3(\lambda(t),\lambda'(t),\lambda''(t))
\end{aligned}$$

We emphasize that the leading behavior of λ is independent of the small parameter $\alpha>0$, associated to v_3, and the crucial source term

$$\frac{4b}{t^2\log^b(t)}$$

comes from v_2, and is a consequence of the particular choice of data for v_2. The other terms on the left-hand side of the above equation come from $v_1, v_3, Q_{\frac{1}{\lambda(t)}}$. In particular, v_4 and v_5 do not contribute to the leading order part of the equation for $\lambda(t)$ because the terms contained in f_3 are subleading, for all λ close (in a C^2 sense) to $\frac{1}{\log^b(t)}$.

To the knowledge of the author, a modulation equation of the above form is quite different from that of previous works. Moreover, in the context of our ansatz, the Volterra form of the modulation equation seems to be related to the fact that $\frac{d}{d\lambda}\Big|_{\lambda=1} Q_\lambda(r) \notin L^2(rdr)$. Indeed, the integral operators acting on λ'' arising in the principal part of the modulation equation come from v_1 and v_3, which were introduced to correct soliton related error terms for large r.

As motivated by the discussion of the simpler ansatz $u_{a,1}$ above, we find an approximate solution to the modulation equation (3.63) for λ of the form

$$\lambda_{0,0}(t) = \frac{1}{\log^b(t)}$$

The equation for λ is then exactly solved around $\lambda_{0,0}$ with a fixed point argument in a weighted C^2 space, and we obtain an exact solution

$$\lambda(t) = \frac{1}{\log^b(t)} + O\left(\frac{1}{\log^b(t)\sqrt{\log(\log(t))}}\right)$$

Afterwards, we show that λ is in fact a C^4 function, and prove quantitative estimates on its derivatives. Along the way, we thus obtain estimates on higher time derivatives of the corrections v_k, whose right-hand sides depend on $\lambda(t)$ (for all $k \neq 2$).

Part 2: Construction of the exact solution

Once we construct u_{ansatz}, we substitute $u = u_{ansatz} + v_6$ into (1.1), thereby obtaining an equation for v_6. Our goal is to solve this equation for v_6 perturbatively, with 0 Cauchy data at infinity. We achieve this, by studying the distorted Fourier transform (in the sense of [**12**]), of v_6. In particular, we (formally) derive the equation for

$$y(t,\omega) = \mathcal{F}(\sqrt{\cdot}v_6(t,\cdot\lambda(t)))(\omega\lambda(t)^2)$$

where we recall that $\mathcal{F}$ denotes the distorted Fourier transform of [**12**]. The particular choice of the rescaling used in the definition of y is explained by noting that the equation for y takes the form

$$\begin{aligned}&\partial_{tt}y + \omega y\\ &\quad= -\mathcal{F}(\sqrt{\cdot}F(t,\cdot\lambda(t)))(\omega\lambda(t)^2) + F_2(y)(t,\omega) - \mathcal{F}(\sqrt{\cdot}F_3(v_6(y))(t,\cdot\lambda(t)))(\omega\lambda(t)^2)\end{aligned}$$

for appropriate F_2, F_3, where

$$F = F_4 + F_5 + F_6$$

is the sum of the error terms of u_{ansatz}. In deriving the y equation, a few properties about the distorted Fourier transform, most importantly, the transferrence identity from [**12**] are used. The equation for y is solved by showing that the map

$$\begin{aligned}T(y)(t,\omega) = -\int_t^\infty \frac{\sin((t-x)\sqrt{\omega})}{\sqrt{\omega}}&\left(F_2(y)(x,\omega) - \mathcal{F}(\sqrt{\cdot}F(x,\cdot\lambda(x)))(\omega\lambda(x)^2)\right.\\ &\left.-\mathcal{F}(\sqrt{\cdot}F_3(u(y))(x,\cdot\lambda(x)))(\omega\lambda(x)^2)\right)dx\end{aligned}$$

has a fixed point in an appropriate Banach space (whose norm is roughly the sum of weighted $L^\infty_t L^2_\omega$ norms of y and $\partial_t y$, see (4.46) for the precise definition) via the contraction mapping theorem. The most delicate term to estimate is

$$-\int_t^\infty \frac{\sin((t-x)\sqrt{\omega})}{\sqrt{\omega}}\left(-\mathcal{F}(\sqrt{\cdot}F_4(x,\cdot\lambda(x)))(\omega\lambda(x)^2)\right)dx$$

In order to obtain sufficient estimates on this term (and its time derivative) in appropriate norms, we must utilize the previously discussed orthogonality condition on F_4. The orthogonality condition on F_4 is utilized by noting that it implies that $\mathcal{F}(\sqrt{\cdot}F_4(x,\cdot\lambda(x)))(\omega\lambda(x)^2)$ has a certain degree of vanishing at small frequencies, and this allows us to integrate by parts in the x variable in the integral above. This, combined with the symbol-type nature of the pointwise estimates on F_4 turn out to provide sufficient decay of the integral above in all norms required by the fixed point argument. Since the density of the spectral measure associated to the distorted Fourier transform of [**12**] (which appears in the weighted norms of our iteration space) has a singularity at low frequencies, such integration by parts would be impossible without the orthogonality condition. In particular, we can not integrate by parts for the analogous integrals involving $F_5 + F_6$. The faster time decay of the L^2 and $\dot{H}^1_e$ norms of this term, relative to F_4 is what allows it to be perturbative, despite the non-orthogonality to $\phi_0(\frac{\cdot}{\lambda(t)})$.

CHAPTER 3

Construction of the ansatz

Fix $b>0$, $0<\alpha<\min\{\frac{1}{b(1040!)},\frac{1}{1040!}\}$, and $N>(5000!)(b+1)$. We consider (1.1) for $t\geq T_0$, where

$$T_0>2e^{e^{1000(b+1)}}+T_{0,1} \tag{3.1}$$

and is otherwise arbitrary for now, and where $T_{0,1}>e^{(128000)^{\frac{4}{b}}}+e^{N^2}$ is such that

$$|\frac{d^j}{dt^j}\left(\frac{\log(\log(t))}{\log^{b+1}(t)}\right)|+|\frac{d^j}{dt^j}\left(\frac{1}{b\log^b(t)\sqrt{\log(\log(t))}}\right)|\leq\frac{1}{200}|\frac{d^j}{dt^j}\left(\frac{1}{\log^b(t)}\right)|$$
$$,t\geq T_{0,1},\quad j=0,1,2$$

$\lambda:[T_0,\infty)\to(0,\infty)$ denotes a $C^2([T_0,\infty))$ map satisfying, for some $C>0$, *independent* of T_0

$$\lambda'(t)<0,\quad|\lambda''(t)|\leq\frac{C}{t^2\log^{b+1}(t)},\quad\text{and}\quad\frac{1}{C\log^C(t)}<\lambda(t)<\frac{1}{2},\quad t\geq T_0 \tag{3.2}$$

and is otherwise arbitrary for now. (Note that the first and third requirements above are not strictly necessary for the validity of most of our procedure. However, for this work, there is no loss of generality in assuming them, since λ will later on be restricted to a class of functions of the form $\lambda=\lambda_0+e$, where λ_0 is some explicit function to be specified later, and e belongs to a certain space of functions such that (among other things) (3.2) holds for λ_0+e).

Also for all estimates appearing in the entire paper, we use the convention that C will always denote a positive constant *independent* of T_0, whose value may change from line to line. Although we have already summarized the ansatz construction in the overview of the proof, we now provide an outline which clarifies the logical structure of the process.

3.0.1. Outline. Step 1, Definitions of the corrections, v_k: For all T_0 and λ as above, and $t\geq T_0$, we define functions v_k, $1\leq k\leq 5$, which were roughly described in the overview of the proof, thereby obtaining

$$u_{ansatz}(t,r)=Q_{\frac{1}{\lambda(t)}}(r)+\sum_{k=1}^{5}v_k(t,r)$$

Step 2, Splitting of the error term of u_{ansatz}, and setup of the modulation equation for λ: We define functions F_4,F_5, and F_6, which split the error term of u_{ansatz} into $-(F_4+F_5+F_6)$. Then, we consider the modulation equation for $\lambda(t)$ resulting from setting

$$\langle F_4(t,\cdot),\phi_0(\frac{\cdot}{\lambda(t)})\rangle=0,\quad t\geq T_0$$

(Note that F_4 depends on $\lambda(t)$). As described (to a certain extent) in the overview of the proof, this modulation equation takes the form of a Volterra equation of the second kind in the variable $\lambda''(t)$ (with kernel and coefficients weakly depending on λ), and our choice of Cauchy data for v_2 gives rise to a leading order solution

$$\lambda_0 = \frac{1}{\log^b(t)} + \int_t^\infty \int_{t_1}^\infty \frac{-b^2 \log(\log(t_2))}{t_2^2 \log^{b+2}(t_2)} dt_2 dt_1$$

to this equation. We then introduce a weighted C^2 space, X, with norm (3.30). From this point on, we further restrict $\lambda(t)$ to be of the form

$$\lambda(t) = \lambda_0(t) + e(t), \quad e \in B$$

where

$$B = \overline{B_1(0)} \subset X$$

(Note that $\lambda_0 + B$ is contained within the class of $\lambda(t)$ we initially started with (which is described by (3.2))).

Step 3, Solving the modulation equation for $\lambda(t)$: Writing $F_4(t,r) = F_4^{\lambda(t)}(t,r)$ to emphasize the λ dependence of F_4, we show that there exists $T_3 > 0$ such that, for all $T_0 \geq T_3$, the equation

$$\langle F_4^{\lambda_0(t)+e(t)}(t,\cdot), \phi_0(\frac{\cdot}{\lambda_0(t)+e(t)})\rangle = 0, \quad t \geq T_0$$

can be solved for $e(t) \in B$, using the contraction mapping principle. An important part of this procedure is that the kernel appearing in the Volterra equation for e (which is independent of e, modulo error terms which can be treated perturbatively) satisfies the structural condition (3.69). From here on, we work under the constraint $T_0 \geq T_3$.

Step 4, Estimates on higher derivatives of λ'': Denoting the solution to the above equation for e by $e_0(t)$, we fix $\lambda(t) = \lambda_0(t) + e_0(t)$. Then, we prove that $\lambda(t)$, which is apriori only in $C^2([T_0,\infty))$, is actually in $C^4([T_0,\infty))$, and establish quantitative estimates on λ''' and λ''''.

Step 5, Estimates on F_k: At this stage, we prove pointwise estimates on F_4, as well as estimates on $||F_k||_{L^2(rdr)}$, $||F_k||_{\dot{H}^1_e}$ for $k = 5, 6$. This completes the ansatz construction. More precisely, our main result of this section is

THEOREM 3.1 (Approximate solution to (1.1)). *For all $b > 0$, there exists $T_3 > 0$ such that, for all $T_0 \geq T_3$, there exists $v_{corr} \in C^4([T_0,\infty); C^2((0,\infty)))$, and $\lambda \in C^4([T_0,\infty))$ such that, if $u = Q_{\frac{1}{\lambda(t)}} + v_{corr}$, then*

$$E_{WM}(u, \partial_t u) < \infty$$

$$-\partial_{tt}u + \partial_{rr}u + \frac{1}{r}\partial_r u - \frac{\sin(2u)}{2r^2} = -\left(F_4 + F_5 + F_6\right)$$

where

$$\frac{1}{\lambda(t)^2}||\left(F_5 + F_6\right)(t,r)||_{L^2(rdr)} \leq \frac{C}{t^4 \log^{3b+2N-1}(t)} \tag{3.3}$$

$$\frac{||\left(F_5 + F_6\right)(t)||_{\dot{H}^1_e}}{\lambda(t)} \leq \frac{C \log^{6+b}(t)}{t^{35/8}} \tag{3.4}$$

$$\langle F_4(t,\cdot), \phi_0(\frac{\cdot}{\lambda(t)})\rangle = 0$$

For $0 \leq k \leq 2, \quad 0 \leq j \leq 1, \quad j+k \leq 2$, *we have*

$$t^j r^k |\partial_r^k \partial_t^j F_4(t,r)| \leq \frac{C \mathbb{1}_{\{r \leq \log^N(t)\}} r}{t^2 \log^{3b+1-2\alpha b}(t)(r^2+\lambda(t)^2)^2} + \frac{C \mathbb{1}_{\{r \leq \frac{t}{2}\}} r}{t^2 \log^{5b+2N-2}(t)(r^2+\lambda(t)^2)^2}$$

In addition, we have

$$|\partial_t^2 F_4(t,r)| \leq \frac{C \mathbb{1}_{\{r \leq \log^N(t)\}} r}{t^4 \log^{3b+1-2\alpha b}(t)(r^2+\lambda(t)^2)^2} + \frac{C \mathbb{1}_{\{r \leq \frac{t}{2}\}} r}{t^4 \log^{5b+2N-2}(t)(r^2+\lambda(t)^2)^2}$$
$$+ \frac{C \mathbb{1}_{\{r \leq \frac{t}{2}\}}}{t^4 \log^{5b+N-2}(t)(r^2+\lambda(t)^2)^2}$$

We also have the following estimates on v_{corr}.

$$||\frac{v_{corr}(x, R\lambda(x))}{R\lambda(x)}||_{L^\infty}^2 + ||\frac{v_{corr}(x, R\lambda(x))}{R\lambda(x)^2(1+R^2)}||_{L^\infty} \leq \frac{C \log(\log(x))}{x^2 \log(x)} \tag{3.5}$$

$$1 + ||\frac{v_{corr}(x, R\lambda(x))}{R}||_{L^\infty} + ||\partial_R(v_{corr}(x, R\lambda(x)))||_{L^\infty} \leq C \tag{3.6}$$

(3.7)
$$||\frac{v_{corr}(x, R\lambda(x))\partial_R(v_{corr}(x, R\lambda(x)))}{R\lambda(x)^2}||_{L^\infty_R((0,1))}$$
$$+ ||\frac{v_{corr}(x, R\lambda(x))\partial_R(v_{corr}(x, R\lambda(x)))}{R^2\lambda(x)^2}||_{L^\infty_R((1,\infty))} + ||\frac{\partial_R(v_{corr}(x, R\lambda(x)))}{(1+R^2)\lambda(x)^2}||_{L^\infty}$$
$$\leq \frac{C \log(\log(x))}{x^2 \log(x)}$$

Finally, λ *is described by*

$$\lambda(t) = \frac{1}{\log^b(t)} + e(t), \quad |e(t)| \leq \frac{C}{\log^b(t)\sqrt{\log(\log(t))}}$$

$$|e^{(j)}(t)| \leq \begin{cases} \frac{C}{t^j \log^{b+1}(t)\sqrt{\log(\log(t))}}, & j = 1,2 \\ \frac{C}{t^j \log^{b+1}(t)}, & j = 3,4 \end{cases}$$

3.0.2. Correcting the large r behavior of $Q_{\frac{1}{\lambda(t)}}$. The error term $\partial_t^2 Q_{\frac{1}{\lambda(t)}}$, which arises from inserting $Q_{\frac{1}{\lambda(t)}}$ into (1.1), does not decay quickly enough for its inner product with $\phi_0(\frac{\cdot}{\lambda(t)})$ to be defined. The first term in the ansatz, v_1, is designed to correct this problem. Note that we do not choose v_1 to solve an equation whose right-hand side is equal to $\partial_t^2 Q_{\frac{1}{\lambda(t)}}$, as doing so would result in a much more difficult equation to solve when we eventually choose $\lambda(t)$. Instead, v_1 is defined as the solution to the equation

$$-\partial_{tt} v_1 + \partial_{rr} v_1 + \frac{1}{r}\partial_r v_1 - \frac{v_1}{r^2} = -2\lambda''(t)\frac{r}{1+r^2}$$

with 0 Cauchy data at infinity. I.e., by Duhamel's principle, we have

$$v_1(t,r) = \int_t^\infty v_s(t,r) ds$$

where v_s is the solution to the following Cauchy problem

$$\begin{cases} -\partial_{tt}v_s + \partial_{rr}v_s + \frac{1}{r}\partial_r v_s - \frac{v_s}{r^2} = 0 \\ v_s(s,r) = 0 \\ \partial_t v_s(s,r) = -2\lambda''(s)\frac{r}{1+r^2} \end{cases} \tag{3.8}$$

We can determine a fairly explicit formula for v_s. The procedure used to determine the formula for v_s may be slightly formal, but the final expression obtained can be seen to be the solution to the Cauchy problem (3.8). Firstly, note that if u is a solution to

$$-\partial_{tt}u + \partial_{rr}u + \frac{1}{r}\partial_r u = 0$$

then, $w = \partial_r u$ is formally a solution to

$$-\partial_{tt}w + \partial_{rr}w + \frac{1}{r}\partial_r w - \frac{w}{r^2} = 0$$

So, we will first write down the spherical means representation formula for the problem

$$\begin{cases} -\partial_{tt}u_1 + \Delta u_1 = 0 \\ u_1(s) = 0 \\ \partial_t u_1(s) = -\lambda''(s)\log(1+|x|^2) \end{cases}$$

then define u by $u(t,|x|) = u_1(t,x)$ (which is possible, since u_1 is radially symmetric). Then,

$$v_s(t,r) = \partial_r u(t,r)$$

will be seen to be the solution to (3.8). From the spherical means representation formula, for $t > s$, we have

$$\begin{aligned} u_1(t,x) &= -\frac{\lambda''(s)}{2\pi}\int_{B(0,t-s)} \frac{\log(1+|y+x|^2)}{((t-s)^2-|y|^2)^{1/2}}dy \\ &= -\frac{\lambda''(s)}{2\pi}\int_0^{t-s}\rho\int_0^{2\pi}\frac{\log(1+|x|^2+2|x|\rho\cos(\theta)+\rho^2)}{((t-s)^2-\rho^2)^{1/2}}d\theta d\rho \end{aligned}$$

Recalling that u is the radial coordinate representative of u_1, we have

(3.9)

$$\partial_{|x|}u = -\frac{\lambda''(s)}{2\pi}\int_0^{t-s}\rho\int_0^{2\pi}\frac{2(|x|+\rho\cos(\theta))}{(1+|x|^2+2|x|\rho\cos(\theta)+\rho^2)((t-s)^2-\rho^2)^{1/2}}d\theta d\rho$$

To do the integral over the angular coordinate, we can regard it as

$$\int_C \frac{2(|x|+\frac{\rho}{2}(z+\frac{1}{z}))}{(1+|x|^2+|x|\rho(z+\frac{1}{z})+\rho^2)((t-s)^2-\rho^2)^{1/2}}\frac{dz}{iz}$$

where C is the boundary of the unit circle in the complex plane, traversed in the counterclockwise direction.

The integrand is a meromorphic function on $\mathbb{C}$, with poles at

$$z = 0, -\left(\frac{1+|x|^2+\rho^2 \pm \sqrt{-4|x|^2\rho^2+(1+|x|^2+\rho^2)^2}}{2|x|\rho}\right)$$

Note that

$$-4|x|^2\rho^2+(1+|x|^2+\rho^2)^2 = 1+2(|x|^2+\rho^2)+(|x|^2-\rho^2)^2 \geq 1$$

So,

$$\frac{1+|x|^2+\rho^2+\sqrt{-4|x|^2\rho^2+(1+|x|^2+\rho^2)^2}}{2|x|\rho} > \frac{1+|x|^2+\rho^2}{2|x|\rho} > \frac{|x|^2+\rho^2}{2|x|\rho} \geq 1$$

On the other hand, we have

$$\sqrt{-4|x|^2\rho^2+(1+|x|^2+\rho^2)^2} \leq (1+|x|^2+\rho^2)$$

so that

$$\frac{1+|x|^2+\rho^2-\sqrt{-4|x|^2\rho^2+(1+|x|^2+\rho^2)^2}}{2|x|\rho} > 0$$

and

$$\begin{aligned}&\frac{1+|x|^2+\rho^2-\sqrt{-4|x|^2\rho^2+(1+|x|^2+\rho^2)^2}}{2|x|\rho}\\ &= \frac{2|x|\rho}{1+|x|^2+\rho^2+\sqrt{-4|x|^2\rho^2+(1+|x|^2+\rho^2)^2}} \leq 1\end{aligned}$$

So, the only two poles of the integrand located inside the unit disk in the complex plane are $z=0$ and $z=z_1=-\left(\frac{1+|x|^2+\rho^2-\sqrt{-4|x|^2\rho^2+(1+|x|^2+\rho^2)^2}}{2|x|\rho}\right)$. The corresponding residues are

$$Res_0 = \frac{1}{i|x|\sqrt{(s-t)^2-\rho^2}}$$

$$Res_{z_1} = \frac{-1+|x|^2-\rho^2}{i|x|\sqrt{-4|x|^2\rho^2+(1+|x|^2+\rho^2)^2}\sqrt{(s-t)^2-\rho^2}}$$

Returning to (3.9), we get, for $t>s$,

$$(\partial_{|x|}u)(t,r) = -\frac{\lambda''(s)}{|x|}\int_0^{t-s}\frac{\rho}{\sqrt{(t-s)^2-\rho^2}}\left(1+\frac{|x|^2-1-\rho^2}{\sqrt{(1+|x|^2+\rho^2)^2-4|x|^2\rho^2}}\right)d\rho$$

By substitution, we see that $\partial_{|x|}u(t,r)$ solves (3.8) for $t>s$. We can extend the solution to $t<s$ with the same Cauchy data at $t=s$ by defining

$$v_s(t,r) = (-\partial_{|x|}u)(s-(t-s),r), \quad t<s$$

so that, for $t<s$, we have

$$v_s(t,r) = \frac{\lambda''(s)}{r}\int_0^{s-t}\frac{\rho}{\sqrt{(s-t)^2-\rho^2}}\left(1+\frac{r^2-1-\rho^2}{\sqrt{(1+r^2+\rho^2)^2-4r^2\rho^2}}\right)d\rho$$

We will only need to use pointwise estimates on v_1, which will be proven shortly, but to give the reader some idea of the behavior of v_s, we note that

$$v_s(t,r) \sim \frac{2\lambda''(s)\left(1-\sqrt{1-a^2}\right)}{a}, \quad 0<a=\frac{r}{(s-t)}<1 \ll s-t$$

We have
(3.10)

$$v_1(t,r) = \int_t^\infty \frac{\lambda''(s)}{r}\int_0^{s-t}\frac{\rho}{\sqrt{(s-t)^2-\rho^2}}\left(1+\frac{r^2-1-\rho^2}{\sqrt{(1+r^2+\rho^2)^2-4r^2\rho^2}}\right)d\rho ds$$

3.0.2.1. *Pointwise estimates on* $\partial_r^j v_1$. In this section, we will prove the following

LEMMA 3.2.

$$v_1(t,r) = r\int_t^\infty \frac{\lambda''(s)}{1+s-t}ds + \text{Err}(t,r) \tag{3.11}$$

where

$$|\text{Err}(t,r)| \leq Cr\log(3+2r)\sup_{x\geq t}|\lambda''(x)|, \quad r>0$$

In addition,

$$|v_1(t,r)| \leq \frac{C}{r}\int_t^\infty |\lambda''(s)|(s-t)ds, \quad r>0 \tag{3.12}$$

Moreover, similar results are true for $\partial_r v_1$*:*

$$\partial_r v_1(t,r) = \int_t^\infty \frac{\lambda''(s)}{1+s-t}ds + E_{\partial_r v_1}(t,r) \tag{3.13}$$

with

$$|E_{\partial_r v_1}(t,r)| \leq C\sup_{x\geq t}(|\lambda''(x)|)\log(3+2r), \quad r>0$$

and

$$|\partial_r v_1(t,r)| \leq \frac{C}{r^2}\left(\sup_{x\geq t}(|\lambda''(x)|(1+(x-t)^2)) + \int_t^\infty |\lambda''(s)|(s-t)ds\right), \quad r>2$$

PROOF. We start with

(3.14)

$$\begin{aligned}
&v_1(t,r)\\
&= \int_t^\infty \frac{\lambda''(s)}{r}\int_0^{s-t}\rho\left(\frac{1}{\sqrt{(s-t)^2-\rho^2}}-\frac{1}{s-t}\right)\left(1+\frac{r^2-1-\rho^2}{\sqrt{(r^2-1-\rho^2)^2+4r^2}}\right)d\rho ds\\
&+\int_t^\infty \frac{\lambda''(s)}{r}\int_0^{s-t}\frac{\rho}{(s-t)}\left(1+\frac{r^2-1-\rho^2}{\sqrt{(r^2-1-\rho^2)^2+4r^2}}\right)d\rho ds
\end{aligned}$$

The first line of (3.14) can be estimated as follows

$$\begin{aligned}
&|\int_t^\infty \frac{\lambda''(s)}{r}\int_0^{s-t}\rho\left(\frac{1}{\sqrt{(s-t)^2-\rho^2}}-\frac{1}{s-t}\right)\left(1+\frac{r^2-1-\rho^2}{\sqrt{(r^2-1-\rho^2)^2+4r^2}}\right)d\rho ds|\\
&\leq \frac{\sup_{x\geq t}(|\lambda''(x)|)}{r}\int_0^\infty \rho\int_{\rho+t}^\infty\left(\frac{1}{\sqrt{(s-t)^2-\rho^2}}-\frac{1}{s-t}\right)\left(1+\frac{r^2-1-\rho^2}{\sqrt{(r^2-1-\rho^2)^2+4r^2}}\right)dsd\rho\\
&\leq \frac{\sup_{x\geq t}(|\lambda''(x)|)}{r}\int_0^\infty \rho\log(2)\left(1+\frac{r^2-1-\rho^2}{\sqrt{(r^2-1-\rho^2)^2+4r^2}}\right)d\rho\\
&\leq C\frac{\sup_{x\geq t}(|\lambda''(x)|)}{r}\cdot r^2\\
&\leq Cr\sup_{x\geq t}(|\lambda''(x)|)
\end{aligned}$$

The second line of (3.14) is split into the following two terms

$$
\begin{aligned}
&\int_t^\infty \frac{\lambda''(s)}{r} \int_0^{s-t} \frac{\rho}{(s-t)} \left(1 + \frac{r^2-1-\rho^2}{\sqrt{(r^2-1-\rho^2)^2+4r^2}}\right) d\rho ds \\
&= \int_t^{t+2(r+1)} \frac{\lambda''(s)}{r} \int_0^{s-t} \frac{\rho}{(s-t)} \left(1 + \frac{r^2-1-\rho^2}{\sqrt{(r^2-1-\rho^2)^2+4r^2}}\right) d\rho ds \\
&+ \int_{t+2(r+1)}^\infty \frac{\lambda''(s)}{r} \int_0^{s-t} \frac{\rho}{(s-t)} \left(1 + \frac{r^2-1-\rho^2}{\sqrt{(r^2-1-\rho^2)^2+4r^2}}\right) d\rho ds
\end{aligned} \tag{3.15}
$$

For the second line of (3.15), we have

$$
\begin{aligned}
&\int_t^{t+2(r+1)} \frac{\lambda''(s)}{r} \int_0^{s-t} \frac{\rho}{(s-t)} \left(1 + \frac{r^2-1-\rho^2}{\sqrt{(r^2-1-\rho^2)^2+4r^2}}\right) d\rho ds \\
&= \int_t^{t+2(r+1)} \frac{\lambda''(s)}{r(s-t)} \frac{\left(1+r^2+(s-t)^2 - \sqrt{(1+(r+s-t)^2)(1+(r-(s-t))^2)}\right)}{2} ds \\
&= \int_t^{t+2(r+1)} \frac{\lambda''(s)}{r(s-t)} \frac{2r^2(s-t)^2}{(1+r^2+(s-t)^2+\sqrt{(1+(r+s-t)^2)(1+(r-(s-t))^2)})} ds
\end{aligned}
$$

So, we have

$$
\begin{aligned}
&\left|\int_t^{t+2(r+1)} \frac{\lambda''(s)}{r} \int_0^{s-t} \frac{\rho}{(s-t)} \left(1 + \frac{r^2-1-\rho^2}{\sqrt{(r^2-1-\rho^2)^2+4r^2}}\right) d\rho ds\right| \\
&\leq Cr \sup_{x\geq t} (|\lambda''(x)|) \int_t^{t+2(r+1)} \frac{(s-t)}{1+r^2} ds \\
&\leq Cr \sup_{x\geq t} (|\lambda''(x)|)
\end{aligned}
$$

For the third line of (3.15), we have

$$
\begin{aligned}
&\int_{t+2(r+1)}^\infty \lambda''(s) \frac{2r(s-t)}{(1+r^2+(s-t)^2+\sqrt{(1+(r+s-t)^2)(1+(r-(s-t))^2)})} ds \\
&= \int_{t+2(r+1)}^\infty \lambda''(s) \frac{2r}{(s-t)} \left(\frac{1}{2} + O\left(\frac{(1+r)^2}{(s-t)^2}\right)\right) ds \\
&= \int_{t+2(r+1)}^\infty \lambda''(s) \frac{r}{(s-t)} ds + E_1(t,r)
\end{aligned}
$$

where

$$
\begin{aligned}
|E_1(t,r)| &\leq \sup_{x\geq t} (|\lambda''(x)|)\, r(1+r)^2 \int_{t+2(r+1)}^\infty \frac{ds}{(s-t)^3} \\
&\leq Cr \sup_{x\geq t} (|\lambda''(x)|)
\end{aligned}
$$

Next,

$$
\begin{aligned}
\int_{t+2(r+1)}^\infty \frac{\lambda''(s) r}{(s-t)} ds &= \int_{t+2(r+1)}^\infty \frac{\lambda''(s) r}{1+s-t} ds \\
&+ \int_{t+2(r+1)}^\infty \lambda''(s) r \left(\frac{1}{s-t} - \frac{1}{1+s-t}\right) ds
\end{aligned}
$$

and

$$|\int_{t+2(r+1)}^{\infty} \lambda''(s) r \left(\frac{1}{s-t} - \frac{1}{1+s-t}\right) ds| \leq r \sup_{x \geq t} (|\lambda''(x)|) \log(1 + \frac{1}{2+2r})$$
$$\leq C \sup_{x \geq t} (|\lambda''(x)|) r$$

Finally,

$$\int_{t+2(r+1)}^{\infty} \frac{\lambda''(s) r}{1+s-t} ds = \int_{t}^{\infty} \frac{\lambda''(s) r}{1+s-t} ds - \int_{t}^{t+2(r+1)} \frac{\lambda''(s) r}{1+s-t} ds$$

with

$$|\int_{t}^{t+2(r+1)} \frac{\lambda''(s) r}{1+s-t} ds| \leq C \sup_{x \geq t} (|\lambda''(x)|) r \log(3+2r)$$

Thus, we obtain the decomposition (3.11), with the desired estimate on Err.

Next, we have

$$|v_1(t,r)| = |\int_{t}^{\infty} \frac{\lambda''(s)}{r} \int_{0}^{s-t} \frac{\rho}{\sqrt{(s-t)^2 - \rho^2}} \left(1 + \frac{r^2 - 1 - \rho^2}{\sqrt{(r^2-1-\rho^2)^2 + 4r^2}}\right) d\rho ds|$$
$$\leq \frac{C}{r} \int_{t}^{\infty} |\lambda''(s)|(s-t) ds$$

Now we treat $\partial_r v_1$:

(3.16)

$$\partial_r v_1(t,r) = \int_{t}^{\infty} \frac{-\lambda''(s)}{r^2} \int_{0}^{s-t} \frac{\rho}{\sqrt{(s-t)^2 - \rho^2}} \left(1 + \frac{r^2 - 1 - \rho^2}{\sqrt{(r^2-1-\rho^2)^2 + 4r^2}}\right) d\rho ds$$
$$+ \int_{t}^{\infty} \frac{\lambda''(s)}{r} \int_{0}^{s-t} \frac{\rho}{\sqrt{(s-t)^2 - \rho^2}} \frac{4r\left(\rho^2 + r^2 + 1\right)}{\left((\rho^2 - r^2 + 1)^2 + 4r^2\right)^{3/2}} d\rho ds$$

Even though the first line of (3.16) is equal to $\frac{-v_1}{r}$, the principal contribution to $\partial_r v_1$ near the origin actually comes from a combination of effects from appropriate parts of both the first and second lines of (3.16). So, we do not simply divide the previous v_1 estimates by $-r$ to treat the first line of (3.16). Instead, we split the first line of (3.16) as

$$\int_{t}^{\infty} \frac{-\lambda''(s)}{r^2} \int_{0}^{s-t} \frac{\rho}{\sqrt{(s-t)^2 - \rho^2}} \left(1 + \frac{r^2 - 1 - \rho^2}{\sqrt{(r^2-1-\rho^2)^2 + 4r^2}}\right) d\rho ds$$
$$= \int_{t}^{\infty} \frac{-\lambda''(s)}{r^2} \int_{0}^{s-t} \rho \left(\frac{1}{\sqrt{(s-t)^2 - \rho^2}} - \frac{1}{(s-t)}\right) \left(1 + \frac{r^2 - 1 - \rho^2}{\sqrt{(r^2-1-\rho^2)^2 + 4r^2}}\right) d\rho ds$$
$$+ \int_{t}^{\infty} \frac{-\lambda''(s)}{r^2} \int_{0}^{s-t} \frac{\rho}{(s-t)} \left(1 + \frac{r^2 - 1 - \rho^2}{\sqrt{(r^2-1-\rho^2)^2 + 4r^2}}\right) d\rho ds$$

and

$$
\begin{aligned}
&|\int_t^\infty \frac{-\lambda''(s)}{r^2}\int_0^{s-t}\rho\left(\frac{1}{\sqrt{(s-t)^2-\rho^2}}-\frac{1}{(s-t)}\right)\left(1+\frac{r^2-1-\rho^2}{\sqrt{(r^2-1-\rho^2)^2+4r^2}}\right)d\rho ds|\\
&\leq C\frac{\sup_{x\geq t}(|\lambda''(x)|)}{r^2}\int_0^\infty \rho\left(1+\frac{r^2-1-\rho^2}{\sqrt{(r^2-1-\rho^2)^2+4r^2}}\right)\int_{\rho+t}^\infty\left(\frac{1}{\sqrt{(s-t)^2-\rho^2}}-\frac{1}{s-t}\right)dsd\rho\\
&\leq \frac{C\sup_{x\geq t}(|\lambda''(x)|)}{r^2}\int_0^\infty \rho\left(1+\frac{r^2-1-\rho^2}{\sqrt{(r^2-1-\rho^2)^2+4r^2}}\right)d\rho\\
&\leq C\sup_{x\geq t}(|\lambda''(x)|)
\end{aligned}
$$

The second line of (3.16) is split in the same way

$$
\begin{aligned}
&\int_t^\infty \frac{\lambda''(s)}{r}\int_0^{s-t}\frac{\rho}{\sqrt{(s-t)^2-\rho^2}}\frac{4r\left(\rho^2+r^2+1\right)}{\left(\left(\rho^2-r^2+1\right)^2+4r^2\right)^{3/2}}d\rho ds\\
&=\int_t^\infty \frac{\lambda''(s)}{r}\int_0^{s-t}\rho\left(\frac{1}{\sqrt{(s-t)^2-\rho^2}}-\frac{1}{(s-t)}\right)\frac{4r\left(\rho^2+r^2+1\right)}{\left(\left(\rho^2-r^2+1\right)^2+4r^2\right)^{3/2}}d\rho ds\\
&+\int_t^\infty \frac{\lambda''(s)}{r}\int_0^{s-t}\frac{\rho}{(s-t)}\frac{4r\left(\rho^2+r^2+1\right)}{\left(\left(\rho^2-r^2+1\right)^2+4r^2\right)^{3/2}}d\rho ds
\end{aligned}
$$

and we again have

$$
\begin{aligned}
&|\int_t^\infty \frac{\lambda''(s)}{r}\int_0^{s-t}\rho\left(\frac{1}{\sqrt{(s-t)^2-\rho^2}}-\frac{1}{(s-t)}\right)\frac{4r\left(\rho^2+r^2+1\right)}{\left((\rho^2-r^2+1)^2+4r^2\right)^{3/2}}|d\rho ds\\
&\leq C\sup_{x\geq t}(|\lambda''(x)|)\int_0^\infty \rho\frac{(\rho^2+r^2+1)}{((\rho^2-r^2+1)^2+4r^2)^{3/2}}\int_{\rho+t}^\infty\left(\frac{1}{\sqrt{(s-t)^2-\rho^2}}-\frac{1}{s-t}\right)dsd\rho\\
&\leq C\sup_{x\geq t}(|\lambda''(x)|)
\end{aligned}
$$

So far, we have

$$
\begin{aligned}
\partial_r v_1(t,r)=&\int_t^\infty \frac{-\lambda''(s)}{r^2}\int_0^{s-t}\frac{\rho}{(s-t)}\left(1+\frac{r^2-1-\rho^2}{\sqrt{(r^2-1-\rho^2)^2+4r^2}}\right)d\rho ds\\
&+\int_t^\infty \frac{\lambda''(s)}{r}\int_0^{s-t}\frac{\rho}{(s-t)}\frac{4r\left(\rho^2+r^2+1\right)}{\left(\left(\rho^2-r^2+1\right)^2+4r^2\right)^{3/2}}d\rho ds\\
&+E_{0,\partial_r v_1}
\end{aligned}
\tag{3.17}
$$

where

$$|E_{0,\partial_r v_1}(t,r)|\leq C\sup_{x\geq t}(|\lambda''(x)|)$$

Now, we combine the first and second lines of (3.17), to get

$$\int_t^\infty \frac{-\lambda''(s)}{r^2}\int_0^{s-t}\frac{\rho}{(s-t)}\left(1+\frac{r^2-1-\rho^2}{\sqrt{(r^2-1-\rho^2)^2+4r^2}}\right)d\rho ds$$
$$+\int_t^\infty \frac{\lambda''(s)}{r}\int_0^{s-t}\frac{\rho}{(s-t)}\frac{4r\left(\rho^2+r^2+1\right)}{\left(\left(\rho^2-r^2+1\right)^2+4r^2\right)^{3/2}}d\rho ds$$
$$=-2\int_t^\infty \lambda''(s)(s-t)\left(\frac{(r^2-1-(s-t)^2)}{\sqrt{\beta}(1+r^2+(s-t)^2+\sqrt{\beta})}\right)ds$$

where

$$\beta = 4r^2+(r^2-1-(s-t)^2)^2$$

Now, we proceed as in the estimates for v_1.

$$|-2\int_t^{t+2(r+1)}\lambda''(s)(s-t)\left(\frac{(r^2-1-(s-t)^2)}{\sqrt{\beta}(1+r^2+(s-t)^2+\sqrt{\beta})}\right)ds|$$
$$\leq C\sup_{x\geq t}\left(|\lambda''(x)|\right)\int_t^{t+2(r+1)}\frac{|s-t|}{1+r^2}ds$$
$$\leq C\sup_{x\geq t}\left(|\lambda''(x)|\right)$$

Next, note that

$$\left(\frac{(r^2-1-(s-t)^2)}{\sqrt{\beta}(1+r^2+(s-t)^2+\sqrt{\beta})}\right)=\frac{-\left(1+\frac{1-r^2}{(s-t)^2}\right)}{(s-t)^2}\left(\frac{1}{\sqrt{1+q}}\cdot\frac{1}{1+y+\sqrt{1+q}}\right)$$

where

$$q=\frac{2(1-r^2)}{(s-t)^2}+\frac{(r^2+1)^2}{(s-t)^4}$$

$$y=\frac{1+r^2}{(s-t)^2}$$

So, for $s-t\geq 2(r+1)$,

$$|q|\leq\frac{9}{16},\quad |y|\leq\frac{1}{4}$$

Using this, we have

$$\left(\frac{(r^2-1-(s-t)^2)}{\sqrt{\beta}(1+r^2+(s-t)^2+\sqrt{\beta})}\right)=\frac{-1}{2(s-t)^2}\left(1+O\left(\frac{1+r^2}{(s-t)^2}\right)\right)$$

So,

$$-2\int_{t+2(r+1)}^\infty \lambda''(s)(s-t)\left(\frac{(r^2-1-(s-t)^2)}{\sqrt{\beta}(1+r^2+(s-t)^2+\sqrt{\beta})}\right)ds$$
$$=\int_{t+2(r+1)}^\infty \frac{\lambda''(s)}{(s-t)}ds+E_{1,\partial_r v_1}(t,r)$$

where

$$|E_{1,\partial_r v_1}(t,r)|\leq C\int_{t+2(r+1)}^\infty \frac{|\lambda''(s)|}{(s-t)}\frac{(1+r^2)}{(s-t)^2}ds\leq C\sup_{x\geq t}\left(|\lambda''(x)|\right)$$

Then,

$$\int_{t+2(r+1)}^{\infty} \frac{\lambda''(s)}{(s-t)} ds = \int_{t+2(r+1)}^{\infty} \frac{\lambda''(s)}{1+s-t} ds + E_{2,\partial_r v_1}(t,r)$$

with

$$|E_{2,\partial_r v_1}(t,r)| \leq \int_{t+2(r+1)}^{\infty} |\lambda''(s)| \left(\frac{1}{s-t} - \frac{1}{1+s-t}\right) ds \leq C \sup_{x \geq t} (|\lambda''(x)|)$$

Finally,

$$\begin{aligned}\int_{t+2(r+1)}^{\infty} \frac{\lambda''(s)}{1+s-t} ds &= \int_{t}^{\infty} \frac{\lambda''(s)}{1+s-t} ds - \int_{t}^{t+2(r+1)} \frac{\lambda''(s)}{1+s-t} ds \\ &= \int_{t}^{\infty} \frac{\lambda''(s)}{1+s-t} ds + E_{3,\partial_r v_1}(t,r)\end{aligned}$$

where

$$|E_{3,\partial_r v_1}(t,r)| \leq C \sup_{x \geq t} (|\lambda''(x)|) \log(3+2r)$$

and we get the desired result

$$\partial_r v_1(t,r) = \int_{t}^{\infty} \frac{\lambda''(s)}{1+s-t} ds + E_{\partial_r v_1}(t,r)$$

with

$$|E_{\partial_r v_1}(t,r)| \leq C \sup_{x \geq t} (|\lambda''(x)|) \log(3+2r)$$

It remains to prove the last estimate in the lemma statement: The first term of (3.16) is estimated by

$$\begin{aligned}&|\int_{t}^{\infty} \frac{-\lambda''(s)}{r^2} \int_{0}^{s-t} \frac{\rho}{\sqrt{(s-t)^2-\rho^2}} \left(1 + \frac{r^2-1-\rho^2}{\sqrt{(r^2-1-\rho^2)^2+4r^2}}\right) d\rho ds| \\ &\leq \frac{C}{r^2} \int_{t}^{\infty} |\lambda''(s)|(s-t) ds\end{aligned}$$

Turning to the second term of (3.16), we have

$$\begin{aligned}&|\int_{t}^{\infty} \frac{\lambda''(s)}{r} \int_{0}^{s-t} \frac{\rho}{\sqrt{(s-t)^2-\rho^2}} \frac{4r\left(\rho^2+r^2+1\right)}{\left(\left(\rho^2-r^2+1\right)^2+4r^2\right)^{3/2}} d\rho ds| \\ &\leq C \int_{0}^{\infty} \frac{\rho(\rho^2+r^2+1)}{((\rho^2-r^2+1)^2+4r^2)^{3/2}} \int_{\rho+t}^{\infty} \frac{|\lambda''(s)|(1+(s-t)^2)}{\sqrt{(s-t)^2-\rho^2}(1+(s-t)^2)} ds d\rho \\ &\leq C \sup_{x \geq t} \left(|\lambda''(x)|(1+(x-t)^2)\right) \\ &\cdot \int_{0}^{\infty} \frac{\rho(\rho^2+r^2+1)}{((\rho^2-r^2+1)^2+4r^2)^{3/2}} \int_{\rho+t}^{\infty} \frac{1}{\sqrt{(s-t)^2-\rho^2}} \frac{1}{(1+(s-t)^2)} ds d\rho \\ &\leq C \sup_{x \geq t} \left(|\lambda''(x)|(1+(x-t)^2)\right) \int_{0}^{\infty} \frac{(\rho^2+r^2+1)}{((\rho^2-r^2+1)^2+4r^2)^{3/2}} \frac{d\rho}{\sqrt{1+\rho^2}}\end{aligned}$$

Let us first note that, for $\rho \leq \frac{r}{2}$, we have

$$r^2 - 1 - \rho^2 > Cr^2, \quad \text{for } r > 2$$

so,

$$\frac{(\rho^2+r^2+1)}{((\rho^2-r^2+1)^2+4r^2)^{3/2}} \leq \frac{C}{r^4}$$

and

$$\int_0^{r/2} \frac{(\rho^2+r^2+1)}{((\rho^2-r^2+1)^2+4r^2)^{3/2}} \frac{d\rho}{\sqrt{1+\rho^2}} \leq \frac{C}{r^4} \sinh^{-1}\left(\frac{r}{2}\right)$$

On the other hand, when $\rho \geq r/2$, we have

$$\begin{aligned}
&\int_{r/2}^{\infty} \frac{(\rho^2+r^2+1)}{((\rho^2-r^2+1)^2+4r^2)^{3/2}} \frac{d\rho}{\sqrt{1+\rho^2}} \\
&\leq C \int_{r/2}^{\infty} \frac{(\rho^2+r^2+1)}{\rho((\rho^2-r^2+1)^2+4r^2)^{3/2}} d\rho \\
&= \frac{r^2 + \frac{3r^4}{\sqrt{9r^4+40r^2+16}} - \frac{9r^2}{\sqrt{9r^4+40r^2+16}} - \frac{12}{\sqrt{9r^4+40r^2+16}}}{4\left(r^2+1\right)^2} \\
&+ \frac{2\log\left(3r^4+\left(\sqrt{9r^4+40r^2+16}+9\right)r^2+\sqrt{9r^4+40r^2+16}+4\right) - 4\log(r) + 1 - \log(4)}{4\left(r^2+1\right)^2} \\
&\leq \frac{C}{r^2}
\end{aligned}$$

So, we conclude that

$$\begin{aligned}
&\left|\int_t^{\infty} \frac{\lambda''(s)}{r} \int_0^{s-t} \frac{\rho}{\sqrt{(s-t)^2-\rho^2}} \frac{4r\left(\rho^2+r^2+1\right)}{\left(\left(\rho^2-r^2+1\right)^2+4r^2\right)^{3/2}} d\rho ds\right| \\
&\leq C \sup_{x\geq t}\left(|\lambda''(x)|(1+(x-t)^2)\right)\frac{1}{r^2}
\end{aligned}$$

Combining these estimates, we get

$$|\partial_r v_1(t,r)| \leq \frac{C}{r^2}\left(\sup_{x\geq t}\left(|\lambda''(x)|(1+(x-t)^2)\right) + \int_t^{\infty} |\lambda''(s)|(s-t)ds\right), \quad r>2$$

□

3.0.3. The free wave correction. The next correction is designed to produce a crucial source term in the equation that we will eventually use to choose a specific λ. In fact, it is this correction which ultimately determines the asymptotics of the λ which we will choose. Define v_2 to be the solution to

$$\begin{cases} -\partial_{tt}v_2 + \partial_{rr}v_2 + \frac{1}{r}\partial_r v_2 - \frac{v_2}{r^2} = 0 \\ v_2(0) = 0 \\ \partial_t v_2(0) = v_{2,0} \end{cases}$$

where

$$\widehat{v_{2,0}}(\xi) = \int_0^{\infty} v_{2,0}(r)J_1(r\xi)rdr = \begin{cases} \frac{4b}{\pi(b-1)} \frac{\chi_{\leq\frac{1}{4}}(\xi)}{\log^{b-1}(\frac{1}{\xi})}, & b\neq 1 \\ \frac{-4}{\pi}\chi_{\leq\frac{1}{4}}(\xi)\log(\log(\frac{1}{\xi})), & b=1 \end{cases}$$

where $\chi_{\leq\frac{1}{4}} \in C_c^{\infty}([0,\infty))$, $0 \leq \chi_{\leq\frac{1}{4}}(x) \leq 1$ for all x, and $\chi_{\leq\frac{1}{4}}$ satisfies

$$\chi_{\leq\frac{1}{4}}(x) = \begin{cases} 1, & 0\leq x\leq\frac{1}{8} \\ 0, & x>\frac{1}{4} \end{cases}$$

and is otherwise arbitrary. For ease of notation, let

$$c_b = \begin{cases} \frac{4b}{\pi(b-1)}, & b\neq 1 \\ \frac{-4}{\pi}, & b=1 \end{cases}$$

Note that this particular form of c_b is due to the fact that part of $\widehat{v_{2,0}}(\xi)$ involves an antiderivative of $\frac{1}{\xi \log^b(\frac{1}{\xi})}$. We have a formula for v_2, namely:

$$v_2(t,r) = c_b \int_0^\infty \sin(t\xi) J_1(r\xi) \chi_{\leq \frac{1}{4}}(\xi) \cdot \begin{cases} \frac{1}{\log^{b-1}(\frac{1}{\xi})}, & b \neq 1 \\ \log(\log(\frac{1}{\xi})), & b = 1 \end{cases} d\xi \tag{3.18}$$

The significance of this particular choice of Cauchy data will be seen later on, when we identify the v_2-related contribution to the equation we use to choose λ.

We will prove pointwise estimates on v_2 later on, but, to give the reader some idea of the behavior of v_2, we note that (for instance, for $b > 1$)

$$v_2(t,r) = \frac{c_b(1-\text{sgn}(t-r))}{2\sqrt{2r}\sqrt{|t-r|}\log^{b-1}(|t-r|)} + E_2(t,r)$$

with

$$|E_2(t,r)| \leq C\left(\frac{1}{\sqrt{r}\sqrt{|t-r|}\log^b(|t-r|)}\right) + \frac{C}{\sqrt{r}\sqrt{t+r}\log^{b-1}(t+r)} + \frac{C}{r\log^{b-1}(r)}$$

$$r > \frac{t}{2}, \quad |t-r| > 5$$

This can be established by a procedure similar to the one which we use later on to compute the inner product of the v_2 linear error term with $\phi_0(\frac{\cdot}{\lambda(t)})$.

3.0.4. Further improvement of the soliton error term. If we substitute $u = Q_{\frac{1}{\lambda(t)}} + v_1 + v_2 + u_3$ into the wave maps equation

$$-\partial_{tt}u + \partial_{rr}u + \frac{1}{r}\partial_r u - \frac{\sin(2u)}{2r^2} = 0$$

we get

$$\begin{aligned} & -\partial_{tt}u_3 + \partial_{rr}u_3 + \frac{1}{r}\partial_r u_3 - \frac{\cos(2Q_{\frac{1}{\lambda(t)}})}{r^2}u_3 \\ & = \partial_{tt}Q_{\frac{1}{\lambda(t)}} + \frac{2\lambda''(t)r}{1+r^2} + \left(\frac{\cos(2Q_{\frac{1}{\lambda(t)}})-1}{r^2}\right)v_1 + \left(\frac{\cos(2Q_{\frac{1}{\lambda(t)}})-1}{r^2}\right)v_2 \\ & + N_2(v_1+v_2) + N(u_3) + L(u_3) \end{aligned} \tag{3.19}$$

where

$$N(f) = \left(\frac{\sin(2f)-2f}{2r^2}\right)\cos(2Q_{\frac{1}{\lambda(t)}}) + \left(\frac{\cos(2f)-1}{2r^2}\right)\sin(2(Q_{\frac{1}{\lambda(t)}} + v_1 + v_2))$$

$$L(f) = \frac{\sin(2f)}{2r^2}\cos(2Q_{\frac{1}{\lambda(t)}})(\cos(2(v_1+v_2))-1) - \frac{\sin(2f)}{2r^2}\sin(2Q_{\frac{1}{\lambda(t)}})\sin(2(v_1+v_2))$$

$$N_2(f) = \frac{\sin(2Q_{\frac{1}{\lambda(t)}})}{2r^2}(\cos(2f)-1) + \frac{\cos(2Q_{\frac{1}{\lambda(t)}})}{2r^2}(\sin(2f)-2f)$$

Note that

$$F_0(t,r) = \partial_{tt}Q_{\frac{1}{\lambda(t)}} + \frac{2\lambda''(t)r}{1+r^2}$$

appears on the right-hand side of (3.19). When the spatial coordinate is renormalized, via

$$r = R\lambda(t)$$

one term arising in the large R expansion of F_0 has insufficient decay in time. To remedy this, we will add another correction, to be denoted v_3. On one hand, choosing v_3 to solve an equation whose right-hand side is exactly equal to F_0 would eventually lead to a much more difficult equation that we use to determine $\lambda(t)$. On the other hand, the error terms remaining after adding the correction v_3 should no longer have insufficient decay in time for large values of R. We therefore proceed as follows: recall that α has been introduced just before (3.1), and satisfies

$$0 < \alpha < \min\{\frac{1}{b(1040!)}, \frac{1}{1040!}\}$$

and let

$$F_{0,1}(t,r) = \frac{2r\lambda''(t)}{(\lambda(t)^2+r^2)}\left(\frac{-1+\lambda(t)^2}{1+r^2} + \frac{1-\lambda(t)^{2\alpha}}{1+r^2\lambda(t)^{2\alpha-2}}\right)$$

Next, we consider v_3, defined as the solution (with 0 Cauchy data at infinity) to the equation

$$-\partial_{tt}v_3 + \partial_{rr}v_3 + \frac{1}{r}\partial_r v_3 - \frac{v_3}{r^2} = F_{0,1}(t,r)$$

Following the same steps as for v_1, we get

$$v_3(t,r) = -\frac{1}{2\pi}\int_t^\infty \int_0^{s-t} \frac{\rho}{\sqrt{(s-t)^2-\rho^2}} \int_0^{2\pi} \frac{F_{0,1}(s,\sqrt{r^2+\rho^2+2r\rho\cos(\theta)})(r+\rho\cos(\theta))}{\sqrt{r^2+\rho^2+2r\rho\cos(\theta)}} d\theta d\rho ds$$

which gives

$$v_3(t,r) = -\frac{1}{r}\int_t^\infty \int_0^{s-t} \frac{\rho}{\sqrt{(s-t)^2-\rho^2}}\lambda''(s)\left(\frac{-1-\rho^2+r^2}{\sqrt{(1+\rho^2-r^2)^2+4r^2}} + \frac{\lambda(s)^2-(r^2-\rho^2)\lambda(s)^{2\alpha}}{\lambda(s)^2\sqrt{1+2(\rho^2+r^2)\lambda(s)^{2\alpha-2}+(\rho^2-r^2)^2\lambda(s)^{4\alpha-4}}}\right) d\rho ds \tag{3.20}$$

The main result of this section is a decomposition of v_3 which will be useful in understanding its contribution to the equation for λ:

LEMMA 3.3.

$$v_3(t,r) = -r\int_{t+6r}^\infty \lambda''(s)(s-t)\left(\frac{1}{1+(s-t)^2} - \frac{1}{\lambda(t)^{2-2\alpha}+(s-t)^2}\right) ds + E_5(t,r) \tag{3.21}$$

where

$$\begin{aligned}|E_5(t,r)| \leq\ & C\left(\sup_{x\geq t}|\lambda''(x)|\right)\cdot \sup_{x\geq t}\left(\frac{|\lambda'(x)|\lambda(t)^\alpha}{\lambda(x)^\alpha}\right)\lambda(t)^{1-2\alpha}\\ &+ Cr\sup_{x\geq t}\left(|\lambda''(x)|\lambda(x)^{\alpha-1}(\lambda(x)^{\alpha-1}-\lambda(t)^{\alpha-1})\right)\lambda(t)^{2-2\alpha}\\ &+ Cr\sup_{x\geq t}|\lambda''(x)|\end{aligned}$$

PROOF.

$$v_3(t,r) = -\frac{1}{r}\int_t^\infty \int_0^{s-t} \frac{\rho}{\sqrt{(s-t)^2-\rho^2}}\lambda''(s)\left(\frac{-1-\rho^2+r^2}{\sqrt{(1+\rho^2-r^2)^2+4r^2}} + \frac{\lambda(s)^2-(r^2-\rho^2)\lambda(s)^{2\alpha}}{\lambda(s)^2\sqrt{1+2(\rho^2+r^2)\lambda(s)^{2\alpha-2}+(\rho^2-r^2)^2\lambda(s)^{4\alpha-4}}}\right) d\rho ds$$

v_3 is then decomposed as

$$v_3(t,r) = v_{3,1}(t,r) + v_{3,2}(t,r)$$

where

$$\begin{aligned} v_{3,1}(t,r) = -\frac{1}{r}\int_t^\infty \int_0^{s-t} \frac{\rho}{(s-t)}\lambda''(s) \Bigg(&\frac{-1-\rho^2+r^2}{\sqrt{(1+\rho^2-r^2)^2+4r^2}} \\ &+ \frac{\lambda(s)^2-(r^2-\rho^2)\lambda(s)^{2\alpha}}{\lambda(s)^2\sqrt{1+2(\rho^2+r^2)\lambda(s)^{2\alpha-2}+(\rho^2-r^2)^2\lambda(s)^{4\alpha-4}}}\Bigg) d\rho ds \end{aligned} \tag{3.22}$$

$$v_{3,2} = v_3 - v_{3,1}$$

Then, we record some pointwise estimates on $v_{3,2}$. If

$$\begin{aligned} F_3(r,\rho,\lambda(s)) &= \frac{1-(r^2-\rho^2)\lambda(s)^{2\alpha-2}}{\sqrt{1+2(\rho^2+r^2)\lambda(s)^{2\alpha-2}+(\rho^2-r^2)^2\lambda(s)^{4\alpha-4}}} \\ &= \frac{1-(r^2-\rho^2)\lambda(s)^{2\alpha-2}}{\sqrt{4r^2\lambda(s)^{2\alpha-2}+(1-(r^2-\rho^2)\lambda(s)^{2\alpha-2})^2}} \end{aligned}$$

then,

$$\begin{aligned} &v_{3,2}(t,r) \\ &= \frac{-1}{r}\int_t^\infty\int_0^{s-t}\rho\left(\frac{1}{\sqrt{(s-t)^2-\rho^2}}-\frac{1}{(s-t)}\right)\lambda''(s)\left(\frac{-1-\rho^2+r^2}{\sqrt{(1+\rho^2-r^2)^2+4r^2}}+F_3(r,\rho,\lambda(t))\right)d\rho ds \\ &-\frac{1}{r}\int_t^\infty\int_0^{s-t}\rho\left(\frac{1}{\sqrt{(s-t)^2-\rho^2}}-\frac{1}{(s-t)}\right)\lambda''(s)\left(F_3(r,\rho,\lambda(s))-F_3(r,\rho,\lambda(t))\right)d\rho ds \end{aligned} \tag{3.23}$$

For the first line of (3.23), we have

$$\begin{aligned} &|-\frac{1}{r}\int_t^\infty\int_0^{s-t}\rho\left(\frac{1}{\sqrt{(s-t)^2-\rho^2}}-\frac{1}{s-t}\right)\lambda''(s)\left(\frac{-1-\rho^2+r^2}{\sqrt{(-1-\rho^2+r^2)^2+4r^2}}+F_3(r,\rho,\lambda(t))\right)d\rho ds| \\ &\leq \frac{C}{r}\left(\sup_{x\geq t}|\lambda''(x)|\right) \\ &\cdot\int_0^\infty \rho|\frac{-1-\rho^2+r^2}{\sqrt{(-1-\rho^2+r^2)^2+4r^2}}+1-1+F_3(r,\rho,\lambda(t))|\int_{t+\rho}^\infty\left(\frac{1}{\sqrt{(s-t)^2-\rho^2}}-\frac{1}{s-t}\right)dsd\rho \\ &\leq \frac{C}{r}\left(\sup_{x\geq t}|\lambda''(x)|\right)\left(\int_0^\infty\rho\left(1-\frac{1+\rho^2-r^2}{\sqrt{(-1-\rho^2+r^2)^2+4r^2}}\right)d\rho+\int_0^\infty\rho\left(1-F_3(r,\rho,\lambda(t))\right)d\rho\right) \\ &\leq Cr\left(\sup_{x\geq t}|\lambda''(x)|\right) \end{aligned}$$

where we used the facts that

$$\frac{-1-\rho^2+r^2}{\sqrt{(-1-\rho^2+r^2)^2+4r^2}} \geq -1$$

and

$$F_3(r,\rho,\lambda(t)) \leq 1$$

To estimate the second line of (3.23), we first note that

$$\begin{aligned} &F_3(r,\rho,\lambda(s)) - F_3(r,\rho,\lambda(t)) \\ &= \int_0^1 \frac{-4r^2 z_\sigma(1+(r^2-\rho^2)z_\sigma^2)}{(1+2(\rho^2+r^2)z_\sigma^2+(\rho^2-r^2)^2 z_\sigma^4)^{3/2}}\left(\lambda(s)^{\alpha-1}-\lambda(t)^{\alpha-1}\right)d\sigma \end{aligned}$$

where

$$z_\sigma = \sigma\lambda(s)^{\alpha-1} + (1-\sigma)\lambda(t)^{\alpha-1}$$

First, we note that

$$1 + 2(\rho^2 + r^2)z^2 + (\rho^2 - r^2)^2 z^4 = 4\rho^2 z^2 + (1 + (r^2 - \rho^2)z^2)^2$$

So,

$$\begin{aligned}&\left|\frac{-4r^2 z_\sigma(1 + (r^2 - \rho^2)z_\sigma^2)}{(1 + 2(\rho^2 + r^2)z_\sigma^2 + (\rho^2 - r^2)^2 z_\sigma^4)^{3/2}}\left(\lambda(s)^{\alpha-1} - \lambda(t)^{\alpha-1}\right)\right| \\ &\leq C\frac{r^2|z_\sigma|\left(\lambda(s)^{\alpha-1} - \lambda(t)^{\alpha-1}\right)}{(1 + 2(\rho^2 + r^2)z_\sigma^2 + (\rho^2 - r^2)^2 z_\sigma^4)}\end{aligned}$$

Since λ is a decreasing function, and $0 < \alpha < \frac{1}{8}$, we have

$$\lambda(t)^{\alpha-1} \leq |z_\sigma| = |\sigma\lambda(s)^{\alpha-1} + (1-\sigma)\lambda(t)^{\alpha-1}| \leq \lambda(s)^{\alpha-1}, \quad 0 \leq \sigma \leq 1, \quad s \geq t$$

So,

$$\begin{aligned}&|F_3(r, \rho, \lambda(s)) - F_3(r, \rho, \lambda(t))| \\ &\leq C\frac{r^2\lambda(s)^{\alpha-1}|\lambda(s)^{\alpha-1} - \lambda(t)^{\alpha-1}|}{(1 + 2(\rho^2 + r^2)\lambda(t)^{2\alpha-2} + (\rho^2 - r^2)^2\lambda(t)^{4\alpha-4})}, \quad s \geq t\end{aligned}$$

This gives

$$\begin{aligned}&\left|-\frac{1}{r}\int_t^\infty \int_0^{s-t} \rho\left(\frac{1}{\sqrt{(s-t)^2 - \rho^2}} - \frac{1}{(s-t)}\right)\lambda''(s)\left(F_3(r,\rho,\lambda(s)) - F_3(r,\rho,\lambda(t))\right)d\rho ds\right| \\ &\leq Cr\int_0^\infty \rho\int_{\rho+t}^\infty \left(\frac{1}{\sqrt{(s-t)^2 - \rho^2}} - \frac{1}{(s-t)}\right)|\lambda''(s)|\frac{\lambda(s)^{\alpha-1}|\lambda(s)^{\alpha-1} - \lambda(t)^{\alpha-1}|dsd\rho}{(1 + 2(\rho^2 + r^2)\lambda(t)^{2\alpha-2} + (\rho^2 - r^2)^2\lambda(t)^{4\alpha-4})} \\ &\leq Cr\sup_{x\geq t}\left(|\lambda''(x)|\lambda(x)^{\alpha-1}|\lambda(x)^{\alpha-1} - \lambda(t)^{\alpha-1}|\right)\int_0^\infty \frac{\rho d\rho}{(1 + 2(\rho^2 + r^2)\lambda(t)^{2\alpha-2} + (\rho^2 - r^2)^2\lambda(t)^{4\alpha-4})}\end{aligned}$$

Then, if $2r > \lambda(t)^{1-\alpha}$, we have

$$\begin{aligned}&\int_0^\infty \frac{\rho d\rho}{(1 + 2(\rho^2 + r^2)\lambda(t)^{2\alpha-2} + (\rho^2 - r^2)^2\lambda(t)^{4\alpha-4})} \\ &\leq C\left(\int_0^{\lambda(t)^{1-\alpha}} \rho d\rho + \int_{\lambda(t)^{1-\alpha}}^{2r} \frac{\rho d\rho}{r^2\lambda(t)^{2\alpha-2}} + \int_{2r}^\infty \frac{\rho d\rho}{\rho^4\lambda(t)^{4\alpha-4}}\right) \\ &\leq C\lambda(t)^{2-2\alpha}\end{aligned}$$

On the other hand, if $2r \leq \lambda(t)^{1-\alpha}$, we have

$$\begin{aligned}&\int_0^\infty \frac{\rho d\rho}{(1 + 2(\rho^2 + r^2)\lambda(t)^{2\alpha-2} + (\rho^2 - r^2)^2\lambda(t)^{4\alpha-4})} \\ &\leq \int_0^{\lambda(t)^{1-\alpha}} \rho d\rho + \int_{\lambda(t)^{1-\alpha}}^\infty \frac{d\rho}{\rho^3\lambda(t)^{4\alpha-4}} \\ &\leq C\lambda(t)^{2-2\alpha}\end{aligned}$$

In total, we get

$$\begin{aligned}&\left|-\frac{1}{r}\int_t^\infty \int_0^{s-t} \rho\left(\frac{1}{\sqrt{(s-t)^2 - \rho^2}} - \frac{1}{(s-t)}\right)\lambda''(s)\left(F_3(r,\rho,\lambda(s)) - F_3(r,\rho,\lambda(t))\right)d\rho ds\right| \\ &\leq Cr\sup_{x\geq t}\left(|\lambda''(x)|\lambda(x)^{\alpha-1}|\lambda(x)^{\alpha-1} - \lambda(t)^{\alpha-1}|\right)\lambda(t)^{2-2\alpha}\end{aligned}$$

It then suffices to study $v_{3,1}$. Firstly, we have

$$|-\frac{1}{r}\int_t^{t+6r}\int_0^{s-t}\frac{\rho}{(s-t)}\lambda''(s)\left(\frac{-1-\rho^2+r^2}{\sqrt{(-1-\rho^2+r^2)^2+4r^2}}+F_3(r,\rho,\lambda(s))\right)d\rho ds|$$
$$\leq \frac{1}{r}\int_t^{t+6r}\int_0^{s-t}\frac{\rho}{(s-t)}|\lambda''(s)|\cdot 2d\rho ds$$
$$\leq Cr\sup_{x\geq t}|\lambda''(x)|$$

Next, we have

$$-\frac{1}{r}\int_{t+6r}^{\infty}\int_0^{s-t}\frac{\rho}{(s-t)}\lambda''(s)\left(\frac{-1-\rho^2+r^2}{\sqrt{(-1-\rho^2+r^2)^2+4r^2}}+F_3(r,\rho,\lambda(s))\right)d\rho ds$$
$$=-2r\int_{6r}^{\infty}\lambda''(t+w)w\left(\frac{1}{(1+w^2)\left(\frac{r^2}{1+w^2}+1+\sqrt{(\frac{r^2}{1+w^2}+1)^2-4\frac{r^2w^2}{(1+w^2)^2}}\right)}\right.$$
$$\left.+\frac{-1}{(\lambda(t+w)^{2-2\alpha}+w^2)\left(\frac{r^2}{\lambda(t+w)^{2-2\alpha}+w^2}+1+\sqrt{(\frac{r^2}{\lambda(t+w)^{2-2\alpha}+w^2}+1)^2-\frac{4r^2w^2}{(\lambda(t+w)^{2-2\alpha}+w^2)^2}}\right)}\right)dw$$

Using the fact that $w=s-t\geq 6r$ in the integral below, we get

$$-\frac{1}{r}\int_{t+6r}^{\infty}\int_0^{s-t}\frac{\rho}{(s-t)}\lambda''(s)\left(\frac{-1-\rho^2+r^2}{\sqrt{(-1-\rho^2+r^2)^2+4r^2}}+F_3(r,\rho,\lambda(s))\right)d\rho ds$$
$$=-2r\int_{6r}^{\infty}\lambda''(t+w)w\left(\frac{1}{2(1+w^2)}-\frac{1}{2(\lambda(t+w)^{2-2\alpha}+w^2)}\right)dw+E_4$$

where

$$|E_4|\leq Cr\sup_{x\geq t}|\lambda''(x)|\int_{6r}^{\infty}\frac{r^2w}{w^2}\left(\frac{1}{1+w^2}+\frac{1}{\lambda(t+w)^{2-2\alpha}+w^2}\right)dw$$
$$\leq Cr\sup_{x\geq t}|\lambda''(x)|$$

So, it suffices to study

$$-2r\int_{6r}^{\infty}\lambda''(t+w)w\left(\frac{1}{2(1+w^2)}-\frac{1}{2(\lambda(t+w)^{2-2\alpha}+w^2)}\right)dw$$

We will make one more reduction, which is to replace $\lambda(t+w)$ in the above expression with $\lambda(t)$. The error in doing this replacement is

$$-r\int_{t+6r}^{\infty}\lambda''(s)(s-t)\left(\frac{1}{(\lambda(t)^{2-2\alpha}+(s-t)^2)}-\frac{1}{(\lambda(s)^{2-2\alpha}+(s-t)^2)}\right)ds$$

But, if

$$F_4(x,s-t)=\frac{1}{x^2+(s-t)^2}$$

then,

$$|F_4(\lambda(s)^{1-\alpha},s-t)-F_4(\lambda(t)^{1-\alpha},s-t)|\leq C\frac{\lambda(t)^{1-\alpha}}{(s-t)^4}\cdot|\lambda(t)^{1-\alpha}-\lambda(s)^{1-\alpha}|$$
$$\leq C\frac{\lambda(t)^{1-2\alpha}}{(s-t)^3}\cdot\sup_{x\geq t}\left(\frac{|\lambda'(x)|\lambda(t)^{\alpha}}{\lambda(x)^{\alpha}}\right)$$

where we use the fact that λ is a decreasing function. So,

$$\begin{aligned}&|-r\int_{t+6r}^{\infty}\lambda''(s)(s-t)\left(\frac{1}{(\lambda(t)^{2-2\alpha}+(s-t)^2)}-\frac{1}{(\lambda(s)^{2-2\alpha}+(s-t)^2)}\right)ds|\\&\leq Cr\left(\sup_{x\geq t}|\lambda''(x)|\right)\cdot\sup_{x\geq t}\left(\frac{|\lambda'(x)|\lambda(t)^{\alpha}}{\lambda(x)^{\alpha}}\right)\lambda(t)^{1-2\alpha}\int_{6r}^{\infty}\frac{w}{w^3}dw\\&\leq C\left(\sup_{x\geq t}|\lambda''(x)|\right)\cdot\sup_{x\geq t}\left(\frac{|\lambda'(x)|\lambda(t)^{\alpha}}{\lambda(x)^{\alpha}}\right)\lambda(t)^{1-2\alpha}\end{aligned}$$

This finally gives (3.21). □

3.0.5. The linear error terms for large r. Despite the decay of

$$\frac{1-\cos(2Q_1(\frac{r}{\lambda(t)}))}{r^2}$$

for large r, we will still need to add another correction which improves the linear error terms of v_1, v_2, v_3, as well as $F_{0,2}=-F_{0,1}+\frac{2\lambda''(t)r}{1+r^2}+\partial_t^2 Q_{\frac{1}{\lambda(t)}}$, for large r. The addition of this correction will not change the leading order contribution of these error terms to the modulation equation, but, will improve the overall error term of the final ansatz for large r. Let $\chi_{\geq 1}\in C^{\infty}(\mathbb{R})$ satisfy

$$\chi_{\geq 1}(x)=\begin{cases}1, & x\geq 2\\ 0, & x<1\end{cases}$$

and

$$0\leq\chi_{\geq 1}(x)\leq 1,\quad x\in\mathbb{R}$$

Then, we recall that N has been defined just before (3.1), let

$$v_{4,c}(t,r)=\chi_{\geq 1}(\frac{2r}{\log^N(t)})\left(\left(\frac{\cos(2Q_1(\frac{r}{\lambda(t)}))-1}{r^2}\right)(v_1+v_2+v_3)+F_{0,2}(t,r)\right)$$

and define v_4 as the solution to

$$-\partial_{tt}v_4+\partial_{rr}v_4+\frac{1}{r}\partial_r v_4-\frac{v_4}{r^2}=v_{4,c}(t,r)$$

with 0 Cauchy data at infinity. In other words, we have

$$v_4(t,r)=\int_t^{\infty}v_{4,s}(t,r)ds$$

where $v_{4,s}$ solves

$$\begin{cases}-\partial_{tt}v_{4,s}+\partial_{rr}v_{4,s}+\frac{1}{r}\partial_r v_{4,s}-\frac{v_{4,s}}{r^2}=0\\ v_{4,s}(s,r)=0\\ \partial_t v_{4,s}(s,r)=v_{4,c}(s,r)\end{cases}$$

So,

$$v_{4,s}(t,r)=\partial_r w_{4,s}(t,r)$$

where

$$u_{4,s}(t,y)=w_{4,s}(t,|y|)$$

and $u_{4,s} : [T_0, \infty) \times \mathbb{R}^2 \to \mathbb{R}$ solves

$$\begin{cases} \partial_{tt} u_{4,s} - \Delta u_{4,s} = 0 \\ u_{4,s}(s,x) = 0 \\ \partial_t u_{4,s}(s,x) = -\int_{|x|}^{\infty} v_{4,c}(s,q)dq \end{cases}$$

We get, for $s \geq t$,

$$u_{4,s}(t,x) = \frac{1}{2\pi} \int_{B_{s-t}(x)} \frac{\left(\int_{|y|}^{\infty} v_{4,c}(s,q)dq\right)}{\sqrt{(s-t)^2 - |y-x|^2}} dy = \frac{1}{2\pi} \int_{B_{s-t}(0)} \frac{\left(\int_{|z+x|}^{\infty} v_{4,c}(s,q)dq\right)}{\sqrt{(s-t)^2 - |z|^2}} dz$$

We use polar coordinates in the z variable, with origin 0, and polar axis $\hat{x}$ for $x \neq 0$. Then, we obtain, apriori for $r \neq 0$,

$$v_4(t,r) = \frac{-1}{2\pi} \int_t^{\infty} \int_0^{s-t} \frac{\rho}{\sqrt{(s-t)^2 - \rho^2}} \int_0^{2\pi} \frac{v_{4,c}(s, \sqrt{r^2 + 2r\rho\cos(\theta) + \rho^2})}{\sqrt{r^2 + 2r\rho\cos(\theta) + \rho^2}} (r + \rho\cos(\theta))\, d\theta d\rho ds$$

If we let

$$G(s,r,\rho) = \int_0^{2\pi} \frac{v_{4,c}(s, \sqrt{r^2 + 2r\rho\cos(\theta) + \rho^2})}{\sqrt{r^2 + 2r\rho\cos(\theta) + \rho^2}} (r + \rho\cos(\theta))\, d\theta \tag{3.24}$$
$$s \geq t, \quad r \geq 0, \quad s - t \geq \rho \geq 0$$

Then,

$$G(s,0,\rho) = 0$$

and

$$\begin{aligned} G(s,r,\rho) &= r \int_0^1 \partial_2 G(s, r\beta, \rho) d\beta \\ &= r \int_0^1 \int_0^{2\pi} \left(\partial_2 v_{4,c}(s, \sqrt{\beta^2 r^2 + 2\beta r\rho\cos(\theta) + \rho^2}) \frac{(\beta r + \rho\cos(\theta))^2}{\beta^2 r^2 + 2\beta r\rho\cos(\theta) + \rho^2} \right. \\ &\quad - v_{4,c}(s, \sqrt{\beta^2 r^2 + 2\beta r\rho\cos(\theta) + \rho^2}) \frac{(\beta r + \rho\cos(\theta))^2}{(\beta^2 r^2 + 2\beta r\rho\cos(\theta) + \rho^2)^{3/2}} \\ &\quad \left. + \frac{v_{4,c}(s, \sqrt{\beta^2 r^2 + 2\beta r\rho\cos(\theta) + \rho^2})}{\sqrt{\beta^2 r^2 + 2\beta r\rho\cos(\theta) + \rho^2}} \right) d\theta d\beta \end{aligned}$$

So, $v_4(t, \cdot)$ is (for instance) continuous on $[0, \infty)$ and we have, for all $r \geq 0$ (including $r = 0$)

$$v_4(t,r) = \frac{-r}{2\pi} \int_t^{\infty} \int_0^{s-t} \frac{\rho}{\sqrt{(s-t)^2 - \rho^2}} \int_0^1 \partial_2 G(s, r\beta, \rho) d\beta d\rho ds \tag{3.25}$$

We will use this formula to prove estimates on v_4, but this will be done later on, once we further restrict the class of functions λ under consideration (which will be done once we introduce an iteration space in which to solve the eventual equation for λ).

3.0.6. The nonlinear error terms involving v_1, v_2, v_3, v_4. Let

$$f_{v_5} = v_1 + v_2 + v_3 + v_4$$

and

$$N_2(f_{v_5})(t,r) = \frac{\sin(2Q_{\frac{1}{\lambda(t)}}(r))}{2r^2}(\cos(2f_{v_5})-1) + \frac{\cos(2Q_{\frac{1}{\lambda(t)}}(r))}{2r^2}(\sin(2f_{v_5})-2f_{v_5})$$

Then, we consider v_5, defined as the solution with 0 Cauchy data at infinity, to the problem

$$-\partial_{tt}v_5 + \partial_{rr}v_5 + \frac{1}{r}\partial_r v_5 - \frac{v_5}{r^2} = N_2(f_{v_5})(t,r)$$

Following the same steps used to obtain (3.25), we obtain the analogous formula for v_5:

$$v_5(t,r) = \frac{-r}{2\pi}\int_t^\infty \int_0^{s-t} \frac{\rho}{\sqrt{(s-t)^2-\rho^2}} \int_0^1 \partial_2 G_5(s, r\beta, \rho) d\beta d\rho ds$$

where

$$\begin{aligned}
&G_5(s,r,\rho)\\
&= r\int_0^1 \partial_2 G_5(s, r\beta, \rho) d\beta\\
&= r\int_0^1\int_0^{2\pi} \Bigg(\partial_2 N_2(f_{v_5})(s, \sqrt{\beta^2r^2+2\beta r\rho\cos(\theta)+\rho^2})\frac{(\beta r+\rho\cos(\theta))^2}{\beta^2r^2+2\beta r\rho\cos(\theta)+\rho^2}\\
&\quad - N_2(f_{v_5})(s, \sqrt{\beta^2r^2+2\beta r\rho\cos(\theta)+\rho^2})\frac{(\beta r+\rho\cos(\theta))^2}{(\beta^2r^2+2\beta r\rho\cos(\theta)+\rho^2)^{3/2}}\\
&\quad + \frac{N_2(f_{v_5})(s, \sqrt{\beta^2r^2+2\beta r\rho\cos(\theta)+\rho^2})}{\sqrt{\beta^2r^2+2\beta r\rho\cos(\theta)+\rho^2}}\Bigg) d\theta d\beta
\end{aligned}$$

As was the case for v_4, we will prove estimates on v_5 later on, once we further restrict the class of functions λ under consideration.

3.0.7. The equation resulting from u_{ansatz}. If we substitute

$$u = Q_{\frac{1}{\lambda(t)}} + v_1 + v_2 + v_3 + v_4 + v_5 + v_6$$

into the wave maps equation

$$-\partial_{tt}u + \partial_{rr}u + \frac{1}{r}\partial_r u - \frac{\sin(2u)}{2r^2} = 0$$

and use the equations solved by v_1, v_2, v_3, v_4, v_5, then, we obtain

(3.26)

$$\begin{aligned}
&-\partial_{tt}v_6+\partial_{rr}v_6+\frac{1}{r}\partial_r v_6-\frac{\cos(2Q_{\frac{1}{\lambda(t)}})}{r^2}v_6\\
&=F_{0,2}(t,r)+N(v_6)+L_1(v_6)+N_2(v_5)\\
&+\frac{\sin(2(v_1+v_2+v_3+v_4))}{2r^2}\left(\cos(2Q_{\frac{1}{\lambda(t)}}+2v_5)-\cos(2Q_{\frac{1}{\lambda(t)}})\right)\\
&+\left(\frac{\cos(2(v_1+v_2+v_3+v_4))-1}{2r^2}\right)\left(\sin(2Q_{\frac{1}{\lambda(t)}}+2v_5)-\sin(2Q_{\frac{1}{\lambda(t)}})\right)\\
&+\left(\frac{\cos(2Q_{\frac{1}{\lambda(t)}})-1}{r^2}\right)(v_1+v_2+v_3+v_4+v_5)\\
&-\chi_{\geq 1}(\frac{2r}{\log^N(t)})\left(\frac{\cos(2Q_{\frac{1}{\lambda(t)}})-1}{r^2}\right)(v_1+v_2+v_3)-\chi_{\geq 1}(\frac{2r}{\log^N(t)})F_{0,2}(t,r)
\end{aligned}$$

where

$$\begin{aligned}
N(f)&=\left(\frac{\sin(2f)-2f}{2r^2}\right)\cos(2Q_{\frac{1}{\lambda(t)}})\\
&+\left(\frac{\cos(2f)-1}{2r^2}\right)\sin(2(Q_{\frac{1}{\lambda(t)}}+v_1+v_2+v_3+v_4+v_5))
\end{aligned}$$

$$\begin{aligned}
L_1(f)&=\frac{\sin(2f)}{2r^2}\cos(2Q_{\frac{1}{\lambda(t)}})(\cos(2(v_1+v_2+v_3+v_4+v_5))-1)\\
&-\frac{\sin(2f)}{2r^2}\sin(2Q_{\frac{1}{\lambda(t)}})\sin(2(v_1+v_2+v_3+v_4+v_5))
\end{aligned}$$

$$N_2(f)=\frac{\sin(2Q_{\frac{1}{\lambda(t)}})}{2r^2}(\cos(2f)-1)+\frac{\cos(2Q_{\frac{1}{\lambda(t)}})}{2r^2}(\sin(2f)-2f)$$

$$F_{0,2}(t,r)=-F_{0,1}(t,r)+\frac{2\lambda''(t)r}{1+r^2}+\partial_t^2 Q_{\frac{1}{\lambda(t)}}$$

Note that we do not combine the terms involving $\chi_{\geq 1}$ with analogous terms having coefficient 1 because the terms involving $\chi_{\geq 1}$ will turn out to have a subleading contribution to the modulation equation for λ. When we solve the final equation, after choosing λ, we will of course make use of the fact that $1-\chi_{\geq 1}(x)$ is supported on the set $x\leq 2$.

We can re-write the right-hand side of (3.26) as $F+F_3$ where

$$F=F_4+F_5+F_6$$

with

(3.27)

$$\begin{aligned}
F_4(t,r)=F_{0,2}(t,r)&+\left(\frac{\cos(2Q_{\frac{1}{\lambda(t)}})-1}{r^2}\right)\left(v_1+v_2+v_3+\left(1-\chi_{\geq 1}(\frac{4r}{t})\right)(v_4+v_5)\right)\\
&-\chi_{\geq 1}(\frac{2r}{\log^N(t)})\left(\frac{\cos(2Q_{\frac{1}{\lambda(t)}})-1}{r^2}\right)(v_1+v_2+v_3)-\chi_{\geq 1}(\frac{2r}{\log^N(t)})F_{0,2}(t,r)
\end{aligned}$$

(3.28)

$$F_5(t,r) = N_2(v_5) + \frac{\sin(2(v_1+v_2+v_3+v_4))}{2r^2}\left(\cos(2Q_{\frac{1}{\lambda(t)}}+2v_5)-\cos(2Q_{\frac{1}{\lambda(t)}})\right) + \left(\frac{\cos(2(v_1+v_2+v_3+v_4))-1}{2r^2}\right)\left(\sin(2Q_{\frac{1}{\lambda(t)}}+2v_5)-\sin(2Q_{\frac{1}{\lambda(t)}})\right)$$

$$F_6(t,r) = \chi_{\geq 1}(\frac{4r}{t})\left(\frac{\cos(2Q_{\frac{1}{\lambda(t)}})-1}{r^2}\right)(v_4(t,r)+v_5(t,r)) \tag{3.29}$$

and

$$F_3 = N(v_6) + L_1(v_6)$$

3.0.8. Choosing $\lambda(t)$. λ will be chosen so that the term

$$\begin{aligned} &F_4(t,r) \\ &= F_{0,2}(t,r) \\ &+ \left(\frac{\cos(2Q_{\frac{1}{\lambda(t)}}(r))-1}{r^2}\right)\left(v_1+v_2+v_3+\left(1-\chi_{\geq 1}(\frac{4r}{t})\right)(v_4+v_5)\right) \\ &- \chi_{\geq 1}(\frac{2r}{\log^N(t)})\left(F_{0,2}(t,r)+(v_1(t,r)+v_2(t,r)+v_3(t,r))\left(\frac{\cos(2Q_{\frac{1}{\lambda(t)}}(r))-1}{r^2}\right)\right) \end{aligned}$$

which appears on the right-hand side of (3.26), is orthogonal to $\phi_0\left(\frac{\cdot}{\lambda(t)}\right)$.

Define the space $(X, ||\cdot||_X)$ by

$$X = \{f \in C^2([T_0,\infty))|||f||_X < \infty\}$$

where

(3.30)

$$||f||_X = \sup_{t\geq T_0}\left(|f(t)|b\log^b(t)\sqrt{\log(\log(t))}+|f'(t)|t\log^{b+1}(t)\sqrt{\log(\log(t))} + |f''(t)|t^2\log^{b+1}(t)\sqrt{\log(\log(t))}\right)$$

In this section, we will first prove the following proposition

PROPOSITION 3.1. *There exists $T_3 > 0$ such that, for all $T_0 \geq T_3$, there exists $\lambda \in C^2([T_0,\infty))$ which solves*

$$\langle F_4(t), \phi_0\left(\frac{\cdot}{\lambda(t)}\right)\rangle = 0, \quad t \geq T_0$$

Moreover,

$$\lambda(t) = \lambda_0(t) + e_0(t), \quad ||e_0||_X \leq 1$$

and

$$\lambda_0(t) = \frac{1}{\log^b(t)} + \int_t^\infty \int_{t_1}^\infty \frac{-b^2\log(\log(t_2))}{t_2^2\log^{b+2}(t_2)}dt_2dt_1$$

After fixing λ as in the above proposition, we will then show that λ, apriori only in $C^2([T_0,\infty))$, is actually in $C^4([T_0,\infty))$, with the estimates

$$|\lambda^{(2+k)}(t)| \leq \frac{C}{t^{2+k}\log^{b+1}(t)}, \quad t \geq T_0, \quad k = 1,2$$

To prove the proposition, we will first show that the equation

$$\langle F_4(t), \phi_0\left(\frac{\cdot}{\lambda(t)}\right)\rangle = 0$$

is equivalent to

$$\begin{aligned}
&-4\int_t^\infty \frac{\lambda''(s)}{1+s-t}ds + \frac{4b}{t^2\log^b(t)} + 4\alpha\log(\lambda(t))\lambda''(t)\\
&-4\int_t^\infty \frac{\lambda''(s)}{(\lambda(t)^{1-\alpha}+s-t)(1+s-t)^3}ds\\
&= -\lambda(t)E_{0,1}(\lambda(t),\lambda'(t),\lambda''(t))\\
&-16\int_t^\infty \lambda''(s)\left(K_3(s-t,\lambda(t)) - K_{3,0}(s-t,\lambda(t))\right)ds\\
&+\frac{16}{\lambda(t)^2}\int_t^\infty K(s-t,\lambda(t))\lambda''(s)ds - \lambda(t)E_{v_2,ip}(t,\lambda(t))\\
&+\frac{16}{\lambda(t)^2}\int_t^\infty \lambda''(s)\left(K_1(s-t,\lambda(t)) - \frac{\lambda(t)^2}{4(1+s-t)}\right)ds\\
&-\lambda(t)\langle\left(\frac{\cos(2Q_{\frac{1}{\lambda(t)}})-1}{r^2}\right)\left((v_4+v_5)\left(1-\chi_{\geq 1}(\frac{4r}{t})\right)+E_5\right)|_{r=R\lambda(t)},\phi_0\rangle\\
&+\lambda(t)\langle\left(\frac{\cos(2Q_{\frac{1}{\lambda(t)}})-1}{r^2}\right)\chi_{\geq 1}(\frac{2r}{\log^N(t)})\left(v_1+v_2+v_3\right)|_{r=R\lambda(t)},\phi_0\rangle\\
&+\lambda(t)\langle\chi_{\geq 1}(\frac{2r}{\log^N(t)})F_{0,2}(t,r)|_{r=R\lambda(t)},\phi_0\rangle\\
&:= G(t,\lambda(t))
\end{aligned} \tag{3.31}$$

Then, we will substitute $\lambda(t) = \lambda_0(t) + e_0(t)$, for $e_0 \in \overline{B_1(0)} \subset X$, and solve the resulting equation for e_0 with a fixed point argument.

We start by studying the relevant inner products of the v_1, v_2 and v_3 terms above.

3.0.8.1. *The inner product of the (rescaled) v_1 linear error term with ϕ_0.* In this section, we will prove

LEMMA 3.4. *For v_1 defined by (3.10), we have*

$$\begin{aligned}
&\langle\left(\frac{\cos(2Q_{\frac{1}{\lambda(t)}})-1}{r^2}\right)v_1|_{r=R\lambda(t)},\phi_0\rangle\\
&=\frac{-16}{\lambda(t)^3}\int_t^\infty \lambda''(s)K(s-t,\lambda(t))ds + \frac{-16}{\lambda(t)^3}\int_t^\infty \lambda''(s)K_1(s-t,\lambda(t))ds
\end{aligned}$$

where

$$\int_t^\infty |K(s-t,\lambda(t))|ds \leq C\lambda(t)^2$$

$$|K_1(x,\lambda(t))| \leq \frac{C\lambda(t)^2 x}{1+x^2}$$

$$\int_t^\infty |K_1(s-t,\lambda(t)) - \frac{\lambda(t)^2}{4(1+s-t)}|ds \leq C\lambda(t)^2 \tag{3.32}$$

$$|K_1(s-t,\lambda(t)) - \frac{\lambda(t)^2}{4(s-t)}| \leq \frac{C\lambda(t)^2(1+\lambda(t)^2)}{(s-t)(1+(s-t)^2)}, \quad s-t \geq 1$$

PROOF. We have

$$\langle \frac{(\cos(2Q_{\frac{1}{\lambda(t)}})-1)}{r^2} v_1\Big|_{r=R\lambda(t)}, \phi_0\rangle = \frac{-1}{\lambda(t)^2}\int_0^\infty v_1(t,R\lambda(t))\frac{8}{(1+R^2)^2}\frac{2R}{1+R^2}RdR$$
$$= -\frac{16}{\lambda(t)^2}\int_0^\infty v_1(t,R\lambda(t))\frac{R^2}{(1+R^2)^3}dR$$

(3.33)

$$-\frac{16}{\lambda(t)^2}\int_0^\infty v_1(t,R\lambda(t))\frac{R^2}{(1+R^2)^3}dR$$
$$= -\frac{16}{\lambda(t)^3}\int_t^\infty \lambda''(s)\int_0^\infty \frac{R}{(1+R^2)^3}\int_0^{s-t}\frac{\rho}{\sqrt{(s-t)^2-\rho^2}}\left(1+\frac{R^2\lambda(t)^2-1-\rho^2}{\sqrt{(R^2\lambda(t)^2-1-\rho^2)^2+4R^2\lambda(t)^2}}\right)d\rho dR ds$$
$$= -\frac{16}{\lambda(t)^3}\int_t^\infty \lambda''(s)\int_0^\infty \frac{R}{(1+R^2)^3}\int_0^{s-t}\frac{\rho}{s-t}\left(1+\frac{R^2\lambda(t)^2-1-\rho^2}{\sqrt{(R^2\lambda(t)^2-1-\rho^2)^2+4R^2\lambda(t)^2}}\right)d\rho dR ds$$
$$-\frac{16}{\lambda(t)^3}\int_t^\infty \lambda''(s)\int_0^\infty \frac{R}{(1+R^2)^3}\int_0^{s-t}\rho\left(\frac{1}{\sqrt{(s-t)^2-\rho^2}}-\frac{1}{s-t}\right)\left(1+\frac{R^2\lambda(t)^2-1-\rho^2}{\sqrt{(R^2\lambda(t)^2-1-\rho^2)^2+4R^2\lambda(t)^2}}\right)d\rho dR ds$$

The last line of (3.33) is of the form

$$-\frac{16}{\lambda(t)^3}\int_t^\infty \lambda''(s)K(s-t,\lambda(t))ds$$

where

$$K(x,\lambda(t))$$
$$= \int_0^\infty \frac{R}{(1+R^2)^3}\int_0^x \rho\left(\frac{1}{\sqrt{x^2-\rho^2}}-\frac{1}{x}\right)\left(1+\frac{R^2\lambda(t)^2-1-\rho^2}{\sqrt{(R^2\lambda(t)^2-1-\rho^2)^2+4R^2\lambda(t)^2}}\right)d\rho dR \geq 0$$

and

(3.34)

$$\int_t^\infty K(s-t,\lambda(t))ds$$
$$= \int_t^\infty\int_0^\infty \frac{R}{(1+R^2)^3}\int_0^{s-t}\rho\left(\frac{1}{\sqrt{(s-t)^2-\rho^2}}-\frac{1}{s-t}\right)\left(1+\frac{R^2\lambda(t)^2-1-\rho^2}{\sqrt{(R^2\lambda(t)^2-1-\rho^2)^2+4R^2\lambda(t)^2}}\right)d\rho dR ds$$
$$= \int_0^\infty \rho\int_0^\infty \frac{R}{(1+R^2)^3}\int_{\rho+t}^\infty\left(\frac{1}{\sqrt{(s-t)^2-\rho^2}}-\frac{1}{s-t}\right)\left(1+\frac{R^2\lambda(t)^2-1-\rho^2}{\sqrt{(R^2\lambda(t)^2-1-\rho^2)^2+4R^2\lambda(t)^2}}\right)ds dR d\rho$$
$$= \int_0^\infty \rho\int_0^\infty \frac{R}{(1+R^2)^3}\left(1+\frac{R^2\lambda(t)^2-1-\rho^2}{\sqrt{(R^2\lambda(t)^2-1-\rho^2)^2+4R^2\lambda(t)^2}}\right)\log(2)dR d\rho$$
$$= \int_0^\infty \frac{R}{(1+R^2)^3}\log(2)R^2\lambda(t)^2 dR$$
$$= \frac{\log(2)}{4}\lambda(t)^2$$

Hence, it remains to calculate

$$-\frac{16}{\lambda(t)^3}\int_t^\infty \lambda''(s)\int_0^\infty \frac{R}{(1+R^2)^3}\int_0^{s-t}\frac{\rho}{s-t}\left(1+\frac{R^2\lambda(t)^2-1-\rho^2}{\sqrt{(R^2\lambda(t)^2-1-\rho^2)^2+4R^2\lambda(t)^2}}\right)d\rho dR ds$$
$$= -\frac{16}{\lambda(t)^3}\int_t^\infty \lambda''(s)\int_0^\infty \frac{R}{(1+R^2)^3}\left(\frac{r^2-\sqrt{((r+s-t)^2+1)\left((r-s+t)^2+1\right)}+(s-t)^2+1}{2(s-t)}\right)|_{r=R\lambda(t)}dR ds$$
$$= -\frac{16}{\lambda(t)^3}\int_t^\infty \lambda''(s)K_1(s-t,\lambda(t))ds$$

where

$$
\begin{aligned}
(3.35)\quad & K_1(w,\lambda(t)) \\
&= \int_0^\infty \frac{R}{(1+R^2)^3}\left(\frac{R^2\lambda(t)^2+w^2+1-\sqrt{(1+R^2\lambda(t)^2+w^2)^2-4R^2\lambda(t)^2w^2}}{2w}\right)dR \\
&= \frac{w\lambda(t)^2\left(\lambda(t)^2+w^2+1\right)}{4(y^2+4w^2)} + \frac{w\lambda(t)^4\log\left(\lambda(t)^2\left(\sqrt{y^2+4w^2}-y\right)\right)}{2\left(y^2+4w^2\right)^{3/2}} \\
&\quad - w\lambda(t)^4\frac{\log\left(4w^2+y^2-\lambda(t)^2y+\left(w^2+1\right)\sqrt{y^2+4w^2}\right)}{2\left(y^2+4w^2\right)^{3/2}}
\end{aligned}
$$

where

$$y=\lambda(t)^2+w^2-1$$

and, for the first line of (3.35), we used the fact that

$$((r+w)^2+1)((r-w)^2+1)=(r^2+w^2+1)^2-4r^2w^2 \tag{3.36}$$

Now, we will prove a pointwise estimate on K_1. Using (3.36), we have

$$
\begin{aligned}
0 &\leq K_1(w,\lambda(t)) \\
&= \int_0^\infty \frac{R}{2w(1+R^2)^3}\left(\frac{4R^2\lambda(t)^2w^2}{1+R^2\lambda(t)^2+w^2+\sqrt{(1+(R\lambda(t)+w)^2)(1+(R\lambda(t)-w)^2)}}\right)dR \\
&\leq C\int_0^\infty \frac{R^3}{(1+R^2)^3}\frac{\lambda(t)^2w}{1+w^2}dR \leq C\lambda(t)^2\frac{w}{1+w^2}
\end{aligned}
$$

So,

$$|K_1(x,\lambda(t))| \leq C\lambda(t)^2\frac{x}{1+x^2} \tag{3.37}$$

For use later on, we will also need to estimate

$$\int_t^\infty |K_1(s-t,\lambda(t))-\frac{\lambda(t)^2}{4(1+s-t)}|ds$$

Note that $K_1(w,\lambda(t))\geq 0$, by its definition as the integral of a non-negative function. We start with

$$
\begin{aligned}
\int_t^{t+1}|K_1(s-t,\lambda(t))-\frac{\lambda(t)^2}{4(1+s-t)}|ds &\leq \int_t^{t+1}K_1(s-t,\lambda(t))ds+\int_t^{t+1}\frac{\lambda(t)^2}{4(1+s-t)}ds \\
&\leq C\lambda(t)^2
\end{aligned}
$$

where we used (3.37).

Now, we consider the region $s-t\geq 1$. Returning to (3.35), we see that

$$
\begin{aligned}
& K_1(w,\lambda(t)) \\
&= \int_0^\infty \frac{R}{(1+R^2)^3}\frac{R^2\lambda(t)^2}{w}dR \\
&\quad - \int_0^\infty \frac{R}{(1+R^2)^3}\frac{4R^2\lambda(t)^2}{2w\left(\sqrt{(-R^2\lambda(t)^2+w^2+1)^2+4R^2\lambda(t)^2}-R^2\lambda(t)^2+w^2+1\right)}dR
\end{aligned}
\tag{3.38}
$$

The right-hand side of the first line of (3.38) is equal to

$$\int_0^\infty \frac{R}{(1+R^2)^3}\frac{R^2\lambda(t)^2}{w}dR=\frac{\lambda(t)^2}{4w}$$

So,

$$K_1(w,\lambda(t)) - \frac{\lambda(t)^2}{4w}$$
$$= -\int_0^\infty \frac{R}{(1+R^2)^3} \frac{4R^2\lambda(t)^2}{2w\left(\sqrt{(1+w^2-R^2\lambda(t)^2)^2+4R^2\lambda(t)^2}+1+w^2-R^2\lambda(t)^2\right)} dR$$

First, we note that

$$1+w^2-R^2\lambda(t)^2 \geq C(1+w^2), \quad \text{if } R\lambda(t) \leq \frac{w}{2}$$

So,

$$\int_0^{\frac{w}{2\lambda(t)}} \frac{R}{(1+R^2)^3} \frac{4R^2\lambda(t)^2}{2w\left(\sqrt{(1+w^2-R^2\lambda(t)^2)^2+4R^2\lambda(t)^2}+1+w^2-R^2\lambda(t)^2\right)} dR$$
$$\leq C\int_0^\infty \frac{R}{(1+R^2)^3} \frac{R^2\lambda(t)^2}{w(1+w^2)} dR \leq \frac{C\lambda(t)^2}{w(1+w^2)}, \quad w \geq 1$$

Next,

$$\frac{1}{\sqrt{(1+w^2-R^2\lambda(t)^2)^2+4R^2\lambda(t)^2}+1+w^2-R^2\lambda(t)^2}$$
$$= \frac{1+w^2-R^2\lambda(t)^2-\sqrt{(1+w^2-R^2\lambda(t)^2)^2+4R^2\lambda(t)^2}}{-4R^2\lambda(t)^2}$$

and

$$\left|\frac{1+w^2-R^2\lambda(t)^2-\sqrt{(1+w^2-R^2\lambda(t)^2)^2+4R^2\lambda(t)^2}}{-4R^2\lambda(t)^2}\right| \leq C\frac{R^2\lambda(t)^2}{R^2\lambda(t)^2} \leq C,$$
$$\text{if} \quad R\lambda(t) > \frac{w}{2}, \quad w \geq 1$$

So,

$$\int_{\frac{w}{2\lambda(t)}}^\infty \frac{R}{(1+R^2)^3} \frac{4R^2\lambda(t)^2}{2w\left(\sqrt{(1+w^2-R^2\lambda(t)^2)^2+4R^2\lambda(t)^2}+1+w^2-R^2\lambda(t)^2\right)} dR$$
$$\leq \int_{\frac{w}{2\lambda(t)}}^\infty \frac{1}{R^5}\frac{R^2\lambda(t)^2}{w} dR \leq \frac{C\lambda(t)^4}{w^3}, \quad w \geq 1$$

Combining these, we get

$$|K_1(w,\lambda(t)) - \frac{\lambda(t)^2}{4w}| \leq \frac{C\lambda(t)^2(1+\lambda(t)^2)}{w(1+w^2)}, \quad w \geq 1 \tag{3.39}$$

So,

$$\int_{t+1}^\infty |K_1(s-t,\lambda(t)) - \frac{\lambda(t)^2}{4(1+s-t)}| ds \leq C\int_{t+1}^\infty \frac{\lambda(t)^2}{4}\left(\frac{1}{s-t}-\frac{1}{1+s-t}\right) ds$$
$$+ C\int_{t+1}^\infty \frac{C\lambda(t)^2(1+\lambda(t)^2)}{(s-t)(1+(s-t)^2)} ds$$
$$\leq C\lambda(t)^2(1+\lambda(t)^2)$$

Recalling (3.2), we conclude

$$\int_t^\infty |K_1(s-t,\lambda(t)) - \frac{\lambda(t)^2}{4(1+s-t)}|ds \leq C\lambda(t)^2 \qquad \square$$

3.0.8.2. *The inner product of the (rescaled) v_2 linear error term with ϕ_0.*

LEMMA 3.5. *For all $b>0$, and v_2 defined by (3.18), we have*

$$\int_0^\infty R\left(\frac{\cos(2Q_1(R))-1}{R^2\lambda(t)^2}\right)\phi_0(R)v_2(t,R\lambda(t))dR = \frac{4b}{\lambda(t)t^2\log^b(t)} + E_{v_2,ip}(t,\lambda(t))$$

where

$$|E_{v_2,ip}(t,\lambda(t))| \leq \frac{C}{\lambda(t)t^2\log^{b+1}(t)} \tag{3.40}$$

PROOF. We start with the case $b \neq 1$. Using our formula for v_2, we get

$$\begin{aligned}&\int_0^\infty R\left(\frac{\cos(2Q_1(R))-1}{R^2\lambda(t)^2}\right)\phi_0(R)v_2(t,R\lambda(t))dR\\ &= -2c_b\int_0^\infty \xi^2\sin(t\xi)\frac{\chi_{\leq\frac{1}{4}}(\xi)}{\log^{b-1}(\frac{1}{\xi})}K_1(\xi\lambda(t))d\xi\end{aligned}$$

where K_1 denotes the modified Bessel function of the second kind. (This follows, for instance, from equation (**6.532** 4) of the table of integrals [**6**]). Recalling that

$$K_1(x) = \frac{1}{x} + O(x\log(x)), \quad x\to 0$$

we can integrate by parts two times, and get

$$\begin{aligned}&-2c_b\int_0^\infty \xi^2\sin(t\xi)\frac{\chi_{\leq\frac{1}{4}}(\xi)}{\log^{b-1}(\frac{1}{\xi})}K_1(\xi\lambda(t))d\xi\\ &=2c_b\int_0^\infty \frac{\sin(t\xi)}{t^2}\partial_\xi^2\left(\xi^2\frac{\chi_{\leq\frac{1}{4}}(\xi)}{\log^{b-1}(\frac{1}{\xi})}K_1(\xi\lambda(t))\right)d\xi\end{aligned}$$

Let

$$F_b(\xi) = \partial_\xi^2\left(\xi^2\frac{\chi_{\leq\frac{1}{4}}(\xi)}{\log^{b-1}(\frac{1}{\xi})}K_1(\xi\lambda(t))\right)$$

Note that

$$F_b(\xi) = \chi_{\leq\frac{1}{4}}(\xi)\partial_\xi^2\left(\xi^2\frac{K_1(\xi\lambda(t))}{\log^{b-1}(\frac{1}{\xi})}\right) + \psi_{v_2}(\xi,\lambda(t)) \tag{3.41}$$

where

$$\psi_{v_2}\in C_c^\infty([\frac{1}{8},\frac{1}{4}])$$

So, integration by parts (for instance, once) gives

$$|2c_b\int_0^\infty \frac{\sin(t\xi)}{t^2}\psi_{v_2}(\xi,\lambda(t))d\xi| \leq \frac{C}{t^3\lambda(t)}$$

To study

$$2c_b\int_0^\infty \chi_{\leq\frac{1}{4}}(\xi)\frac{\sin(t\xi)}{t^2}\partial_\xi^2\left(\xi^2\frac{K_1(\xi\lambda(t))}{\log^{b-1}(\frac{1}{\xi})}\right)d\xi$$

we can use the asymptotics of K_1 to write

$$\partial_\xi^2\left(\xi^2\frac{K_1(\xi\lambda(t))}{\log^{b-1}(\frac{1}{\xi})}\right)=\partial_\xi^2\left(\frac{\xi}{\lambda(t)\log^{b-1}(\frac{1}{\xi})}\right)+F_{v_2}(\xi,\lambda(t)),\quad \xi\leq\frac{1}{4}\tag{3.42}$$

where

$$\begin{aligned}F_{v_2}(\xi,y)=&\frac{\xi^2y^2\log\left(\frac{1}{\xi}\right)\left(-3\log\left(\frac{1}{\xi}\right)-2b+2\right)K_0(y\xi)}{\xi y\log^{b+1}(\frac{1}{\xi})}\\&+\frac{(b-1)\left(\log\left(\frac{1}{\xi}\right)+b\right)(\xi yK_1(y\xi)-1)}{\xi y\log^{b+1}(\frac{1}{\xi})}\\&+\frac{\xi^3y^3\log^2\left(\frac{1}{\xi}\right)K_1(y\xi)}{\log^{b+1}(\frac{1}{\xi})\xi y},\quad \xi\leq\frac{1}{4}\end{aligned}$$

Using the simple estimates, valid for all $x>0$,

$$\begin{aligned}|-1+xK_1(x)|&\leq Cx^2(|\log(x)|+1)\\|K_0(x)|&\leq C(|\log(x)|+1)\\|xK_1(x)|&\leq C\end{aligned}\tag{3.43}$$

we get, for any $\lambda(t)>0$,

$$|F_{v_2}(\xi,\lambda(t))|\leq C\frac{\xi\lambda(t)}{\log^{b-1}(\frac{1}{\xi})}\left(|\log(\xi)|+|\log(\lambda(t))|\right),\quad \xi\leq\frac{1}{4}$$

Similarly, we have

$$\begin{aligned}\partial_\xi F_{v_2}(\xi,y)=&\frac{(b-1)\left(-\log^2\left(\frac{1}{\xi}\right)+b^2+b\right)\log^{-b-2}\left(\frac{1}{\xi}\right)(\xi yK_1(y\xi)-1)}{\xi^2y}\\&+y\frac{\left(\xi y\log\left(\frac{1}{\xi}\right)\left(4\log\left(\frac{1}{\xi}\right)+3b-3\right)K_1(y\xi)\right)}{\log^{b+1}\left(\frac{1}{\xi}\right)}\\&-yK_0(y\xi)\frac{\left(6(b-1)\log\left(\frac{1}{\xi}\right)+3(b-1)b+\log^2\left(\frac{1}{\xi}\right)(\xi^2y^2+3)\right)}{\log^{b+1}(\frac{1}{\xi})}\end{aligned}$$

Again, using (3.43), we get, for all $\lambda(t)>0$,

$$|\partial_\xi F_{v_2}(\xi,\lambda(t))|\leq\frac{C\lambda(t)\left(|\log(\xi)|+|\log(\lambda(t))|\right)}{\log^{b-1}(\frac{1}{\xi})}\left(1+\lambda(t)^2\xi^2\right),\quad \xi\leq\frac{1}{4}$$

So, we can integrate by parts (for instance) once, and recall (3.2), to get

$$|2c_b\int_0^\infty\chi_{\leq\frac{1}{4}}(\xi)\frac{\sin(t\xi)}{t^2}F_{v_2}(\xi,\lambda(t))d\xi|\leq\frac{C\lambda(t)|\log(\lambda(t))|}{t^3}$$

Next, we need to consider

$$2c_b\int_0^\infty \chi_{\leq\frac{1}{4}}(\xi)\frac{\sin(t\xi)}{t^2}\partial_\xi^2\left(\frac{\xi}{\lambda(t)\log^{b-1}(\frac{1}{\xi})}\right)d\xi$$
$$=\frac{2c_b}{\lambda(t)}\int_0^{1/2}\chi_{\leq\frac{1}{4}}(\xi)\frac{\sin(t\xi)}{t^2}\left(\frac{b-1}{\xi\log^b(\frac{1}{\xi})}+\frac{b(b-1)}{\xi\log^{b+1}(\frac{1}{\xi})}\right)d\xi$$

From one integration by parts we see that

$$|\frac{2c_b}{\lambda(t)}\int_0^{1/2}\left(\chi_{\leq\frac{1}{4}}(\xi)-1\right)\frac{\sin(t\xi)}{t^2}\left(\frac{b-1}{\xi\log^b(\frac{1}{\xi})}+\frac{b(b-1)}{\xi\log^{b+1}(\frac{1}{\xi})}\right)d\xi|\leq\frac{C}{t^3\lambda(t)}$$

So, we need only consider

$$\begin{aligned}&\frac{2c_b}{\lambda(t)}\int_0^{1/2}\frac{\sin(t\xi)}{t^2}\left(\frac{b-1}{\xi\log^b(\frac{1}{\xi})}+\frac{b(b-1)}{\xi\log^{b+1}(\frac{1}{\xi})}\right)d\xi\\&=\frac{2c_b}{\lambda(t)}\int_0^{t/2}\frac{\sin(u)}{t^2}\left(\frac{b-1}{u\log^b(\frac{t}{u})}+\frac{b(b-1)}{u\log^{b+1}(\frac{t}{u})}\right)du\end{aligned}\tag{3.44}$$

Let us start by studying, for $a>0$,

$$\int_0^{t/2}\frac{\sin(u)}{u\log^a(t/u)}du$$

Let

$$f(z)=\frac{e^{iz}}{z\log^a(t/z)}$$

where we use the principal branch of log and $(\cdot)^a$ Then, f is analytic on (for instance) D given by

$$D=\mathbb{C}\setminus((-\infty,0]\cup[2t/3,\infty))$$

For $0<\epsilon\leq\frac{1}{2}$, consider the contour $C_{t,\epsilon}$ in D given by:

$$C_{t,\epsilon}=\{\frac{t}{2}e^{i\theta}|0\leq\theta\leq\frac{\pi}{2}\}\cup\{iy|\epsilon\leq y\leq\frac{t}{2}\}\cup\{\epsilon e^{i\theta}|0\leq\theta\leq\frac{\pi}{2}\}\cup[\epsilon,t/2]$$

traversed in the counter-clockwise direction. By Cauchy's residue theorem,

$$\int_{C_{t,\epsilon}}f(z)dz=0$$

On the other hand,

$$\begin{aligned}\int_{C_{t,\epsilon}}f(z)dz=&\int_0^{\frac{\pi}{2}}i\frac{e^{\frac{it}{2}\cos(\theta)}e^{-\frac{t}{2}\sin(\theta)}}{\log^a(2e^{-i\theta})}d\theta-\int_\epsilon^{t/2}\frac{e^{-y}}{y\log^a(\frac{t}{iy})}dy-\int_0^{\frac{\pi}{2}}i\frac{e^{i\epsilon e^{i\theta}}}{\log^a(\frac{te^{-i\theta}}{\epsilon})}d\theta\\&+\int_\epsilon^{t/2}\frac{e^{iu}}{u\log^a(t/u)}du\end{aligned}$$

So,

$$\begin{aligned}\int_\epsilon^{t/2}\frac{\sin(u)}{u\log^a(t/u)}du=&\operatorname{Im}\left(\int_\epsilon^{t/2}\frac{e^{-y}}{y\log^a(\frac{t}{iy})}dy\right)+\operatorname{Im}\left(\int_0^{\frac{\pi}{2}}i\frac{e^{i\epsilon e^{i\theta}}}{\log^a(\frac{te^{-i\theta}}{\epsilon})}d\theta\right)\\&-\operatorname{Im}\left(\int_0^{\frac{\pi}{2}}i\frac{e^{\frac{it}{2}\cos(\theta)}e^{-\frac{t}{2}\sin(\theta)}}{\log^a(2e^{-i\theta})}d\theta\right)\end{aligned}$$

Letting $\epsilon \to 0$, we have

$$\begin{aligned}\int_0^{t/2} \frac{\sin(u)}{u\log^a(t/u)}du &= \lim_{\epsilon\to 0} \operatorname{Im}\left(\int_\epsilon^{t/2} \frac{e^{-y}}{y\log^a(\frac{t}{iy})}dy\right) \\ &\quad - \operatorname{Im}\left(\int_0^{\frac{\pi}{2}} i\frac{e^{\frac{it}{2}\cos(\theta)}e^{-\frac{t}{2}\sin(\theta)}}{\log^a(2e^{-i\theta})}d\theta\right)\end{aligned} \tag{3.45}$$

Note that

$$\operatorname{Im}\left(\int_\epsilon^{t/2} \frac{e^{-y}}{y\log^a(\frac{t}{iy})}dy\right) = \int_\epsilon^{t/2} \frac{e^{-y}\sin(a\tan^{-1}(\frac{\pi}{2(\log(t)-\log(y))}))}{y((\log(t)-\log(y))^2+\frac{\pi^2}{4})^{a/2}}dy$$

So,

$$\lim_{\epsilon\to 0}\operatorname{Im}\left(\int_\epsilon^{t/2} \frac{e^{-y}}{y\log^a(\frac{t}{iy})}dy\right) = \int_0^{t/2} \frac{e^{-y}\sin(a\tan^{-1}(\frac{\pi}{2(\log(t)-\log(y))}))}{y((\log(t)-\log(y))^2+\frac{\pi^2}{4})^{a/2}}dy$$

We have

$$\frac{e^{-y}\sin(a\tan^{-1}(\frac{\pi}{2(\log(t)-\log(y))}))}{y((\log(t)-\log(y))^2+\frac{\pi^2}{4})^{a/2}} = a\frac{\pi}{2}\frac{e^{-y}}{y\log^{a+1}(t/y)} + \text{Err}, \quad 0 \le y \le t/2$$

where

$$|\text{Err}| \le C\frac{e^{-y}}{y}\frac{1}{\log^{3+a}(t/y)}, \quad 0 \le y \le t/2$$

So,

$$\begin{aligned}\int_0^{t/2} \frac{e^{-y}\sin(a\tan^{-1}(\frac{\pi}{2(\log(t)-\log(y))}))}{y((\log(t)-\log(y))^2+\frac{\pi^2}{4})^{a/2}}dy &= a\frac{\pi}{2}\int_0^{t/2}\frac{e^{-y}}{y\log^{a+1}(t/y)}dy \\ &\quad + \int_0^{t/2}\text{Err}\,dy\end{aligned} \tag{3.46}$$

Let us start by considering

$$\int_0^1 \frac{1}{y}\frac{e^{-y}dy}{\log^{a+1}(t/y)} = \int_0^1 \frac{dy}{y\log^{a+1}(t/y)} + \int_0^1 \frac{(e^{-y}-1)}{y\log^{a+1}(\frac{t}{y})}dy$$

Treating each term individually, we have:

$$\int_0^1 \frac{dy}{y\log^{a+1}(t/y)} = \frac{1}{a\log^a(t)}$$

$$|\int_0^1 \frac{(e^{-y}-1)}{y\log^{a+1}(\frac{t}{y})}dy| \le \frac{C}{\log^{a+1}(t)}\int_0^1\frac{|e^{-y}-1|}{y}dy \le \frac{C}{\log^{a+1}(t)}$$

where we used the fact that

$$y \mapsto \frac{1}{\log^{a+1}(\frac{t}{y})}, \text{ is increasing on } (0,1]$$

So,

$$\int_0^1 \frac{1}{y}\frac{e^{-y}dy}{\log^{a+1}(t/y)} = \frac{1}{a\log^a(t)} + O\left(\frac{1}{\log^{a+1}(t)}\right)$$

Also,

$$\int_1^{\sqrt{t}} \frac{1}{y} \frac{e^{-y} dy}{(\log(t) - \log(y))^{a+1}} \leq 2^{a+1} \int_1^{\sqrt{t}} \frac{1}{y} \frac{e^{-y} dy}{\log^{a+1}(t)} = O(\frac{1}{\log^{a+1}(t)})$$

Finally,

$$\int_{\sqrt{t}}^{t/2} \frac{1}{y} \frac{e^{-y} dy}{(\log(t) - \log(y))^{a+1}} \leq \frac{1}{\sqrt{t}} \int_{\sqrt{t}}^{t/2} \frac{e^{-y}}{\log^{a+1}(2)} dy = O(\frac{e^{-\sqrt{t}}}{\sqrt{t}})$$

So,

$$\int_0^{t/2} \frac{1}{y} \frac{e^{-y} dy}{\log^{a+1}(t/y)} = \frac{1}{a \log^a(t)} + O\left(\frac{1}{\log^{a+1}(t)}\right) \tag{3.47}$$

Treating the Err term in (3.46) in the same way as was used to obtain (3.47), we have

$$\begin{aligned} \int_0^{t/2} \frac{e^{-y} \sin(a \tan^{-1}(\frac{\pi}{2(\log(t)-\log(y))}))}{y((\log(t) - \log(y))^2 + \frac{\pi^2}{4})^{a/2}} dy &= a\frac{\pi}{2} \left(\frac{1}{a \log^a(t)} + O\left(\frac{1}{\log^{a+1}(t)}\right)\right) \\ &\quad + O\left(\frac{1}{\log^{a+2}(t)}\right) \\ &= \frac{\pi}{2 \log^a(t)} + O\left(\frac{1}{\log^{a+1}(t)}\right) \end{aligned}$$

Lastly, let us estimate

$$\begin{aligned} |-\text{Im}\left(\int_0^{\frac{\pi}{2}} i \frac{e^{\frac{it}{2}\cos(\theta)} e^{-\frac{t}{2}\sin(\theta)}}{\log^a(2e^{-i\theta})} d\theta\right)| &\leq C \int_0^{\frac{\pi}{2}} \frac{e^{-\frac{t}{2}\sin(\theta)}}{\log^a(2)} d\theta \\ &\leq C \int_0^{\frac{\pi}{2}} e^{-\frac{t}{2}\frac{2}{\pi}\theta} d\theta = O\left(\frac{1}{t}\right) \end{aligned}$$

So, (3.45) can be written as

$$\int_0^{t/2} \frac{\sin(u)}{u \log^a(t/u)} du = \frac{\pi}{2 \log^a(t)} + O\left(\frac{1}{\log^{a+1}(t)}\right)$$

Since (3.45) is valid for all $a > 0$, we can repeat the identical steps done in this section for the case $b = 1$, and apply the above result to (3.44) (and its analog for $b = 1$), thereby obtaining, for all $b > 0$:

$$\int_0^\infty R \left(\frac{\cos(2Q_1(R)) - 1}{R^2 \lambda(t)^2}\right) \phi_0(R) v_2(t, R\lambda(t)) dR = \frac{4b}{\lambda(t) t^2 \log^b(t)} + E_{v_2,ip}(t, \lambda(t))$$

where, for $b \neq 1$:

(3.48)

$$\begin{aligned}E_{v_2,ip}(t,\lambda(t)) &= 2c_b \int_0^\infty \frac{\sin(t\xi)}{t^2}\psi_{v_2}(\xi,\lambda(t))d\xi \\ &+ 2c_b \int_0^\infty \chi_{\leq\frac{1}{4}}(\xi)\frac{\sin(t\xi)}{t^2}F_{v_2}(\xi,\lambda(t))d\xi \\ &+ 2\frac{c_b}{\lambda(t)}\int_0^{\frac{1}{2}}\left(\chi_{\leq\frac{1}{4}}(\xi)-1\right)\frac{\sin(t\xi)}{t^2}\left(\frac{b-1}{\xi\log^b(\frac{1}{\xi})}+\frac{b(b-1)}{\xi\log^{b+1}(\frac{1}{\xi})}\right)d\xi \\ &+ \frac{2c_b}{\lambda(t)}\left(\int_0^{\frac{t}{2}}\frac{\sin(u)(b-1)}{t^2u\log^b(\frac{t}{u})}du - \frac{(b-1)\pi}{2t^2\log^b(t)} + \int_0^{\frac{t}{2}}\frac{\sin(u)b(b-1)}{t^2u\log^{b+1}(\frac{t}{u})}du\right)\end{aligned}$$

When $b=1$, $E_{v_2,ip}$ has the same form as (3.48), except the third and fourth lines changed to

$$\begin{aligned}&\frac{2c_1}{\lambda(t)}\int_0^{\frac{1}{2}}\left(\chi_{\leq\frac{1}{4}}(\xi)-1\right)\frac{\sin(t\xi)}{\xi t^2}\left(-\frac{1}{\log^2(\frac{1}{\xi})}-\frac{1}{\log(\frac{1}{\xi})}\right)d\xi \\ &+\frac{2c_1}{\lambda(t)}\left(\int_0^{\frac{1}{2}}\frac{\sin(t\xi)}{\xi t^2}\left(\frac{-1}{\log(\frac{1}{\xi})}-\frac{1}{\log^2(\frac{1}{\xi})}\right)d\xi+\frac{\pi}{2t^2\log(t)}\right)\end{aligned}$$

For all $b>0$,

$$|E_{v_2,ip}(t,\lambda(t))| \leq \frac{C}{\lambda(t)t^2\log^{b+1}(t)}$$ □

3.0.8.3. *Pointwise estimates on $\partial_t^i\partial_r^j v_2$.*

LEMMA 3.6. *For any $b>0$, there exists $C>0$ such that For $0\leq j\leq 2$, $0\leq k\leq 1$, $j+k\leq 2$,*

$$\partial_t^j\partial_r^k v_2(t,r) = \frac{\partial_t^j\partial_r^k\left(\frac{-br}{t^2}\right)}{\log^b(t)} + E_{\partial_t^j\partial_r^k v_2}(t,r) \tag{3.49}$$

where

$$|E_{\partial_t^j\partial_r^k v_2}(t,r)| \leq C\partial_t^j\partial_r^k\left(\frac{r}{t^2\log^{b+1}(t)}\right)+C\partial_t^j\partial_r^k\left(\frac{r^2}{t^3\log^b(t)}\right), \quad r\leq\frac{t}{2}$$

For $0\leq j\leq 2$, $0\leq k\leq 2$, $j+k\leq 2$,

$$|\partial_t^j\partial_r^k v_2(t,r)| \leq \frac{C}{\sqrt{r}}, \quad r\geq\frac{t}{2} \tag{3.50}$$

For $0\leq j\leq 1$, $0\leq k\leq 1$, $j+k\leq 2$,

$$|\partial_t^j\partial_r^k v_2(t,r)| \leq \frac{C\log(r)}{|t-r|^{1+j+k}}, \quad r\geq\frac{t}{2}, \quad r\neq t \tag{3.51}$$

Finally, we also have the estimates

$$|\partial_r^2 v_2(t,r)| \leq \begin{cases}\frac{Cr}{t^4\log^b(t)}, & r\leq\frac{t}{2}\\ \frac{C\log(r)}{|t-r|^3}, & \frac{t}{2}<r\neq t\end{cases} \tag{3.52}$$

PROOF. In the course of proving these estimates, we will occasionally make use of the following formula, which can be found, for example, in Appendix B of [**10**]: for $n > -\frac{1}{2}$

$$J_n(x) = \frac{1}{\Gamma(n+\frac{1}{2})\sqrt{\pi}} \left(\frac{x}{2}\right)^n \int_0^{\pi} \cos(x\cos(\theta))\sin^{2n}(\theta)d\theta \tag{3.53}$$

Similarly, we will need, Lemma 8.1 of [**10**], which states

$$J_{\frac{d-2}{2}}(x) = \frac{\left(\frac{x}{2}\right)^{\frac{d-2}{2}}}{\pi^{\frac{d-1}{2}}} \mathrm{Re}\left(e^{-ix}\Phi_{\frac{d-2}{2}}(x)\right), \quad x>0 \tag{3.54}$$

where

$$|\Phi^{(k)}_{\frac{d-2}{2}}|(x) \leq C_k x^{-\frac{(d-1)}{2}-k}$$

We will first prove the lemma for the case $b \neq 1$, and then remark what the identical procedures give, for the case $b=1$.

We will start with the region $0 \leq r \leq \frac{t}{2}$, and use (3.53) for $n=1$ in the formula for v_2:

$$\begin{aligned}
&v_2(t,r)\\
&= c_b \int_0^{\infty} \sin(t\xi) J_1(r\xi) \frac{\chi_{\leq \frac{1}{4}}(\xi)}{\log^{b-1}(\frac{1}{\xi})} d\xi\\
&= \frac{c_b}{\pi} \int_0^{\pi} \sin^2(\theta) \int_0^{\infty} \sin(t\xi) r\xi \cos(r\xi\cos(\theta)) \frac{\chi_{\leq \frac{1}{4}}(\xi)}{\log^{b-1}(\frac{1}{\xi})} d\xi d\theta\\
&= \frac{c_b}{2\pi} \int_0^{\pi} \sin^2(\theta) \int_0^{\infty} r\xi \left(\sin(\xi(t+r\cos(\theta))) + \sin(\xi(t-r\cos(\theta)))\right) \frac{\chi_{\leq \frac{1}{4}}(\xi)}{\log^{b-1}(\frac{1}{\xi})} d\xi d\theta
\end{aligned}$$

Integrating by parts twice, we get

$$v_2(t,r) = -\frac{c_b r}{2\pi} \int_0^{\pi} \sin^2(\theta) \int_0^{\infty} \partial_{\xi}^2 \left(\frac{\xi \chi_{\leq \frac{1}{4}}(\xi)}{\log^{b-1}(\frac{1}{\xi})}\right) \left(\frac{\sin(\xi t_+)}{t_+^2} + \frac{\sin(\xi t_-)}{t_-^2}\right) d\xi d\theta$$

where

$$t_{+/-} = t \pm r\cos(\theta)$$

We divide the terms analogously to how similar terms were treated in the previous subsection. In particular, we have

$$\begin{aligned}
v_2(t,r) = \frac{-c_b r}{2\pi} \int_0^{\pi} \sin^2(\theta) \int_0^{\infty} &\left(\chi_{\leq \frac{1}{4}}(\xi) \left(\frac{b-1}{\xi \log^b(\frac{1}{\xi})} + \frac{b(b-1)}{\xi \log^{b+1}(\frac{1}{\xi})}\right) + \psi(\xi)\right)\\
&\left(\frac{\sin(\xi t_+)}{t_+^2} + \frac{\sin(\xi t_-)}{t_-^2}\right) d\xi d\theta
\end{aligned}$$

where $\psi \in C_c^{\infty}([\frac{1}{8}, \frac{1}{4}])$. So, we can integrate by parts (for example, once) to treat the ψ term:

$$\begin{aligned}
&|\frac{-c_b r}{2\pi} \int_0^{\pi} \sin^2(\theta) \int_0^{\infty} \psi(\xi) \left(\frac{\sin(\xi t_+)}{t_+^2} + \frac{\sin(\xi t_-)}{t_-^2}\right) d\xi d\theta| \leq Cr \int_0^{\pi} \int_0^{\infty} \frac{|\psi'(\xi)| d\xi d\theta}{t^3}\\
&\leq \frac{Cr}{t^3}, \quad r \leq \frac{t}{2}
\end{aligned}$$

where we use

$$\frac{1}{|t_{\pm}|} \leq \frac{C}{t}, \quad r \leq \frac{t}{2}$$

Then, it remains to study

(3.55)

$$\frac{-c_b r}{2\pi} \int_0^{\pi} \sin^2(\theta) \int_0^{\infty} \left(\chi_{\leq \frac{1}{4}}(\xi) \left(\frac{b-1}{\xi \log^b(\frac{1}{\xi})} + \frac{b(b-1)}{\xi \log^{b+1}(\frac{1}{\xi})}\right)\right) \left(\frac{\sin(\xi t_+)}{t_+^2} + \frac{\sin(\xi t_-)}{t_-^2}\right) d\xi d\theta$$
$$= \frac{-c_b r}{2\pi} \int_0^{\pi} \sin^2(\theta) \int_0^{\frac{1}{2}} \left(\frac{b-1}{\xi \log^b(\frac{1}{\xi})} + \frac{b(b-1)}{\xi \log^{b+1}(\frac{1}{\xi})}\right) \left(\frac{\sin(\xi t_+)}{t_+^2} + \frac{\sin(\xi t_-)}{t_-^2}\right) d\xi d\theta$$
$$+ \frac{-c_b r}{2\pi} \int_0^{\pi} \sin^2(\theta) \int_0^{\frac{1}{2}} \left(\chi_{\leq \frac{1}{4}}(\xi) - 1\right) \left(\frac{b-1}{\xi \log^b(\frac{1}{\xi})} + \frac{b(b-1)}{\xi \log^{b+1}(\frac{1}{\xi})}\right) \left(\frac{\sin(\xi t_+)}{t_+^2} + \frac{\sin(\xi t_-)}{t_-^2}\right) d\xi d\theta$$

We start with the second line of (3.55). Since all other terms will use the identical argument, we first consider the term involving $\frac{1}{\log^b(t)}$ and t_+, namely

(3.56)
$$\begin{aligned} &\frac{-c_b r}{2\pi} \int_0^{\pi} \sin^2(\theta) \left((b-1) \int_0^{\frac{t_+}{2}} \frac{\sin(u) du}{u \log^b(\frac{t_+}{u}) t_+^2}\right) d\theta \\ &= -\frac{c_b r}{2\pi} \int_0^{\pi} \sin^2(\theta)(b-1) \left(\frac{\pi}{2 \log^b(t_+) t_+^2} + \text{Err}(t, r, \theta)\right) d\theta \end{aligned}$$

where

$$|\text{Err}(t, r, \theta)| \leq \frac{C}{t_+^2 \log^{b+1}(t_+)}$$

and we use our calculation of (3.45) from the previous subsection. To treat this last integral, we start with the first term:

$$\begin{aligned} &\frac{-c_b r}{2\pi} \cdot (b-1) \frac{\pi}{2} \int_0^{\pi} \frac{\sin^2(\theta) d\theta}{\log^b(t_+) t_+^2} \\ &= \frac{-c_b r(b-1)}{4} \frac{1}{t^2 \log^b(t)} \int_0^{\pi} \frac{\sin^2(\theta)}{(1 + \frac{r\cos(\theta)}{t})^2} \frac{d\theta}{\left(1 + \frac{\log(1 + \frac{r\cos(\theta)}{t})}{\log(t)}\right)^b} \\ &= \frac{-c_b r(b-1)}{4} \left(\frac{\pi}{2t^2 \log^b(t)} + E_{v_2,1}(t, r)\right), \quad r \leq \frac{t}{2} \end{aligned}$$

with

$$|E_{v_2,1}(t, r)| \leq C \frac{r}{t^3 \log^b(t)}, \quad r \leq \frac{t}{2}$$

For the second term of (3.56), we have

$$\begin{aligned} &|\frac{-c_b r}{2\pi} \int_0^{\pi} \sin^2(\theta)(b-1) \text{Err}(t, r, \theta) d\theta| \leq Cr \int_0^{\pi} \frac{d\theta}{t^2 \log^{b+1}(t)} \\ &\leq \frac{Cr}{t^2 \log^{b+1}(t)}, \quad r \leq \frac{t}{2} \end{aligned}$$

where we used

$$\frac{1}{t_+^2 \log^{b+1}(t_+)} \leq \frac{C}{t^2 \log^{b+1}(t)}, \quad r \leq \frac{t}{2}$$

Combining these, we get

$$\frac{-c_b r}{2\pi}\int_0^\pi \sin^2(\theta)\left((b-1)\int_0^{\frac{t_+}{2}} \frac{\sin(u)du}{u\log^b(\frac{t_+}{u})t_+^2}\right)d\theta = \frac{-br}{2t^2\log^b(t)} + E_{v_2,2}(t,r), \quad r\leq \frac{t}{2}$$

$$|E_{v_2,2}(t,r)| \leq C\left(\frac{r^2}{t^3\log^b(t)} + C\frac{r}{t^2\log^{b+1}(t)}\right), \quad r\leq\frac{t}{2}$$

We use the identical procedure to treat all the other terms in the second line of (3.55), and get

$$\frac{-c_b r}{2\pi}\int_0^\pi \sin^2(\theta)\int_0^{\frac{1}{2}}\left(\frac{b-1}{\xi\log^b(\frac{1}{\xi})}+\frac{b(b-1)}{\xi\log^{b+1}(\frac{1}{\xi})}\right)\left(\frac{\sin(\xi t_+)}{t_+^2}+\frac{\sin(\xi t_-)}{t_-^2}\right)d\xi d\theta$$
$$= \frac{-br}{t^2\log^b(t)} + E_{v_2,3}(t,r)$$

with

$$|E_{v_2,3}(t,r)| \leq C\left(\frac{r}{t^2\log^{b+1}(t)}+\frac{r^2}{t^3\log^b(t)}\right), \quad r\leq \frac{t}{2}$$

Finally, for the third line of (3.55), we integrate by parts, identically to how a similar term was treated in the last subsection, and get

$$|\frac{-c_b r}{2\pi}\int_0^\pi \sin^2(\theta)\int_0^{\frac{1}{2}}\left(\chi_{\leq\frac{1}{4}}(\xi)-1\right)\left(\frac{b-1}{\xi\log^b(\frac{1}{\xi})}+\frac{b(b-1)}{\xi\log^{b+1}(\frac{1}{\xi})}\right)\left(\frac{\sin(\xi t_+)}{t_+^2}+\frac{\sin(\xi t_-)}{t_-^2}\right)d\xi d\theta|$$
$$\leq \frac{Cr}{t^3}, \quad r\leq\frac{t}{2}$$

This completes the proof of (3.49) (for v_2).

We will need two more estimates on v_2. One is given by using the simple estimate

$$|J_1(x)| \leq \frac{C}{\sqrt{x}}$$

which yields

$$|v_2(t,r)| \leq \frac{C}{\sqrt{r}}$$

We will obtain another estimate in the region $r\geq \frac{t}{2}$, by first decomposing v_2 as follows:

$$\begin{aligned} v_2(t,r) &= c_b\int_0^\infty \sin(t\xi)J_1(r\xi)\frac{\chi_{\leq\frac{1}{4}}(\xi)}{\log^{b-1}(\frac{1}{\xi})}d\xi \\ &= c_b\int_0^\infty \chi_{\leq 1}(r\xi)\sin(t\xi)J_1(r\xi)\frac{\chi_{\leq\frac{1}{4}}(\xi)}{\log^{b-1}(\frac{1}{\xi})}d\xi \\ &+ c_b\int_0^\infty (1-\chi_{\leq 1}(r\xi))\sin(t\xi)J_1(r\xi)\frac{\chi_{\leq\frac{1}{4}}(\xi)}{\log^{b-1}(\frac{1}{\xi})}d\xi \end{aligned} \tag{3.57}$$

where $\chi_{\leq 1}\in C_c^\infty([0,\infty))$, $0\leq\chi_{\leq 1}\leq 1$,

$$\chi_{\leq 1}(x) = \begin{cases} 1, & x\leq\frac{1}{2} \\ 0, & x\geq 1\end{cases}$$

and $\chi_{\leq 1}$ is otherwise arbitrary.

We start with the second line of (3.57). Using the simple estimate

$$|J_1(x)| \le Cx$$

we get

$$\begin{aligned}|\int_0^\infty \chi_{\le 1}(r\xi)\sin(t\xi)J_1(r\xi)\frac{\chi_{\le\frac14}(\xi)}{\log^{b-1}(\frac1\xi)}d\xi| &\le Cr\int_0^\infty \chi_{\le 1}(r\xi)\xi\frac{\chi_{\le\frac14}(\xi)}{\log^{b-1}(\frac1\xi)}d\xi\\ &\le \frac{C}{r\log^{b-1}(r)}, \quad r\ge\frac t2>4\end{aligned}$$

We need only estimate the remaining integral in (3.57), namely,

$$\int_0^\infty (1-\chi_{\le1}(r\xi))\sin(t\xi)J_1(r\xi)\frac{\chi_{\le\frac14}(\xi)}{\log^{b-1}(\frac1\xi)}d\xi$$

Note that this integral vanishes if $r<2$, by the support properties of $\chi_{\le 1}$ and $\chi_{\le\frac14}$. We use (3.54), for $d=4$. Then, for $r>4$, we have

$$\begin{aligned}&|\int_0^\infty (1-\chi_{\le1}(r\xi))\sin(t\xi)J_1(r\xi)\frac{\chi_{\le\frac14}(\xi)}{\log^{b-1}(\frac1\xi)}d\xi|\\ &= |\mathrm{Re}\left(\int_0^\infty (1-\chi_{\le1}(r\xi))\sin(t\xi)\frac{r\xi}{2\pi^{3/2}}e^{-ir\xi}\Phi_1(r\xi)\frac{\chi_{\le\frac14}(\xi)}{\log^{b-1}(\frac1\xi)}d\xi\right)| \qquad (3.58)\\ &\le \frac{C}{|t-r|}\int_0^\infty |\partial_\xi\left((1-\chi_{\le1}(r\xi))r\xi\Phi_1(r\xi)\frac{\chi_{\le\frac14}(\xi)}{\log^{b-1}(\frac1\xi)}\right)|d\xi\end{aligned}$$

Note that

$$\begin{aligned}\int_0^\infty |\chi'_{\le1}(r\xi)|r|r\xi\Phi_1(r\xi)|\frac{|\chi_{\le\frac14}(\xi)|}{\log^{b-1}(\frac1\xi)}d\xi &\le Cr\int_0^{1/r}\frac{d\xi}{\log^{b-1}(\frac1\xi)}\\ &\le \frac{C}{\log^{b-1}(r)}, \quad r\ge\frac t2\end{aligned}$$

where we used the support properties of χ and symbol property of Φ_1.

Next,

$$\begin{aligned}\int_0^\infty (1-\chi_{\le1}(r\xi))\,|\partial_\xi(r\xi\Phi_1(r\xi))|\frac{|\chi_{\le\frac14}(\xi)|}{\log^{b-1}(\frac1\xi)}d\xi &\le Cr\int_0^\infty\frac{(1-\chi_{\le1}(r\xi))}{(r\xi)^{3/2}}\frac{|\chi_{\le\frac14}(\xi)|}{\log^{b-1}(\frac1\xi)}d\xi\\ &\le \frac{C}{\sqrt r}\int_{\frac{1}{2r}}^{\frac14}\frac{d\xi}{\xi^{3/2}\log^{b-1}(\frac1\xi)}\\ &\le \frac{C}{\log^{b-1}(r)}, \quad r\ge\frac t2\end{aligned}$$

Finally, we use the same procedure to get

$$\begin{aligned}&\int_0^\infty (1-\chi_{\leq 1}(r\xi))\,|r\xi\Phi_1(r\xi)||\partial_\xi\left(\frac{\chi_{\leq\frac{1}{4}}(\xi)}{\log^{b-1}(\frac{1}{\xi})}\right)|d\xi\\ &\leq \frac{C}{\sqrt{r}}\int_{\frac{1}{2r}}^{\frac{1}{4}}\left(\frac{|\chi'_{\leq\frac{1}{4}}(\xi)|}{\sqrt{\xi}\log^{b-1}(\frac{1}{\xi})}+\frac{1}{\xi^{3/2}\log^b(\frac{1}{\xi})}\right)d\xi\\ &\leq \frac{C}{\log^b(r)},\quad r\geq\frac{t}{2}\end{aligned}$$

Returning to (3.58), we get, for $r\geq\frac{t}{2}$,

$$\frac{C}{|t-r|}\int_0^\infty |\partial_\xi\left((1-\chi_{\leq 1}(r\xi))\,r\xi\Phi_1(r\xi)\frac{\chi_{\leq\frac{1}{4}}(\xi)}{\log^{b-1}(\frac{1}{\xi})}\right)|d\xi\leq\frac{C}{|t-r|\log^{b-1}(r)}$$

Note that, if $\frac{t}{2}\leq r$, then, $|t-r|\leq r$, so

$$\frac{C}{r\log^{b-1}(r)}\leq\frac{C}{|t-r|\log^{b-1}(r)},\quad r\geq\frac{t}{2},\quad r\neq t$$

So, we can combine the previous two estimates to conclude (3.51), thereby concluding the proof of the v_2 estimates (in the case $b\neq 1$).

We now proceed to prove the similar statements about $\partial_r v_2$. We have

$$\partial_r v_2(t,r)=\frac{c_b}{2}\int_0^\infty \sin(t\xi)\xi\left(J_0(r\xi)-J_2(r\xi)\right)\frac{\chi_{\leq\frac{1}{4}}(\xi)}{\log^{b-1}(\frac{1}{\xi})}d\xi$$

and start by treating the J_0 term. Using (3.53) for $n=0$, we have

$$\begin{aligned}&\frac{c_b}{2}\int_0^\infty \sin(t\xi)\xi J_0(r\xi)\frac{\chi_{\leq\frac{1}{4}}(\xi)}{\log^{b-1}(\frac{1}{\xi})}d\xi\\ &=\frac{c_b}{4\pi}\int_0^\pi\int_0^\infty\left(\sin(\xi t_+)+\sin(\xi t_-)\right)\frac{\xi\chi_{\leq\frac{1}{4}}(\xi)}{\log^{b-1}(\frac{1}{\xi})}d\xi d\theta\\ &=\frac{-c_b}{4\pi}\int_0^\pi\int_0^\infty\left(\frac{\sin(\xi t_+)}{t_+^2}+\frac{\sin(\xi t_-)}{t_-^2}\right)\partial_\xi^2\left(\frac{\xi\chi_{\leq\frac{1}{4}}(\xi)}{\log^{b-1}(\frac{1}{\xi})}\right)d\xi d\theta\end{aligned}$$

We follow the identical procedure used to prove (3.49) for v_2, the only major difference being an extra $\sin^2(\theta)$ in the θ integral in the v_2 case. We get

$$\frac{c_b}{2}\int_0^\infty \sin(t\xi)\xi J_0(r\xi)\frac{\chi_{\leq\frac{1}{4}}(\xi)}{\log^{b-1}(\frac{1}{\xi})}d\xi=\frac{-b}{t^2\log^b(t)}+E_{\partial_r v_2,1}(t,r)$$

where

$$|E_{\partial_r v_2}(t,r)|\leq\frac{Cr}{t^3\log^b(t)}+\frac{C}{t^2\log^{b+1}(t)},\quad r\leq\frac{t}{2}$$

Now, we treat the J_2 term, again using (3.53), this time for $n=2$.

$$-\frac{c_b}{2}\int_0^\infty \sin(t\xi)\xi J_2(r\xi)\frac{\chi_{\leq\frac{1}{4}}(\xi)}{\log^{b-1}(\frac{1}{\xi})}d\xi$$

$$=\frac{-c_b r^2}{6\pi}\int_0^\infty \sin(t\xi)\xi^3\int_0^\pi \cos(r\xi\cos(\theta))\sin^4(\theta)\frac{\chi_{\leq\frac{1}{4}}(\xi)}{\log^{b-1}(\frac{1}{\xi})}d\theta d\xi$$

$$=\frac{-c_b r^2}{12\pi}\int_0^\pi \sin^4(\theta)\int_0^\infty \left(\sin(\xi(t+r\cos(\theta)))+\sin(\xi(t-r\cos(\theta)))\right)\frac{\xi^3\chi_{\leq\frac{1}{4}}(\xi)}{\log^{b-1}(\frac{1}{\xi})}d\xi d\theta$$

$$=\frac{-c_b r^2}{12\pi}\int_0^\pi \sin^4(\theta)\int_0^\infty \left(\frac{\sin(\xi(t+r\cos(\theta)))}{(t+r\cos(\theta))^4}+\frac{\sin(\xi(t-r\cos(\theta)))}{(t-r\cos(\theta))^4}\right)\partial_\xi^4\left(\frac{\xi^3\chi_{\leq\frac{1}{4}}(\xi)}{\log^{b-1}(\frac{1}{\xi})}\right)d\xi d\theta$$

Note that

$$\partial_\xi^4\left(\frac{\xi^3\chi_{\leq\frac{1}{4}}(\xi)}{\log^{b-1}(\frac{1}{\xi})}\right)$$

$$=\chi_{\leq\frac{1}{4}}(\xi)\left(\frac{11b(b-1)}{\xi\log^{b+1}(\frac{1}{\xi})}+\frac{6(b-1)}{\xi\log^{b}(\frac{1}{\xi})}+\frac{b(b-1)(b+1)(b+2)}{\xi\log^{b+3}(\frac{1}{\xi})}+\frac{6b(b-1)(b+1)}{\xi\log^{b+2}(\frac{1}{\xi})}\right)$$

$$+\psi(\xi)$$

where

$$\psi\in C_c^\infty([\frac{1}{8},\frac{1}{4}])$$

We treat the ψ term in the same way as in the v_2 estimate, and use the same argument as in the v_2 estimate to handle the other terms. In total, we get

$$|-\frac{c_b}{2}\int_0^\infty \sin(t\xi)\xi J_2(r\xi)\frac{\chi_{\leq\frac{1}{4}}(\xi)}{\log^{b-1}(\frac{1}{\xi})}d\xi|\leq\frac{Cr^2}{t^4\log^b(t)},\quad r\leq\frac{t}{2}$$

Combining our results, we then get (3.49) for $\partial_r v_2$.

Now, we study the region $r\geq\frac{t}{2}$. Again using the simple estimate

$$|J_1'(x)|\leq\frac{C}{\sqrt{x}}$$

we get

$$|\partial_r v_2(t,r)|\leq\frac{C}{\sqrt{r}}$$

Then, we decompose

$$\partial_r v_2(t,r)=I_r+II_r$$

with

$$I_r=\frac{c_b}{2}\int_0^\infty \sin(t\xi)\xi\chi_{\leq 1}(r\xi)\left(J_0(r\xi)-J_2(r\xi)\right)\frac{\chi_{\leq\frac{1}{4}}(\xi)}{\log^{b-1}(\frac{1}{\xi})}d\xi$$

Using the simple estimate

$$|J_k(x)|\leq C,\quad k=0,2$$

we get

$$|I_r|\leq|\frac{c_b}{2}\int_0^\infty \sin(t\xi)\chi_{\leq 1}(r\xi)\xi\left(J_0(r\xi)-J_2(r\xi)\right)\frac{\chi_{\leq\frac{1}{4}}(\xi)}{\log^{b-1}(\frac{1}{\xi})}d\xi|$$

$$\leq C\int_0^{\min\{\frac{1}{r},\frac{1}{4}\}}\frac{\xi d\xi}{\log^{b-1}(\frac{1}{\xi})}\leq\frac{C}{r^2\log^{b-1}(r)},\quad r\geq\frac{t}{2}\geq 4$$

It remains to estimate II_r. We write

$$II_r = II_{r,0} + II_{r,2}$$

with

$$II_{r,0} = \frac{c_b}{2} \int_0^\infty \sin(t\xi)\left(1 - \chi_{\leq 1}(r\xi)\right)\xi J_0(r\xi) \frac{\chi_{\leq \frac{1}{4}}(\xi)}{\log^{b-1}(\frac{1}{\xi})} d\xi$$

$$II_{r,2} = -\frac{c_b}{2} \int_0^\infty \sin(t\xi)\left(1 - \chi_{\leq 1}(r\xi)\right)\xi J_2(r\xi) \frac{\chi_{\leq \frac{1}{4}}(\xi)}{\log^{b-1}(\frac{1}{\xi})} d\xi$$

Again using (3.54) for $d = 2$, we get

$$\begin{aligned}
|II_{r,0}| &\leq C \int_0^\infty |\partial_\xi^2 \left((1 - \chi_{\leq 1}(r\xi))\,\xi \Phi_0(r\xi) \frac{\chi_{\leq \frac{1}{4}}(\xi)}{\log^{b-1}(\frac{1}{\xi})} \right)| \frac{d\xi}{(t-r)^2} \\
&\leq \frac{C}{(t-r)^2} \int_0^\infty \frac{\mathbb{1}_{\{r\xi \geq \frac{1}{2}\}} \mathbb{1}_{\{\xi \leq \frac{1}{4}\}}}{\sqrt{r}\xi^{3/2} \log^{b-1}(\frac{1}{\xi})} d\xi \\
&\leq \frac{C}{(t-r)^2} \int_{\frac{1}{2r}}^{\frac{1}{4}} \frac{d\xi}{\sqrt{r}\xi^{3/2} \log^{b-1}(\frac{1}{\xi})} \leq \frac{C}{(t-r)^2} \frac{1}{\sqrt{r}} \int_{\frac{1}{2r}}^{\frac{1}{4}} \frac{d\xi}{\xi^{3/2} \log^{b-1}(\frac{1}{\xi})} \\
&\leq \frac{C}{(t-r)^2 \log^{b-1}(r)}, \quad r \geq \frac{t}{2}
\end{aligned} \tag{3.59}$$

For $II_{r,2}$, we use (3.54) for $d = 6$. Using the same argument as for $II_{r,0}$, we get

$$\begin{aligned}
|II_{r,2}| &\leq C \int_0^\infty |\partial_\xi^2 \left((1 - \chi_{\leq 1}(r\xi))\,\xi (r\xi)^2 \Phi_2(r\xi) \frac{\chi_{\leq \frac{1}{4}}(\xi)}{\log^{b-1}(\frac{1}{\xi})} \right)| \frac{d\xi}{(t-r)^2} \\
&\leq \frac{C}{(t-r)^2 \log^{b-1}(r)}, \quad r \geq \frac{t}{2}
\end{aligned}$$

Combining these, we have

$$|II_r| \leq \frac{C}{(t-r)^2 \log^{b-1}(r)}, \quad r \geq \frac{t}{2}$$

Using the same reasoning as was used to prove the analogous estimate for v_2, if $r \geq \frac{t}{2}, \quad r \neq t$, we have

$$\frac{1}{r^2 \log^{b-1}(r)} \leq \frac{C}{(t-r)^2 \log^{b-1}(r)}$$

which gives (3.51) for $\partial_r v_2$.

Next, we study $\partial_t v_2$, in the region $0 \leq r \leq \frac{t}{2}$. Using (3.53), we have

$$\begin{aligned}
&\partial_t v_2(t,r)\\
&= c_b \int_0^\infty \cos(t\xi) \frac{\xi J_1(r\xi)\chi_{\leq \frac{1}{4}}(\xi)}{\log^{b-1}(\frac{1}{\xi})} d\xi\\
&= \frac{c_b}{\pi} \int_0^\pi \sin^2(\theta) \int_0^\infty \xi \cos(t\xi) r\xi \cos(r\xi\cos(\theta)) \frac{\chi_{\leq \frac{1}{4}}(\xi)}{\log^{b-1}(\frac{1}{\xi})} d\xi d\theta\\
&= \frac{c_b r}{\pi} \int_0^\pi \sin^2(\theta) \int_0^\infty \frac{\xi^2}{2} \left(\cos(\xi(t+r\cos(\theta))) + \cos(\xi(t-r\cos(\theta)))\right) \frac{\chi_{\leq \frac{1}{4}}(\xi)}{\log^{b-1}(\frac{1}{\xi})} d\xi d\theta\\
&= \frac{c_b r}{2\pi} \int_0^\pi \sin^2(\theta) \int_0^\infty \left(\frac{\sin(\xi(t+r\cos(\theta)))}{(t+r\cos(\theta))^3} + \frac{\sin(\xi(t-r\cos(\theta)))}{(t-r\cos(\theta))^3}\right) \partial_\xi^3 \left(\frac{\xi^2 \chi_{\leq \frac{1}{4}}(\xi)}{\log^{b-1}(\frac{1}{\xi})}\right) d\xi d\theta
\end{aligned}$$

We note that

$$\begin{aligned}
\partial_\xi^3 \left(\frac{\xi^2}{\log^{b-1}(\frac{1}{\xi})} \chi_{\leq \frac{1}{4}}(\xi)\right) &= \left(\frac{(b-1)b(b+1)}{\xi \log^{b+2}(\frac{1}{\xi})} + \frac{3b(b-1)}{\xi \log^{b+1}(\frac{1}{\xi})} + \frac{2(b-1)}{\xi \log^{b}(\frac{1}{\xi})}\right) \chi_{\leq \frac{1}{4}}(\xi)\\
&\quad + \psi(\xi)
\end{aligned}$$

where

$$\psi \in C_c^\infty([\frac{1}{8}, \frac{1}{4}])$$

The integral to estimate is therefore of exactly the same form as that treated in estimating v_2, and repeating this procedure gives (3.49) for $\partial_t v_2$. For the region $r \geq \frac{t}{2}$, we can again use

$$|J_1(x)| \leq \frac{C}{\sqrt{x}}$$

to get

$$|\partial_t v_2(t,r)| \leq \frac{C}{\sqrt{r}}$$

We now consider the region $r \geq \frac{t}{2}$. First, let us write

$$\partial_t v_2(t,r) = I_t(t,r) + II_t(t,r)$$

where

$$I_t = c_b \int_0^\infty \chi_{\leq 1}(r\xi) \xi \cos(t\xi) J_1(r\xi) \frac{\chi_{\leq \frac{1}{4}}(\xi)}{\log^{b-1}(\frac{1}{\xi})} d\xi$$

$$II_t = c_b \int_0^\infty (1 - \chi_{\leq 1}(r\xi))\, \xi \cos(t\xi) J_1(r\xi) \frac{\chi_{\leq \frac{1}{4}}(\xi)}{\log^{b-1}(\frac{1}{\xi})} d\xi$$

Then, we use

$$|J_1(x)| \leq Cx$$

to get

$$|I_t| \leq C \int_0^{\min\{\frac{1}{r}, \frac{1}{4}\}} \frac{r\xi^2}{\log^{b-1}(\frac{1}{\xi})} d\xi \leq \frac{C}{r^2 \log^{b-1}(r)}, \quad r \geq 4$$

As usual, we use (3.54) with $d = 4$ to get

$$II_t = \frac{c_b}{4\pi^{3/2}} \mathrm{Re} \left(\int_0^\infty (1 - \chi_{\leq 1}(r\xi))\, \xi \left(e^{i(t-r)\xi} + e^{-i(t+r)\xi}\right) \frac{r\xi \Phi_1(r\xi) \chi_{\leq \frac{1}{4}}(\xi)}{\log^{b-1}(\frac{1}{\xi})} d\xi\right)$$

and this gives

$$|II_t| \leq C \int_0^\infty \frac{1}{|t-r|^2} |\partial_\xi^2 \left(\frac{(1-\chi_{\leq 1}(r\xi))\, \xi^2 r \Phi_1(r\xi) \chi_{\leq \frac{1}{4}}(\xi)}{\log^{b-1}(\frac{1}{\xi})} \right) | d\xi$$

But, comparing this expression with (3.59), we see that the only difference is

$$\Phi_0(r\xi) \text{ is replaced by } \Phi_1(r\xi) \cdot r\xi$$

Comparing the symbol-type estimates on Φ_d which follow (3.54), we can use the same procedure used to treat (3.59) to get

$$|II_t| \leq \frac{C}{(t-r)^2 \log^{b-1}(r)}, \quad r \geq \frac{t}{2}$$

This gives (3.51) for $\partial_t v_2$.

Next, we obtain estimates on $\partial_r^2 v_2$, using the same procedure as above. In particular, we use

$$J_1''(x) = \frac{1}{4}(J_3(x) - 3J_1(x))$$

to get

$$\partial_r^2 v_2(t,r) = \frac{c_b}{4} \int_0^\infty \sin(t\xi)\xi^2 \left(-3J_1(r\xi) + J_3(r\xi)\right) \frac{\chi_{\leq \frac{1}{4}}(\xi)}{\log^{b-1}(\frac{1}{\xi})} d\xi$$

Then, we use (3.53) for J_k, $\quad k = 1,3$. For the term involving J_k, for $k = 1$ or $k = 3$, we integrate by parts $k+3$ times in ξ and get:

$$\begin{aligned}
&\partial_r^2 v_2(t,r) \\
&= \frac{-3c_b r}{8\pi} \int_0^\pi \sin^2(\theta) \int_0^\infty \left(\frac{\sin(\xi(t + r\cos(\theta)))}{(t + r\cos(\theta))^4} + \frac{\sin((t - r\cos(\theta))\xi)}{(t - r\cos(\theta))^4} \right) \partial_\xi^4 \left(\frac{\xi^3 \chi_{\leq \frac{1}{4}}(\xi)}{\log^{b-1}(\frac{1}{\xi})} \right) d\xi d\theta \\
&\quad - \frac{c_b r^3}{120\pi} \int_0^\pi \sin^6(\theta) \int_0^\infty \left(\frac{\sin(\xi(t + r\cos(\theta)))}{(t + r\cos(\theta))^6} + \frac{\sin(\xi(t - r\cos(\theta)))}{(t - r\cos(\theta))^6} \right) \partial_\xi^6 \left(\frac{\xi^5 \chi_{\leq \frac{1}{4}}(\xi)}{\log^{b-1}(\frac{1}{\xi})} \right) d\xi d\theta
\end{aligned}$$

We will not need as precise of a description of $\partial_r^2 v_2$ here, since such a description for small r will be obtained later using the equation solved by v_2. So, we use the identical procedure used for $\partial_r v_2$, and get

$$|\partial_r^2 v_2(t,r)| \leq \frac{Cr}{t^4 \log^b(t)}, \quad r \leq \frac{t}{2}$$

For the larger r estimates, we can again use

$$|J_k(x)| \leq \frac{C}{\sqrt{x}}, \quad k = 1,3$$

to get

$$|\partial_r^2 v_2(t,r)| \leq C \int_0^\infty \frac{\xi^2}{\sqrt{r\xi}} \frac{\chi_{\leq \frac{1}{4}}(\xi)}{\log^{b-1}(\frac{1}{\xi})} d\xi \leq \frac{C}{\sqrt{r}}$$

Lastly, for another estimate for large r, we first split $\partial_r^2 v_2$ as follows

$$\partial_r^2 v_2(t,r) = I_{rr} + II_{rr}$$

and

$$I_{rr} = \frac{c_b}{4} \int_0^\infty \sin(t\xi)\xi^2 \chi_{\leq 1}(r\xi) \left(-3J_1(r\xi) + J_3(r\xi)\right) \frac{\chi_{\leq \frac{1}{4}}(\xi)}{\log^{b-1}(\frac{1}{\xi})} d\xi$$

$$II_{rr} = \frac{c_b}{4}\int_0^\infty \sin(t\xi)\xi^2\left(1-\chi_{\leq 1}(r\xi)\right)\left(-3J_1(r\xi)+J_3(r\xi)\right)\frac{\chi_{\leq\frac{1}{4}}(\xi)}{\log^{b-1}(\frac{1}{\xi})}d\xi$$

which give

$$|I_{rr}| \leq C\int_0^{\min\{\frac{1}{r},\frac{1}{4}\}} \frac{\xi^2 r\xi}{\log^{b-1}(\frac{1}{\xi})}d\xi \leq \frac{C}{r^3\log^{b-1}(r)}, \quad r\geq 4$$

Next, we again use (3.54) for $J_k, \quad k=1,3$, and integrate by parts 3 times, to get

$$\begin{aligned}|II_{rr}| &\leq \sum_{k\in\{1,3\}}\frac{C}{|t-r|^3}\int_0^\infty \left|\partial_\xi^3\left(\frac{\xi^2\chi_{\leq\frac{1}{4}}(\xi)}{\log^{b-1}(\frac{1}{\xi})}(r\xi)^k\Phi_k(r\xi)\left(1-\chi_{\leq 1}(r\xi)\right)\right)\right|d\xi\\ &\leq \frac{C}{|t-r|^3}\int_0^\infty\left(\frac{\mathbb{1}_{\{r\xi\geq\frac{1}{2}\}}\mathbb{1}_{\{\xi\leq\frac{1}{4}\}}}{\sqrt{r}\xi^{3/2}\log^{b-1}(\frac{1}{\xi})}+\sqrt{\frac{r}{\xi}}\frac{\mathbb{1}_{\{1\geq r\xi\geq\frac{1}{2}\}}\mathbb{1}_{\{\xi\leq\frac{1}{4}\}}}{\log^{b-1}(\frac{1}{\xi})}\right)d\xi\\ &\leq \frac{C}{|t-r|^3\log^{b-1}(r)}, \quad r>\frac{t}{2}\end{aligned}$$

Combining the above estimates, we get (3.52)

Next, we study $\partial_t^2 v_2$ in the region $r\leq\frac{t}{2}$. We proceed just as for v_2:

$$\partial_t^2 v_2(t,r) = \frac{-c_b}{2\pi}\int_0^\pi \sin^2(\theta)\int_0^\infty r\left(\frac{\sin(\xi t_+)}{t_+^4}+\frac{\sin(\xi t_-)}{t_-^4}\right)\partial_\xi^4\left(\frac{\chi_{\leq\frac{1}{4}}(\xi)\cdot\xi^3}{\log^{b-1}(\frac{1}{\xi})}\right)d\xi d\theta$$

We then use

$$\begin{aligned}&\partial_\xi^4\left(\frac{\chi_{\leq\frac{1}{4}}(\xi)\cdot\xi^3}{\log^{b-1}(\frac{1}{\xi})}\right)\\ &=\chi_{\leq\frac{1}{4}}(\xi)\left(\frac{6b\left(b^2-1\right)}{\xi\log^{b+2}\left(\frac{1}{\xi}\right)}+\frac{(b-1)b(b+1)(b+2)}{\xi\log^{b+3}\left(\frac{1}{\xi}\right)}+\frac{11(b-1)b}{\xi\log^{b+1}\left(\frac{1}{\xi}\right)}+\frac{6(b-1)}{\xi\log^{b}\left(\frac{1}{\xi}\right)}\right)\\ &+\psi(\xi)\\ &, \quad \psi\in C_c^\infty([\frac{1}{8},\frac{1}{4}])\end{aligned}$$

The integral is of the same form as that considered during the v_2 estimates, and we get (3.49) for $\partial_t^2 v_2$. (Note that any other estimates on $\partial_t^2 v_2$ which we may need can be obtained using the other estimates of the lemma, and the equation solved by v_2).

Finally, we will need a similar formula for $\partial_{tr}v_2$: We have

$$\partial_{tr}v_2(t,r) = \frac{c_b}{2}\int_0^\infty \cos(t\xi)\xi^2\left(J_0(r\xi)-J_2(r\xi)\right)\frac{\chi_{\leq\frac{1}{4}}(\xi)d\xi}{\log^{b-1}(\frac{1}{\xi})}$$

Applying (3.53), the J_0 term is

$$\frac{c_b}{4\pi}\int_0^\pi\int_0^\infty\left(\frac{\sin(\xi t_+)}{t_+^3}+\frac{\sin(\xi t_-)}{t_-^3}\right)\partial_\xi^3\left(\frac{\xi^2\chi_{\leq\frac{1}{4}}(\xi)}{\log^{b-1}(\frac{1}{\xi})}\right)d\xi d\theta$$

Comparing this to the analogous integral treated while studying $\partial_t v_2$, we get

$$\frac{c_b}{4\pi}\int_0^\pi\int_0^\infty\left(\frac{\sin(\xi t_+)}{t_+^3}+\frac{\sin(\xi t_-)}{t_-^3}\right)\partial_\xi^3\left(\frac{\xi^2\chi_{\leq\frac{1}{4}}(\xi)}{\log^{b-1}(\frac{1}{\xi})}\right)d\xi d\theta$$
$$=\frac{2b}{t^3\log^b(t)}+E_{\partial_{12}v_2,1}(t,r)$$

where

$$|E_{\partial_{12}v_2,1}(t,r)|\leq C\left(\frac{r}{t^4\log^b(t)}+\frac{1}{t^3\log^{b+1}(t)}\right),\quad r\leq\frac{t}{2}$$

For the J_2 term, we use (3.53), and get

$$\frac{c_b}{2}\int_0^\infty\cos(t\xi)\xi^2(-J_2(r\xi))\frac{\chi_{\leq\frac{1}{4}}(\xi)d\xi}{\log^{b-1}(\frac{1}{\xi})}$$
$$=\frac{c_b r^2}{12\pi}\int_0^\pi\sin^4(\theta)\int_0^\infty\left(\frac{\sin(\xi t_+)}{t_+^5}+\frac{\sin(\xi t_-)}{t_-^5}\right)\partial_\xi^5\left(\frac{\xi^4\chi_{\leq\frac{1}{4}}(\xi)}{\log^{b-1}(\frac{1}{\xi})}\right)d\xi d\theta$$

we then treat this integral in exactly the same way as the J_0 term, and get

$$|\frac{c_b}{2}\int_0^\infty\cos(t\xi)\xi^2(-J_2(r\xi))\frac{\chi_{\leq\frac{1}{4}}(\xi)d\xi}{\log^{b-1}(\frac{1}{\xi})}|\leq\frac{Cr^2}{t^5\log^b(t)},\quad r\leq\frac{t}{2}$$

This gives (3.49) for $\partial_{tr}v_2$.

Finally, (3.50) is proven for $\partial_t^2 v_2$ and $\partial_{tr}v_2$, and (3.51) is proven for $\partial_{tr}v_2$ in exactly the same way as for the other derivatives.

The identical procedure shows that (3.49) is still true for $b=1$. In the region $r\geq\frac{t}{2}$, (3.50) is still true for $b=1$, and for the second estimate in the region $r\geq\frac{t}{2}$, we get for $0\leq j\leq 1,\quad 0\leq k\leq 1,\quad j+k\leq 2$

$$|\partial_t^j\partial_r^k v_2(t,r)|\leq\frac{C}{|t-r|^{1+j+k}}\log(\log(r)),\quad t\neq r\geq\frac{t}{2},\quad b=1$$

(Note the difference with the case $b\neq 1$, which had $\frac{C}{|t-r|^{1+j+k}\log^{b-1}(r)}$ on the right hand side). In any case, (3.51) is true for all $b>0$ (for simplicity, we use $\log(r)$ instead of $\begin{cases}\log^{1-b}(r), & b\neq 1\\ \log(\log(r)), & b=1\end{cases}$ in (3.51)). □

3.0.8.4. *The inner product of the (rescaled) v_3 linear error term.*

LEMMA 3.7. *For v_3 defined in (3.20), we have*

$$\int_0^\infty\left(\frac{\cos(2Q_1(R))-1}{R^2\lambda(t)^2}\right)v_3(t,R\lambda(t))\phi_0(R)RdR$$
$$=\frac{16}{\lambda(t)}\int_t^\infty K_3(s-t,\lambda(t))\lambda''(s)ds$$
$$+\int_0^\infty\left(\frac{\cos(2Q_1(R))-1}{R^2\lambda(t)^2}\right)E_5(t,R\lambda(t))\phi_0(R)RdR$$

where E_5 is as in (3.21),

$$K_3(w,\lambda(t))=\left(\frac{w}{1+w^2}-\frac{w}{\lambda(t)^{2-2\alpha}+w^2}\right)\frac{w^4}{4(w^2+36\lambda(t)^2)^2}$$

and

$$\int_0^\infty |K_3(w,\lambda(t)) - K_{3,0}(w,\lambda(t))|dw \leq C \tag{3.60}$$

where

$$K_{3,0}(w,\lambda(t)) = -\frac{1}{4(\lambda(t)^{1-\alpha}+w)(1+w)^3}$$

PROOF. We start with

(3.61)
$$\begin{aligned}
&\int_0^\infty \left(\frac{\cos(2Q_1(R))-1}{R^2\lambda(t)^2}\right) v_3(t,R\lambda(t))\phi_0(R)RdR\\
&= \frac{16}{\lambda(t)}\int_0^\infty \frac{R^3}{(1+R^2)^3}\int_{t+6R\lambda(t)}^\infty \lambda''(s)\left(\frac{s-t}{1+(s-t)^2} - \frac{(s-t)}{\lambda(t)^{2-2\alpha}+(s-t)^2}\right)dsdR\\
&+ \int_0^\infty \left(\frac{\cos(2Q_1(R))-1}{R^2\lambda(t)^2}\right) E_5(t,R\lambda(t))\phi_0(R)RdR
\end{aligned}$$

where we recall the decomposition (3.21).

We study the second line of (3.61) in more detail, since the third line is as in the lemma statement.

$$\begin{aligned}
&\frac{16}{\lambda(t)}\int_0^\infty \frac{R^3}{(1+R^2)^3}\int_{t+6R\lambda(t)}^\infty \lambda''(s)\left(\frac{s-t}{1+(s-t)^2} - \frac{(s-t)}{\lambda(t)^{2-2\alpha}+(s-t)^2}\right)dsdR\\
&= \frac{16}{\lambda(t)}\int_t^\infty K_3(s-t,\lambda(t))\lambda''(s)ds
\end{aligned}$$

where

$$K_3(w,\lambda(t)) = \left(\frac{w}{1+w^2} - \frac{w}{\lambda(t)^{2-2\alpha}+w^2}\right)\frac{w^4}{4(w^2+36\lambda(t)^2)^2}$$

We now estimate

$$\int_0^\infty |K_3(w,\lambda(t)) - K_{3,0}(w,\lambda(t))|dw$$

where

$$K_{3,0}(w,\lambda(t)) = -\frac{1}{4(\lambda(t)^{1-\alpha}+w)(1+w)^3}$$

We start with

$$\begin{aligned}
&\int_0^{\lambda(t)^{1-\alpha}} |K_3(w,\lambda(t))|dw + \int_0^{\lambda(t)^{1-\alpha}} \frac{dw}{4(\lambda(t)^{1-\alpha}+w)(1+w)^3}\\
&\leq C\int_0^{\lambda(t)^{1-\alpha}} \frac{wdw}{(1+w^2)} + C\int_0^{\lambda(t)^{1-\alpha}} \frac{wdw}{\lambda(t)^{2-2\alpha}} + C\int_0^{\lambda(t)^{1-\alpha}} \frac{dw}{\lambda(t)^{1-\alpha}}\\
&\leq C
\end{aligned}$$

Next,

$$\begin{aligned}
&\int_{\lambda(t)^{1-\alpha}}^{\frac{1}{2}} |K_3(w,\lambda(t)) + \frac{1}{4(\lambda(t)^{1-\alpha}+w)(1+w)^3}|dw \\
&\leq \int_{\lambda(t)^{1-\alpha}}^{\frac{1}{2}} |w\left(\frac{1}{1+w^2} - \frac{1}{\lambda(t)^{2-2\alpha}+w^2}\right)\left(\frac{w^4}{4(w^2+36\lambda(t)^2)^2} - \frac{1}{4}\right)|dw \\
&+ \int_{\lambda(t)^{1-\alpha}}^{\frac{1}{2}} |\frac{w}{4}\left(\frac{1}{1+w^2} - \frac{1}{\lambda(t)^{2-2\alpha}+w^2}\right) + \frac{1}{4(\lambda(t)^{1-\alpha}+w)}|dw \\
&+ \int_{\lambda(t)^{1-\alpha}}^{\frac{1}{2}} |\frac{-1}{4(\lambda(t)^{1-\alpha}+w)} + \frac{1}{4(\lambda(t)^{1-\alpha}+w)(1+w)^3}|dw
\end{aligned} \tag{3.62}$$

For the second line of (3.62), we have

$$\begin{aligned}
&\int_{\lambda(t)^{1-\alpha}}^{\frac{1}{2}} |w\left(\frac{1}{1+w^2} - \frac{1}{\lambda(t)^{2-2\alpha}+w^2}\right)\left(\frac{w^4}{4(w^2+36\lambda(t)^2)^2} - \frac{1}{4}\right)|dw \\
&\leq C\int_{\lambda(t)^{1-\alpha}}^{\frac{1}{2}} w\left(\frac{1}{\lambda(t)^{2-2\alpha}+w^2} - \frac{1}{1+w^2}\right)\frac{\lambda(t)^2}{w^2}dw \\
&\leq C\lambda(t)^{2\alpha}|\log(\lambda(t)^{2-2\alpha})| \leq C
\end{aligned}$$

For the third line of (3.62), we have

$$\begin{aligned}
&\int_{\lambda(t)^{1-\alpha}}^{\frac{1}{2}} |\frac{w}{4}\left(\frac{1}{1+w^2} - \frac{1}{\lambda(t)^{2-2\alpha}+w^2}\right) + \frac{1}{4(\lambda(t)^{1-\alpha}+w)}|dw \\
&\leq \int_{\lambda(t)^{1-\alpha}}^{\frac{1}{2}} \frac{wdw}{4(1+w^2)} + \int_{\lambda(t)^{1-\alpha}}^{\frac{1}{2}} |\frac{1}{\lambda(t)^{1-\alpha}+w} - \frac{w}{\lambda(t)^{2-2\alpha}+w^2}|dw \\
&\leq C
\end{aligned}$$

Finally, the last line of (3.62) is estimated as follows:

$$\begin{aligned}
&\int_{\lambda(t)^{1-\alpha}}^{\frac{1}{2}} |\frac{-1}{4(\lambda(t)^{1-\alpha}+w)} + \frac{1}{4(\lambda(t)^{1-\alpha}+w)(1+w)^3}|dw \\
&\leq C\int_{\lambda(t)^{1-\alpha}}^{\frac{1}{2}} \frac{wdw}{(\lambda(t)^{1-\alpha}+w)} \\
&\leq C
\end{aligned}$$

Then, we consider

$$\begin{aligned}
&\int_{\frac{1}{2}}^{\infty} |K_3(w,\lambda(t)) + \frac{1}{4(\lambda(t)^{1-\alpha}+w)(1+w)^3}|dw \\
&\leq \int_{\frac{1}{2}}^{\infty} |K_3(w,\lambda(t))|dw + \int_{\frac{1}{2}}^{\infty} \frac{dw}{(\lambda(t)^{1-\alpha}+w)(1+w)^3} \\
&\leq C\int_{\frac{1}{2}}^{\infty} \frac{w|\lambda(t)^{2-2\alpha}-1|}{w^4}dw + C \\
&\leq C
\end{aligned}$$

Combining the above, we get that

$$\int_0^\infty |K_3(w,\lambda(t)) - K_{3,0}(w,\lambda(t))|dw \leq C \qquad \square$$

3.0.8.5. *Leading behavior of λ and set-up of the modulation equation.* From the previous subsections, we have

$$\begin{aligned}
&\langle F_4(t,\cdot\lambda(t)),\phi_0\rangle\\
&= \frac{-16}{\lambda(t)^3}\int_t^\infty \lambda''(s)K_1(s-t,\lambda(t))ds + \frac{4b}{\lambda(t)t^2\log^b(t)} + \frac{4\alpha\log(\lambda(t))\lambda''(t)}{\lambda(t)}\\
&+ \frac{16}{\lambda(t)}\int_t^\infty \lambda''(s)K_{3,0}(s-t,\lambda(t))ds\\
&+ E_{0,1}(\lambda(t),\lambda'(t),\lambda''(t)) + \frac{16}{\lambda(t)}\int_t^\infty \lambda''(s)\left(K_3(s-t,\lambda(t)) - K_{3,0}(s-t,\lambda(t))\right)ds\\
&+ \frac{-16}{\lambda(t)^3}\int_t^\infty \lambda''(s)K(s-t,\lambda(t))ds + E_{v_2,ip}(t,\lambda(t))\\
&+ \langle\left(\frac{\cos(2Q_{\frac{1}{\lambda(t)}})-1}{r^2}\right)\left((v_4+v_5)\left(1-\chi_{\geq 1}(\frac{4r}{t})\right)+E_5\right)|_{r=R\lambda(t)},\phi_0\rangle\\
&- \langle\left(\frac{\cos(2Q_{\frac{1}{\lambda(t)}})-1}{r^2}\right)\chi_{\geq 1}(\frac{2r}{\log^N(t)})\left(v_1+v_2+v_3\right)|_{r=R\lambda(t)},\phi_0\rangle\\
&- \langle\chi_{\geq 1}(\frac{2r}{\log^N(t)})F_{0,2}(t,r)|_{r=R\lambda(t)},\phi_0\rangle
\end{aligned}$$

where

$$E_{0,1}(\lambda(t),\lambda'(t),\lambda''(t)) = 2\frac{\lambda'(t)^2}{\lambda(t)^2} + \frac{2\lambda''(t)}{\lambda(t)} - \frac{4\alpha\lambda''(t)\log(\lambda(t))}{\lambda(t)}\left(\frac{1}{-1+\lambda(t)^{2\alpha}}+1\right)$$

So, the equation resulting from

$$\langle F_4(t),\phi_0(\frac{\cdot}{\lambda(t)})\rangle = 0$$

is what we recorded at the beginning of this section, namely (3.31). For convenience, we repeat the equation here.

$$
\begin{aligned}
& -4\int_t^\infty \frac{\lambda''(s)}{1+s-t}ds+\frac{4b}{t^2\log^b(t)}+4\alpha\log(\lambda(t))\lambda''(t)\\
& -4\int_t^\infty \frac{\lambda''(s)}{(\lambda(t)^{1-\alpha}+s-t)(1+s-t)^3}ds\\
& = -\lambda(t)E_{0,1}(\lambda(t),\lambda'(t),\lambda''(t)) - 16\int_t^\infty \lambda''(s)\left(K_3(s-t,\lambda(t))-K_{3,0}(s-t,\lambda(t))\right)ds\\
& + \frac{16}{\lambda(t)^2}\int_t^\infty K(s-t,\lambda(t))\lambda''(s)ds - \lambda(t)E_{v_2,ip}(t,\lambda(t))\\
& + \frac{16}{\lambda(t)^2}\int_t^\infty \lambda''(s)\left(K_1(s-t,\lambda(t))-\frac{\lambda(t)^2}{4(1+s-t)}\right)ds \qquad (3.63)\\
& -\lambda(t)\langle\left(\frac{\cos(2Q_{\frac{1}{\lambda(t)}})-1}{r^2}\right)\left((v_4+v_5)\left(1-\chi_{\geq 1}(\frac{4r}{t})\right)+E_5\right)|_{r=R\lambda(t)},\phi_0\rangle\\
& +\lambda(t)\langle\left(\frac{\cos(2Q_{\frac{1}{\lambda(t)}})-1}{r^2}\right)\chi_{\geq 1}(\frac{2r}{\log^N(t)})\left(v_1+v_2+v_3\right)|_{r=R\lambda(t)},\phi_0\rangle\\
& +\lambda(t)\langle\chi_{\geq 1}(\frac{2r}{\log^N(t)})F_{0,2}(t,r)|_{r=R\lambda(t)},\phi_0\rangle\\
& := G(t,\lambda(t))
\end{aligned}
$$

The key point is that there is leading order cancellation between the four terms on the first two lines of (3.63) (which means cancellation of terms of size $\frac{1}{t^2\log^b(t)}$ and terms of size $\frac{\log(\log(t))}{t^2\log^{b+1}(t)}$) when we substitute $\lambda=\lambda_0$ into these terms, where

$$\lambda_0(t)=\frac{1}{\log^b(t)}+\int_t^\infty\int_{t_1}^\infty \frac{-b^2\log(\log(t_2))}{t_2^2\log^{b+2}(t_2)}dt_2dt_1 := \lambda_{0,0}+\lambda_{0,1}$$

In order to show this cancellation, we first let $\lambda(t)=\lambda_0(t)+e(t)$, and re-write the equation for e in the following way:

$$
\begin{aligned}
& -4\int_t^\infty \frac{\lambda_{0,0}''(s)}{1+s-t}ds+\frac{4b}{t^2\log^b(t)}\\
& -4\int_t^\infty \frac{\lambda_{0,1}''(s)}{1+s-t}ds+4\alpha\log(\lambda_{0,0}(t))\lambda_{0,0}''(t)-4\int_t^\infty \frac{\lambda_{0,0}''(s)ds}{(\lambda_{0,0}(t)^{1-\alpha}+s-t)(1+s-t)^3}\\
& -4\int_t^\infty \frac{e''(s)ds}{1+s-t}+4\alpha\log(\lambda_0(t))e''(t)-4\int_t^\infty \frac{e''(s)ds}{(\lambda_0(t)^{1-\alpha}+s-t)(1+s-t)^3}\\
& +4\alpha(\log(\lambda(t))-\log(\lambda_{0,0}(t)))\lambda_{0,0}''(t)+4\alpha\log(\lambda(t))\left(\lambda_0''(t)-\lambda_{0,0}''(t)\right)\\
& +4\alpha e''(t)\left(\log(\lambda(t))-\log(\lambda_0(t))\right) \qquad (3.64)\\
& -4\int_t^\infty \frac{e''(s)}{(1+s-t)^3}\left(\frac{1}{\lambda(t)^{1-\alpha}+s-t}-\frac{1}{\lambda_0(t)^{1-\alpha}+s-t}\right)ds\\
& -4\int_t^\infty \frac{(\lambda_0''(s)-\lambda_{0,0}''(s))ds}{(\lambda_{0,0}(t)^{1-\alpha}+s-t)(1+s-t)^3}\\
& -4\int_t^\infty \frac{\lambda_0''(s)}{(1+s-t)^3}\left(\frac{1}{(\lambda(t)^{1-\alpha}+s-t)}-\frac{1}{(\lambda_{0,0}(t)^{1-\alpha}+s-t)}\right)ds\\
& = G(t,\lambda(t))\,.
\end{aligned}
$$

Now, we will show that the terms in each of the first two lines of (3.64) cancel to leading order. More precisely, this means that both terms in the first line of (3.64) are of size $\frac{1}{t^2\log^b(t)}$, but their sum has size bounded above by $\frac{1}{t^2\log^{b+1}(t)}$. Similarly, each term on the second line of (3.64) is of size $\frac{\log(\log(t))}{t^2\log^{b+1}(t)}$, but their sum has size bounded above by $\frac{1}{t^2\log^{b+1}(t)}$.

$$\begin{aligned}
-4\int_t^\infty \frac{\lambda_{0,0}''(s)}{1+s-t}ds &= -4\int_t^{2t} \frac{\lambda_{0,0}''(s)}{1+s-t}ds - 4\int_{2t}^\infty \frac{\lambda_{0,0}''(s)}{1+s-t}ds\\
&= -4\lambda_{0,0}''(t)\int_t^{2t}\frac{ds}{1+s-t} - 4\int_t^{2t}\frac{\lambda_{0,0}''(s)-\lambda_{0,0}''(t)}{1+s-t}ds\\
&- 4\int_{2t}^\infty \frac{\lambda_{0,0}''(s)}{1+s-t}ds\\
&= -\frac{4b}{t^2\log^{b+1}(t)}\log(1+t) + O(\frac{1}{t^2\log^{b+1}(t)}) + \text{Err}
\end{aligned}$$

where

$$\begin{aligned}
|\text{Err}| &\leq 4\int_t^{2t}\frac{||\lambda_{0,0}'''||_{L^\infty(t,2t)}(s-t)}{1+s-t}ds + \frac{4}{1+t}\int_{2t}^\infty |\lambda_{0,0}''(s)|ds\\
&\leq \frac{C}{t^2\log^{b+1}(t)}
\end{aligned}$$

So, we get

$$-4\int_t^\infty \frac{\lambda_{0,0}''(s)}{1+s-t}ds = -\frac{4b}{t^2\log^b(t)} + E_{\lambda_{0,0}}$$

where

$$|E_{\lambda_{0,0}}| \leq \frac{C}{t^2\log^{b+1}(t)}$$

Next, we have

$$\begin{aligned}
&-4\int_t^\infty \frac{\lambda_{0,0}''(x)dx}{(\lambda_{0,0}(t)^{1-\alpha}+x-t)(1+x-t)^3}\\
&= -4\int_t^\infty \frac{b}{x^2\log^{b+1}(x)}\frac{dx}{(\lambda_{0,0}(t)^{1-\alpha}+x-t)(1+x-t)^3} + E_{v_{3,ip}}
\end{aligned}$$

where

$$|E_{v_{3,ip}}| \leq C\int_t^\infty \frac{dx}{x^2\log^{b+2}(x)(\lambda_{0,0}(t)^{1-\alpha}+x-t)(1+x-t)^3}$$

Then, we have

$$\begin{aligned}
&|-4b\int_t^{t+\lambda_{0,0}(t)^{1-\alpha}} \frac{dx}{x^2\log^{b+1}(x)(\lambda_{0,0}(t)^{1-\alpha}+x-t)(1+x-t)^3}|\\
&\leq C\int_t^{t+\lambda_{0,0}(t)^{1-\alpha}} \frac{dx}{x^2\log^{b+1}(x)}\frac{1}{\lambda_{0,0}(t)^{1-\alpha}}\\
&\leq \frac{C}{t^2\log^{b+1}(t)}
\end{aligned}$$

The second term to consider is

(3.65)

$$
\begin{aligned}
&-4b\int_{t+\log^{(\alpha-1)b}(t)}^{t+\log^{(\alpha-1)b}(t)+\frac{1}{2}}\frac{1}{x^2\log^{b+1}(x)}\frac{1}{(\log^{(\alpha-1)b}(t)+x-t)}\frac{dx}{(1+x-t)^3}\\
&=\frac{-4b}{t^2\log^{b+1}(t)}\int_{t+\log^{(\alpha-1)b}(t)}^{t+\log^{(\alpha-1)b}(t)+\frac{1}{2}}\frac{dx}{(\log^{(\alpha-1)b}(t)+x-t)}\\
&-4b\int_{t+\log^{(\alpha-1)b}(t)}^{t+\log^{(\alpha-1)b}(t)+\frac{1}{2}}\left(\frac{1}{x^2\log^{b+1}(x)}-\frac{1}{t^2\log^{b+1}(t)}\right)\frac{1}{(\log^{(\alpha-1)b}(t)+x-t)}dx\\
&-4b\int_{t+\log^{(\alpha-1)b}(t)}^{t+\log^{(\alpha-1)b}(t)+\frac{1}{2}}\frac{1}{x^2\log^{b+1}(x)}\frac{1}{(\log^{(\alpha-1)b}(t)+x-t)}\left(\frac{1}{(1+x-t)^3}-1\right)dx
\end{aligned}
$$

The second line of (3.65) is treated as follows:

$$
\begin{aligned}
&\frac{-4b}{t^2\log^{b+1}(t)}\int_{t+\log^{(\alpha-1)b}(t)}^{t+\log^{(\alpha-1)b}(t)+\frac{1}{2}}\frac{dx}{(\log^{(\alpha-1)b}(t)+x-t)}\\
&=\frac{-4b}{t^2\log^{b+1}(t)}\left(\log(2\log^{(\alpha-1)b}(t)+\frac{1}{2})-\log(2\log^{(\alpha-1)b}(t))\right)\\
&=\frac{4b^2(\alpha-1)\log(\log(t))}{t^2\log^{b+1}(t)}+E_{v_{3,ip},1b}
\end{aligned}
$$

where

$$|E_{v_{3,ip},1b}|\leq\frac{C}{t^2\log^{b+1}(t)}$$

The third line of (3.65) is estimated by:

$$
\begin{aligned}
&|-4b\int_{t+\log^{(\alpha-1)b}(t)}^{t+\log^{(\alpha-1)b}(t)+\frac{1}{2}}\left(\frac{1}{x^2\log^{b+1}(x)}-\frac{1}{t^2\log^{b+1}(t)}\right)\frac{1}{(\log^{(\alpha-1)b}(t)+x-t)}dx|\\
&\leq\frac{C}{t^3\log^{b+1}(t)}\int_{t+\log^{(\alpha-1)b}(t)}^{t+\log^{(\alpha-1)b}(t)+\frac{1}{2}}\frac{(x-t)dx}{(\log^{(\alpha-1)b}(t)+x-t)}\\
&\leq\frac{C}{t^3\log^{b+1}(t)}
\end{aligned}
$$

The last line of (3.65) is estimated by

$$
\begin{aligned}
&|-4b\int_{t+\log^{(\alpha-1)b}(t)}^{t+\log^{(\alpha-1)b}(t)+\frac{1}{2}}\frac{1}{x^2\log^{b+1}(x)}\frac{1}{(\log^{(\alpha-1)b}(t)+x-t)}\left(\frac{1}{(1+x-t)^3}-1\right)dx|\\
&\leq C\int_{t+\log^{(\alpha-1)b}(t)}^{t+\log^{(\alpha-1)b}(t)+\frac{1}{2}}\frac{1}{x^2\log^{b+1}(x)}\frac{x-t}{(\log^{(\alpha-1)b}(t)+x-t)}dx\\
&\leq\frac{C}{t^2\log^{b+1}(t)}
\end{aligned}
$$

The third term to consider is

$$
\begin{aligned}
&|-4b\int_{t+\log^{(\alpha-1)b}(t)+\frac{1}{2}}^{\infty}\frac{1}{x^2\log^{b+1}(x)}\frac{1}{(\log^{(\alpha-1)b}(t)+x-t)}\frac{dx}{(1+x-t)^3}|\\
&\leq \frac{C}{t^2\log^{b+1}(t)}\int_{t+\log^{(\alpha-1)b}(t)+\frac{1}{2}}^{\infty}\frac{dx}{(x-t)^4}\\
&\leq \frac{C}{t^2\log^{b+1}(t)}
\end{aligned}
$$

Then, we estimate $E_{v_3,ip}$:

$$
\begin{aligned}
|E_{v_3,ip}| &\leq \frac{C}{t^2\log^{b+2}(t)}\int_t^{\infty}\frac{dx}{(\log^{(\alpha-1)b}(t)+x-t)(1+x-t)^3}\\
&\leq \frac{C\log(\log(t))}{t^2\log^{b+2}(t)}
\end{aligned}
$$

In total, we have

$$
-4\int_t^{\infty}\frac{\lambda_{0,0}''(x)dx}{(\log^{(\alpha-1)b}(t)+x-t)(1+x-t)^3} = E_{v_3,ip,f}+\frac{4b^2(\alpha-1)\log(\log(t))}{t^2\log^{b+1}(t)}
$$

where

$$
|E_{v_3,ip,f}| \leq \frac{C}{t^2\log^{b+1}(t)}
$$

By the same procedure used to study the analogous term involving $\lambda_{0,0}$, we have

$$
-4\int_t^{\infty}\frac{\lambda_{0,1}''(s)ds}{1+s-t} = \frac{4b^2\log(\log(t))}{t^2\log^{b+1}(t)}+E_{v_3,ip,01}
$$

where

$$
|E_{v_3,ip,01}| \leq \frac{C\log(\log(t))}{t^2\log^{b+2}(t)}
$$

Combining the above, we can show the cancellation, to leading order, of the terms on each of the first two lines of (3.64):

$$
-4\int_t^{\infty}\frac{\lambda_{0,0}''(s)}{1+s-t}ds+\frac{4b}{t^2\log^b(t)} = E_{\lambda_{0,0}}
$$

with

$$
|E_{\lambda_{0,0}}| \leq \frac{C}{t^2\log^{b+1}(t)}
$$

and

$$
\begin{aligned}
&-4\int_t^{\infty}\frac{\lambda_{0,1}''(s)}{1+s-t}ds+4\alpha\log(\lambda_{0,0}(t))\lambda_{0,0}''(t)-4\int_t^{\infty}\frac{\lambda_{0,0}''(s)ds}{(\lambda_{0,0}(t)^{1-\alpha}+s-t)(1+s-t)^3}\\
&= E_{v_3,ip,01}+E_{v_3,ip,f}+E_{\lambda_{0,1}}
\end{aligned}
$$

with

$$
E_{\lambda_{0,1}}(t) = 4\alpha\log(\lambda_{0,0}(t))\left(\lambda_{0,0}''(t)-\frac{b}{t^2\log^{b+1}(t)}\right)
$$

$$
|E_{v_3,ip,01}|+|E_{v_3,ip,f}|+|E_{\lambda_{0,1}}| \leq \frac{C}{t^2\log^{b+1}(t)}
$$

We recall that we are considering (3.63) for $t\in[T_0,\infty)$, with T_0 satisfying (3.1).

The equation for e can be written as follows:

(3.66)
$$
\begin{aligned}
&-4\int_t^\infty \frac{e''(s)ds}{\log(\lambda_0(s))(1+s-t)}+4\alpha e''(t)-4\int_t^\infty \frac{e''(s)ds}{\log(\lambda_0(s))(\lambda_0(t)^{1-\alpha}+s-t)(1+s-t)^3}\\
&=\frac{1}{\log(\lambda_0(t))}\left(-E_{\lambda_{0,0}}-E_{v_3,ip,01}-E_{v_3,ip,f}-E_{\lambda_{0,1}}\right)\\
&+\frac{1}{\log(\lambda_0(t))}\left(G(t,\lambda(t))-4\alpha(\log(\lambda(t))-\log(\lambda_{0,0}(t)))\lambda_{0,0}''(t)-4\alpha\log(\lambda(t))\left(\lambda_0''(t)-\lambda_{0,0}''(t)\right)\right)\\
&+\frac{1}{\log(\lambda_0(t))}\left(-4\alpha e''(t)\left(\log(\lambda(t))-\log(\lambda_0(t))\right)\right)\\
&+\frac{1}{\log(\lambda_0(t))}\left(4\int_t^\infty \frac{e''(s)}{(1+s-t)^3}\left(\frac{1}{\lambda(t)^{1-\alpha}+s-t}-\frac{1}{\lambda_0(t)^{1-\alpha}+s-t}\right)ds\right)\\
&+\frac{4}{\log(\lambda_0(t))}\int_t^\infty \frac{(\lambda_0''(s)-\lambda_{0,0}''(s))ds}{(\lambda_{0,0}(t)^{1-\alpha}+s-t)(1+s-t)^3}\\
&+\frac{4}{\log(\lambda_0(t))}\int_t^\infty \frac{\lambda_0''(s)}{(1+s-t)^3}\left(\frac{1}{(\lambda(t)^{1-\alpha}+s-t)}-\frac{1}{(\lambda_{0,0}(t)^{1-\alpha}+s-t)}\right)ds\\
&+4\int_t^\infty e''(s)\left(\frac{1}{\log(\lambda_0(t))}-\frac{1}{\log(\lambda_0(s))}\right)\frac{1}{(1+s-t)}ds\\
&+4\int_t^\infty e''(s)\left(\frac{1}{\log(\lambda_0(t))}-\frac{1}{\log(\lambda_0(s))}\right)\frac{1}{(\lambda_0(t)^{1-\alpha}+s-t)(1+s-t)^3}ds\\
&:=RHS(e,t)
\end{aligned}
$$

where $\lambda(t)=\lambda_0(t)+e(t)$. In order to study this equation, let us first consider the problem of solving an equation of the form

$$
-\int_t^\infty \frac{y(s)}{\log(\lambda_0(s))}\left(\frac{1}{1+s-t}+\frac{1}{(\lambda_0(t)^{1-\alpha}+s-t)(1+s-t)^3}\right)ds+\alpha y(t)=F(t)\quad ,t\geq T_0
$$

where

$$
F\in C([T_0,\infty)),\quad |F(x)|\leq \frac{C}{x^2} \tag{3.67}
$$

If

$$
x(t)=y(-t),\quad H(t)=\frac{F(-t)}{\alpha},\quad t\leq -T_0
$$

then our equation becomes

$$
\begin{aligned}
&-\int_{-\infty}^t \frac{x(s)}{\alpha\log(\lambda_0(-s))}\left(\frac{1}{1-s+t}+\frac{1}{(\lambda_0(-t)^{1-\alpha}-s+t)(1-s+t)^3}\right)ds+x(t)\\
&=H(t),\quad t\leq -T_0
\end{aligned}
$$

which can be written as

$$
\int_J x(s)K(t,s)ds+x(t)=H(t),\quad t\in J \tag{3.68}
$$

where

$$
J=(-\infty,-T_0]
$$

and, for $(t,s)\in J^2$,

$$
\begin{aligned}
K(t,s)&=-\frac{\mathbb{1}_{\leq 0}(s-t)}{\alpha\log(\lambda_0(-s))}\left(\frac{1}{1-s+t}+\frac{1}{(\lambda_0(-t)^{1-\alpha}-s+t)(1-s+t)^3}\right)\\
&=\frac{\mathbb{1}_{\leq 0}(s-t)}{\alpha|\log(\lambda_0(-s))|}\left(\frac{1}{1-s+t}+\frac{1}{(\lambda_0(-t)^{1-\alpha}-s+t)(1-s+t)^3}\right)
\end{aligned}
$$

Note that $K(t,s)=0$ when $s>t$, so K is a (non-negative) Volterra kernel on J^2.

Moreover, K is of type $L^\infty_{\rm loc}$ on J. To see this, it suffices to consider $C=[a,d]\subset J$ a compact subinterval, $g\in L^1(C), f\in L^\infty(C)$ with $||g||_{L^1(C)}\le 1, ||f||_{L^\infty(C)}\le 1$ and estimate

$$\begin{aligned}
&\int_C\int_C |g(t)||K(t,s)||f(s)|dsdt\\
&\le \int_C \frac{|g(t)|}{2\alpha}\left(\int_a^t \frac{ds}{1-s+t}+\frac{1}{\lambda_0(-t)^{1-\alpha}}\int_{-\infty}^t \frac{ds}{(1-s+t)^3}\right)dt\\
&\le \frac{\log(1+d-a)+\frac{1}{2\lambda_0(-a)^{1-\alpha}}}{2\alpha}
\end{aligned}$$

where we used the facts (which follow from (3.1) and (3.2))

$$t\mapsto \frac{1}{\lambda_0(-t)^{1-\alpha}} \text{ is decreasing,}\qquad \frac{1}{|\log(\lambda_0(-t))|}\le\frac{1}{2},\qquad t\in J$$

Moreover, if $s\le u\le v\le t$, and $(t,v,u,s)\in J^4$, then,

$$K(v,s)K(t,u)\le K(t,s)K(v,u) \tag{3.69}$$

To verify this, let us note that, by the given conditions on s,u,v,t, we have

$$1=\mathbb{1}_{\le 0}(s-v)=\mathbb{1}_{\le 0}(u-t)=\mathbb{1}_{\le 0}(s-t)=\mathbb{1}_{\le 0}(u-v)$$

and

$$K(t,s)=\frac{\mathbb{1}_{\le 0}(s-t)}{\alpha|\log(\lambda_0(-s))|}k(t,s)$$

for

$$k(t,s)=\frac{1}{1-s+t}+\frac{1}{(\lambda_0(-t)^{1-\alpha}-s+t)(1-s+t)^3}$$

So, it suffices to show that

$$\frac{\partial^2}{\partial s\partial t}\log(k(t,s))\le 0,\quad s\le t$$

Then, we note that

$$\log(k(t,s))=\log(\frac{1}{1-s+t})+\log\left(1+\frac{1}{(\lambda_0(-t)^{1-\alpha}-s+t)(1-s+t)^2}\right)$$

and we have

$$\begin{aligned}
&\partial_{st}\log\left(1+\frac{1}{(\lambda_0(-t)^{1-\alpha}-s+t)(1-s+t)^2}\right)\\
&=\left(\frac{-\left(2(1-s+t)(\lambda_0(-t)^{1-\alpha}-s+t)+(1-s+t)^2(-\lambda_0'(-t)(1-\alpha)\lambda_0(-t)^{-\alpha}+1)\right)}{((1-s+t)^2(\lambda_0(-t)^{1-\alpha}-s+t)+1)^2}\right)\\
&\cdot\left(\frac{2}{1-s+t}+\frac{1}{(\lambda_0(-t)^{1-\alpha}-s+t)}\right)\\
&+\left(\frac{1}{(1-s+t)^2(\lambda_0(-t)^{1-\alpha}-s+t)+1}\right)\left(\frac{-2}{(1-s+t)^2}-\frac{(-(1-\alpha)\lambda_0(-t)^{-\alpha}\lambda_0'(-t)+1)}{(\lambda_0(-t)^{1-\alpha}-s+t)^2}\right)\\
&\le 0
\end{aligned}$$

where we recall $\alpha\le\frac{1}{4}$ and $\lambda_0'(x)\le 0,\quad x\ge T_0$. This completes the verification of (3.69).

Now, by Theorem 8.6 (sec. 9.8, pg. 259) of [**7**], K has a non-negative resolvent, r, which is locally of type L^∞ on J^2. Recall that the resolvent kernel, r, satisfies (a.e) the equations

$$r + K * r = K, \quad r + r * K = K \tag{3.70}$$

where the $*$ operation, (as defined in [**7**], Definition 2.3) between two measurable functions on J^2, or between a measurable function on J^2 and one on J (when the integrands are integrable) is

$$(a * b)(t, s) = \int_J a(t, u) b(u, s) du$$

$$(a * f)(t) = \int_J a(t, u) f(u) du$$

In fact, we also have

$$\frac{K(t, s)}{K(u, s)} \le 2, \quad s \le u \le t \tag{3.71}$$

because

$$\begin{aligned}\frac{K(t,s)}{K(u,s)} &= \frac{\frac{1}{1-s+t} + \frac{1}{(\lambda_0(-t)^{1-\alpha}-s+t)(1-s+t)^3}}{\frac{1}{1-s+u} + \frac{1}{(\lambda_0(-u)^{1-\alpha}-s+u)(1-s+u)^3}} \\ &\le \frac{1-s+u}{1-s+t} + \frac{(\lambda_0(-u)^{1-\alpha}-s+u)(1-s+u)^3}{(\lambda_0(-t)^{1-\alpha}-s+t)(1-s+t)^3} \\ &\le 2, \quad s \le u \le t\end{aligned}$$

Hence, by Theorem 8.5(sec. 9.8, pg. 258) of [**7**] r is in fact a non-negative Volterra kernel of type L^∞, and we have the estimate

$$\int_{-\infty}^t r(t, u) du \le 2, \quad \text{a.e. } t \in J \tag{3.72}$$

By our assumptions on F, namely (3.67), H satisfies the property that (e.g)

$$H(\cdot)\,(\cdot)^2 \in L^\infty(J) \tag{3.73}$$

Then, we have a solution to (3.68) (a.e.) given by the formula

$$x = H - (r * H) \tag{3.74}$$

To see this, we first note that (3.68) is

$$K * x + x = H$$

Then,

$$K * (r * H) = (K * r) * H$$

by Fubini's theorem. Fubini's theorem is applicable because

$$\int_J\int_J |K(t,u)||r(u,s)||H(s)|dsdu = \int_{-\infty}^t\int_{-\infty}^u |K(t,u)||r(u,s)||H(s)|dsdu$$
$$\leq ||H(\cdot)\,(\cdot)^2\,||_{L^\infty(J)}\int_{-\infty}^t\int_{-\infty}^u K(t,u)\frac{r(u,s)}{s^2}dsdu$$
$$\leq ||H(\cdot)\,(\cdot)^2\,||_{L^\infty(J)}\int_{-\infty}^t\int_{-\infty}^u \frac{K(t,u)}{u^2}r(u,s)dsdu$$
$$\leq 2||H(\cdot)\,(\cdot)^2\,||_{L^\infty(J)}\int_{-\infty}^t \frac{K(t,u)}{u^2}du$$
$$\leq \frac{2||H(\cdot)\,(\cdot)^2\,||_{L^\infty(J)}}{\alpha|\log(\lambda_0(T_0))|}\left(1+\frac{1}{\lambda_0(-t)^{1-\alpha}}\right)\int_{-\infty}^t\frac{du}{u^2}$$
$$\leq \frac{2||H(\cdot)\,(\cdot)^2\,||_{L^\infty(J)}}{\alpha|\log(\lambda_0(T_0))|}\left(1+\frac{1}{\lambda_0(-t)^{1-\alpha}}\right)\frac{1}{|t|},\quad t\in J$$

by (3.72), (3.73), and inspection of the formula for K. Now, substituting

$$x = H - (r*H)$$

using Fubini's theorem as above, and using the resolvent equation

$$r + K*r = K$$

we see that (3.74) is a solution to (3.68).

But, this means that we have a solution to the equation

$$\alpha y(t) = F(t) + \int_t^\infty \frac{y(s)}{\log(\lambda_0(s))}\left(\frac{1}{1+s-t}+\frac{1}{(\lambda_0(t)^{1-\alpha}+s-t)(1+s-t)^3}\right)ds$$
$$,\quad \text{a.e. } t\geq T_0$$

Since we are considering this equation for F satisfying (3.67) the right-hand side of the equation above is a continuous function of t. So, y agrees with a continuous function a.e., and hence, we may extend y given a.e. by (3.74) to a continuous function of $t\in[T_0,\infty)$.

(3.74), written in terms of y reads

$$y(t) = \frac{F(t)}{\alpha} - \int_t^\infty \frac{F(s)}{\alpha}r(-t,-s)ds \tag{3.75}$$

and (3.72) implies that

$$\int_t^\infty r(-t,-z)dz \leq 2,\quad \text{a.e. } t\geq T_0 \tag{3.76}$$

Now, we are finally ready to solve (3.66). We recall the complete, normed vector space $(X,||\cdot||_X)$ defined in the beginning of this section by

$$X = \{f\in C^2([T_0,\infty))|||f||_X<\infty\}$$

where

$$||f||_X = \sup_{t\geq T_0}\Big(|f(t)|b\log^b(t)\sqrt{\log(\log(t))} + |f'(t)|t\log^{b+1}(t)\sqrt{\log(\log(t))}$$
$$+|f''(t)|t^2\log^{b+1}(t)\sqrt{\log(\log(t))}\Big)$$

We quickly remark that all previous manipulations done on v_k, including estimates and representations of inner products, are valid for all $\lambda = \lambda_0 + e, \quad e \in \overline{B_1(0)} \subset X$, since (3.2) is valid for all λ of this form. For $e \in \overline{B_1(0)} \subset X$, $RHS(e,t)$ is a continuous function of $t \in [T_0, \infty)$. We will now estimate $RHS(e,t)$ for an arbitrary $e \in \overline{B_1(0)} \subset X$. The estimate we will obtain will then allow us to define a map, T, on $\overline{B_1}(0) \subset X$, and prove some properties about it, using the discussion above. Eventually, we will show that T has a fixed point.

We start by estimating all the terms of $RHS(e,t)$, except for the one involving G, for $e \in \overline{B_1(0)} \subset X$. From our previous calculations, we have

$$|E_{\lambda_{0,0}}| + |E_{v_3,ip,01}| + |E_{v_3,ip,f}| + |E_{\lambda_{0,1}}| \leq \frac{C}{t^2 \log^{b+1}(t)}$$

So, the first line of $RHS(e,t)$ is bounded above in absolute value by

$$\frac{C}{\log(\log(t)) t^2 \log^{b+1}(t)}$$

Next, we note that

$$\lambda_{0,1}(t) = \int_t^\infty \int_{t_1}^\infty \frac{-b^2 \log(\log(t_2))}{t_2^2 \log^{b+2}(t_2)} dt_2 dt_1 = \frac{-b^2 \log(\log(t))}{(b+1) \log^{b+1}(t)} + O\left(\frac{1}{\log^{b+1}(t)}\right)$$

Then, using the fact that $e \in \overline{B_1(0)} \subset X$, we get that the terms in the second line of $RHS(e,t)$, except for the one depending on G are bounded above in absolute value by

$$\frac{C}{(\log(\log(t)))^{3/2} t^2 \log^{b+1}(t)}$$

Similarly, the third, fourth, fifth, sixth, seventh, and eighth lines of $RHS(e,t)$ are bounded above in absolute value (respectively) by

$$\begin{aligned} &\frac{C}{(\log(\log(t)))^2 t^2 \log^{b+1}(t)} + \frac{C}{(\log(\log(t)))^2 \log^{b+1}(t) t^2} + \frac{C \log(\log(t))}{t^2 \log^{b+2}(t)} \\ &+ \frac{C}{(\log(\log(t)))^{3/2} t^2 \log^{b+1}(t)} + \frac{C}{t^2 \log^{b+2}(t) (\log(\log(t))^{5/2}} \\ &+ \frac{C}{(\log(\log(t))^{5/2} t^3 \log^{b+2}(t)} \end{aligned}$$

Now, we proceed to estimate the terms from G. In the below expressions, we note that $\lambda(t) = \lambda_0(t) + e(t)$, and $e \in \overline{B}_1(0) \subset X$ is arbitrary. Recalling the definition of $E_{0,1}$, we have

$$|\lambda(t) E_{0,1}(\lambda(t), \lambda'(t), \lambda''(t))| \leq \frac{C}{t^2 \log^{b+1}(t)}$$

Then, from (3.60), we have

$$|-16 \int_t^\infty \lambda''(s) \left(K_3(s-t, \lambda(t)) - K_{3,0}(s-t, \lambda(t))\right) ds| \leq \frac{C}{t^2 \log^{b+1}(t)}$$

From (3.40), we have

$$|\lambda(t) E_{v_2,ip}(t, \lambda(t))| \leq \frac{C}{t^2 \log^{b+1}(t)}$$

Using (3.34) and (3.32), we get

$$|\frac{16}{\lambda(t)^2}\int_t^\infty \lambda''(s)K(s-t,\lambda(t))ds| \leq \frac{C}{t^2\log^{b+1}(t)}$$
$$|\frac{16}{\lambda(t)^2}\int_t^\infty \lambda''(x)\left(K_1(x-t,\lambda(t))-\frac{\lambda(t)^2}{4(1+x-t)}\right)dx| \leq \frac{C}{t^2\log^{b+1}(t)}$$

In order to proceed, we will use some pointwise estimates on v_3, v_4 and v_5.

3.0.8.6. *Pointwise estimates on* $v_3, \partial_r^j v_3$. Here, we will prove two simple pointwise estimate on v_3 which do not directly follow from (3.21). As with all of our work from now on, every estimate is valid for any λ of the form

$$\lambda(t)=\lambda_0(t)+e(t), \quad e\in\overline{B}_1(0)\subset X$$

LEMMA 3.8. *We have the following pointwise estimates on* $\partial_r^j v_3, \quad j=0,1,2$:

$$|v_3(t,r)| \leq \frac{Cr\log(\log(t))}{t^2\log^{b+1}(t)} \tag{3.77}$$

$$|v_3(t,r)| \leq \frac{C}{r}\int_t^\infty |\lambda''(s)|(s-t)ds \tag{3.78}$$

$$|\partial_r v_3(t,r)| \leq \frac{C}{t^2\log^b(t)} \tag{3.79}$$

PROOF. We again make the decomposition

$$v_3=v_{3,1}+v_{3,2}$$

and use the same estimates on $v_{3,2}$ proven while obtaining (3.21), to get

$$|v_{3,2}(t,r)| \leq \frac{Cr}{t^2\log^{b+1}(t)}$$

For $v_{3,1}$, whose definition is (3.22), we have

$$|v_{3,1}(t,r)| \leq \frac{Cr}{t^2\log^{b+1}(t)}+Cr\int_{6r}^\infty |\lambda''(t+w)|w|\frac{1}{(\lambda(t+w)^{2-2\alpha}+w^2)}-\frac{1}{1+w^2}|dw$$

In order to estimate this integral, we use the fact that $\lambda'(x)\leq 0, \quad x\geq T_0$, and get

$$\begin{aligned}\int_0^1 w\frac{|\lambda''(t+w)|dw}{(\lambda(t+w)^{2-2\alpha}+w^2)} &\leq \frac{C}{t^2\log^{b+1}(t)}\frac{1}{\log^{(2\alpha-2)b}(t)}\int_0^1\frac{wdw}{1+w^2\lambda(t)^{2\alpha-2}}\\ &\leq \frac{C\log(\log(t))}{t^2\log^{b+1}(t)}\end{aligned} \tag{3.80}$$

Also,

$$\int_0^1\frac{w|\lambda''(t+w)|dw}{1+w^2}\leq\frac{C}{t^2\log^{b+1}(t)}$$

On the other hand, if $w\geq 1$, then,

$$|\frac{1}{(\lambda(t+w)^{2-2\alpha}+w^2)}-\frac{1}{1+w^2}|\leq\frac{C}{w^4}$$

which gives

$$\int_1^\infty |\lambda''(t+w)|w|\frac{1}{(\lambda(t+w)^{2-2\alpha}+w^2)}-\frac{1}{1+w^2}|dw\leq\frac{C}{t^2\log^{b+1}(t)}$$

In total, we get

$$|v_{3,1}(t,r)| \leq \frac{Cr\log(\log(t))}{t^2\log^{b+1}(t)}$$

which gives (3.77). For the second pointwise estimate on v_3 in the lemma, we use

$$\begin{aligned}|v_3(t,r)| &\leq \frac{1}{r}\int_t^\infty \int_0^{s-t} \frac{\rho}{\sqrt{(s-t)^2-\rho^2}}|\lambda''(s)|\cdot 2d\rho ds\\ &\leq \frac{C}{r}\int_t^\infty |\lambda''(s)|(s-t)ds\end{aligned}$$

We now prove the estimate on $\partial_r v_3$ in the lemma statement. We recall the definition of v_3

(3.81)

$$\begin{aligned}&v_3(t,r)\\ &= \frac{-1}{r}\int_t^\infty\int_0^{s-t}\frac{\rho}{\sqrt{(s-t)^2-\rho^2}}\lambda''(s)\left(\frac{-1-\rho^2+r^2}{\sqrt{(1+\rho^2-r^2)^2+4r^2}}+F_3(r,\rho,\lambda(s))\right)d\rho ds\end{aligned}$$

Then, we make a decomposition analogous to $v_3 = v_{3,1}+v_{3,2}$, used previously, and treat each term separately:

$$\begin{aligned}&|\int_t^\infty\int_0^{s-t}\rho\left(\frac{1}{\sqrt{(s-t)^2-\rho^2}}-\frac{1}{(s-t)}\right)\frac{\lambda''(s)}{r}\partial_r\left(\frac{-1-\rho^2+r^2}{\sqrt{(1+\rho^2-r^2)^2+4r^2}}\right)d\rho ds|\\ &\leq C\left(\sup_{x\geq t}|\lambda''(x)|\right)\int_0^\infty \rho\left(\frac{1+\rho^2+r^2}{(4r^2+(1+\rho^2-r^2)^2)^{3/2}}\right)d\rho\\ &\leq C\sup_{x\geq t}|\lambda''(x)|\end{aligned}$$

$$\begin{aligned}&|\int_t^\infty\int_0^{s-t}\rho\left(\frac{1}{\sqrt{(s-t)^2-\rho^2}}-\frac{1}{(s-t)}\right)\frac{\lambda''(s)}{r}\partial_r\left(F_3(r,\rho,\lambda(s))\right)d\rho ds|\\ &\leq C\sup_{x\geq t}\left(|\lambda''(x)|\lambda(x)^{4\alpha-4}\right)\int_0^\infty \rho\frac{(\rho^2+r^2+\lambda(t)^{2-2\alpha})}{(1+2(\rho^2+r^2)\lambda(t)^{2\alpha-2}+(\rho^2-r^2)^2\lambda(t)^{4\alpha-4})^{3/2}}d\rho\\ &\leq C\sup_{x\geq t}\left(|\lambda''(x)|\lambda(x)^{4\alpha-4}\right)\lambda(t)^{4-4\alpha}\end{aligned}$$

where we used

$$\frac{|\partial_r F_3|}{r} \leq \frac{C((\rho^2+r^2)\lambda(s)^{4\alpha-4}+\lambda(s)^{2\alpha-2})}{(1+2(\rho^2+r^2)\lambda(s)^{2\alpha-2}+(\rho^2-r^2)^2\lambda(s)^{4\alpha-4})^{3/2}} \tag{3.82}$$

Next, we have the term where the r derivative acts on the $\frac{1}{r}$ factored out of the integrals in (3.81). For this term, we simply note that

$$\begin{aligned}&\frac{1}{r^2}\int_t^\infty\int_0^{s-t}\frac{\rho\lambda''(s)}{\sqrt{(s-t)^2-\rho^2}}\left(\frac{-1-\rho^2+r^2}{\sqrt{(1+\rho^2-r^2)^2+4r^2}}+F_3(r,\rho,\lambda(s))\right)d\rho ds\\ &= \frac{-v_3(t,r)}{r}\end{aligned}$$

Then, using (3.77), we have

$$\begin{aligned}&|\frac{1}{r^2}\int_t^\infty\int_0^{s-t}\frac{\rho\lambda''(s)}{\sqrt{(s-t)^2-\rho^2}}\left(\frac{-1-\rho^2+r^2}{\sqrt{(1+\rho^2-r^2)^2+4r^2}}+F_3(r,\rho,\lambda(s))\right)d\rho ds|\\&\leq C\frac{\log(\log(t))}{t^2\log^{b+1}(t)}\end{aligned}$$

The last term to estimate is

$$-\int_t^\infty\frac{1}{(s-t)}\int_0^{s-t}\rho\frac{\lambda''(s)}{r}\partial_r\left(\frac{-1-\rho^2+r^2}{\sqrt{(1+\rho^2-r^2)^2+4r^2}}+F_3(r,\rho,\lambda(s))\right)d\rho ds$$

If $s-t\leq\frac{1}{2}$, we start with

$$\begin{aligned}&|\int_0^{s-t}\frac{\rho}{r}\partial_r\left(\frac{-1-\rho^2+r^2}{\sqrt{(1+\rho^2-r^2)^2+4r^2}}\right)d\rho|\\&=|\frac{-1-r^2+(s-t)^2}{\sqrt{4(s-t)^2+(1+r^2-(s-t)^2)^2}}+1|\\&=\frac{4(s-t)^2}{\sqrt{4(s-t)^2+(1+r^2-(s-t)^2)^2}(1+r^2-(s-t)^2+\sqrt{(1+r^2-(s-t)^2)^2+4(s-t)^2})}\\&\leq C(s-t)^2,\quad s-t\leq\frac{1}{2}\end{aligned}$$

This gives

$$\begin{aligned}&|-\int_t^{t+\frac{1}{2}}\frac{1}{(s-t)}\int_0^{s-t}\rho\frac{\lambda''(s)}{r}\partial_r\left(\frac{-1-\rho^2+r^2}{\sqrt{(1+\rho^2-r^2)^2+4r^2}}\right)d\rho ds|\\&\leq\frac{C}{t^2\log^{b+1}(t)}\end{aligned}$$

On the other hand, we have

$$\begin{aligned}&|-\int_{t+\frac{1}{2}}^\infty\frac{1}{(s-t)}\int_0^{s-t}\rho\frac{\lambda''(s)}{r}\partial_r\left(\frac{-1-\rho^2+r^2}{\sqrt{(1+\rho^2-r^2)^2+4r^2}}\right)d\rho ds|\\&\leq\int_{t+\frac{1}{2}}^\infty\frac{1}{(s-t)}|\lambda''(s)|\cdot 2ds\\&\leq C\sup_{x\geq t}\left(x|\lambda''(x)|\right)\frac{\log(t)}{t}\end{aligned}$$

Now, we have to treat the F_3 related terms. We start by recalling (3.82).

Then, we use a slightly different procedure:

$$|-\int_t^\infty \frac{1}{(s-t)}\int_0^{s-t}\rho\frac{\lambda''(s)}{r}\partial_r\left(F_3(r,\rho,\lambda(s))\right)d\rho ds|$$
$$\leq C\int_0^\infty \rho\int_{\rho+t}^\infty \frac{|\lambda''(s)|}{(s-t)}\frac{(\rho^2+r^2+\lambda(s)^{2-2\alpha})\lambda(s)^{4\alpha-4}}{(1+2(\rho^2+r^2)\lambda(s)^{2\alpha-2}+(\rho^2-r^2)^2\lambda(s)^{4\alpha-4})^{3/2}}dsd\rho$$
$$\leq C\int_0^\infty \rho\int_{\rho+t}^\infty \frac{|\lambda''(s)|}{(s-t)}\frac{(\rho^2+r^2+\lambda(t)^{2-2\alpha})\lambda(s)^{4\alpha-4}}{(1+2(\rho^2+r^2)\lambda(t)^{2\alpha-2}+(\rho^2-r^2)^2\lambda(t)^{4\alpha-4})^{3/2}}dsd\rho$$
$$\leq C\int_0^\infty \frac{\rho(\rho^2+r^2+\lambda(t)^{2-2\alpha})}{(1+2(\rho^2+r^2)\lambda(t)^{2\alpha-2}+(\rho^2-r^2)^2\lambda(t)^{4\alpha-4})^{3/2}}\frac{1}{\log^{(4\alpha-4)b}(t)t\log^{b+1}(t)}\int_{\rho+t}^\infty\frac{ds}{s(s-t)}d\rho$$
$$\leq C\int_0^t \frac{\rho(\rho^2+r^2+\lambda(t)^{2-2\alpha})}{(1+2(\rho^2+r^2)\lambda(t)^{2\alpha-2}+(\rho^2-r^2)^2\lambda(t)^{4\alpha-4})^{3/2}}\frac{\log(1+\frac{t}{\rho})d\rho}{\log^{(4\alpha-4)b}(t)t^2\log^{b+1}(t)}$$
$$+C\int_t^\infty \frac{\rho(\rho^2+r^2+\lambda(t)^{2-2\alpha})}{(1+2(\rho^2+r^2)\lambda(t)^{2\alpha-2}+(\rho^2-r^2)^2\lambda(t)^{4\alpha-4})^{3/2}}\frac{d\rho}{\log^{(4\alpha-4)b}(t)t^2\log^{b+1}(t)}$$

We finally consider two subsets of the region $r\leq t$ separately:

$$\int_0^{\lambda(t)^{1-\alpha}} \frac{\rho(\rho^2+r^2+\lambda(t)^{2-2\alpha})}{(1+2(\rho^2+r^2)\lambda(t)^{2\alpha-2}+(\rho^2-r^2)^2\lambda(t)^{4\alpha-4})^{3/2}}\frac{(\log(t)+|\log(\rho)|)d\rho}{\log^{(4\alpha-4)b}(t)t^2\log^{b+1}(t)}$$
$$\leq C\int_0^{\lambda(t)^{1-\alpha}} \frac{\rho\lambda(t)^{2-2\alpha}(\log(t)+|\log(\rho)|)d\rho}{(1+2(\rho^2+r^2)\lambda(t)^{2\alpha-2}+(\rho^2-r^2)^2\lambda(t)^{4\alpha-4})^{1/2}}\frac{1}{\log^{(4\alpha-4)b}(t)t^2\log^{b+1}(t)}$$
$$\leq \frac{C}{t^2\log^b(t)}$$

Then,

$$\int_{\lambda(t)^{1-\alpha}}^t \frac{\rho(\rho^2+r^2+\lambda(t)^{2-2\alpha})}{(1+2(\rho^2+r^2)\lambda(t)^{2\alpha-2}+(\rho^2-r^2)^2\lambda(t)^{4\alpha-4})^{3/2}}\frac{(\log(t)+|\log(\rho)|)d\rho}{\log^{(4\alpha-4)b}(t)t^2\log^{b+1}(t)}$$
$$\leq \frac{C\log(t)}{t^2\log^{(4\alpha-4)b}(t)\log^{b+1}(t)}\int_0^\infty \frac{\rho(\rho^2+r^2+\lambda(t)^{2-2\alpha})d\rho}{(1+2(\rho^2+r^2)\lambda(t)^{2\alpha-2}+(\rho^2-r^2)^2\lambda(t)^{4\alpha-4})^{3/2}}$$
$$\leq \frac{C}{t^2\log^b(t)}$$

The final integral to estimate is then

$$\int_t^\infty \frac{\rho(\rho^2+r^2+\lambda(t)^{2-2\alpha})d\rho}{(1+2(\rho^2+r^2)\lambda(t)^{2\alpha-2}+(\rho^2-r^2)^2\lambda(t)^{4\alpha-4})^{3/2}}\frac{1}{\log^{(4\alpha-4)b}(t)t^2\log^{b+1}(t)}$$
$$\leq \frac{C}{t^2\log^{b+1}(t)}$$

This gives (3.79). □

3.0.8.7. *Pointwise estimates on v_4, $\partial_r v_4$.* In this section, we prove

LEMMA 3.9. *For all λ of the form*

$$\lambda(t)=\lambda_0(t)+e(t),\quad e\in\overline{B}_1(0)\subset X$$

we have the pointwise estimates

$$|v_4(t,r)|\leq\begin{cases}\frac{Cr}{t^2\log^{3b+2N-1}(t)}, & r\leq\frac{t}{2}\\ \frac{C}{\sqrt{r}t\log^{\frac{3N}{2}+3b-1}(t)}, & r>\frac{t}{2}\end{cases}\tag{3.83}$$

$$|\partial_r v_4(t,r)|\leq\begin{cases}\frac{C}{t^2\log^{3b+2N-1}(t)}, & r\leq\frac{t}{2}\\ \frac{C}{\sqrt{r}t\log^{3b-1+\frac{5N}{2}}(t)}, & r\geq\frac{t}{2}\end{cases}\tag{3.84}$$

and

$$|\partial_t v_4(t,r)| \leq \frac{C}{\sqrt{r} t \log^{3b-1+\frac{5N}{2}}(t)}, \quad r \geq \frac{t}{2}$$

We also have the L^2 estimates

$$||\partial_t v_4||_{L^2(rdr)} + ||\partial_r v_4||_{L^2(rdr)} + ||\frac{v_4}{r}||_{L^2(rdr)} \leq \frac{C}{t \log^{2N+3b}(t)}$$

PROOF. We start by considering $v_4(t,r)$, for $\frac{t}{2} > r > 0$. In order to ease notation, let $x \in \mathbb{R}^2$ be defined by $x = r\mathbf{e}_1$. Then, we recall (3.25) and get

$$v_4(t,r) = \frac{-r}{2\pi} \int_0^1 \int_t^\infty \int_{B_{s-t}(0)} \frac{1}{\sqrt{(s-t)^2 - |y|^2}} \left(\frac{\partial_2 v_{4,c}(s, |\beta x + y|) \left((\beta x + y) \cdot \hat{x}\right)^2}{|\beta x + y|^2} - \frac{v_{4,c}(s, |\beta x + y|) \left((\beta x + y) \cdot \hat{x}\right)^2}{|\beta x + y|^3} + \frac{v_{4,c}(s, |\beta x + y|)}{|\beta x + y|} \right) dA(y) ds d\beta$$

Note that, for $|x| \leq \frac{t}{2}$, $0 < \beta < 1$, and $|y| \leq s-t$, we have

$$s - |\beta x + y| \geq \frac{t}{2}$$

This means that, for the purposes of estimating v_4 in the region $r \leq \frac{t}{2}$, we can use (3.51) to estimate v_2, for all $r \geq \frac{t}{2}$. We then combine this with the estimates for v_1, v_3, and $F_{0,2}$, to get

$$|v_{4,c}(t,r)| \leq C|\chi_{\geq 1}(\frac{2r}{\log^N(t)})| \begin{cases} \frac{1}{t^2 r^3 \log^{3b}(t)}, & r \leq \frac{t}{2} \\ \frac{\log(r)}{\log^{2b}(t) r^4 |t-r|} + \frac{\log^{2b\alpha}(t)}{t^2 r^3 \log^{3b+1}(t)}, & r \geq \frac{t}{2} \end{cases} \tag{3.85}$$

and similarly, for the derivatives, we have

$$|\partial_r v_{4,c}(t,r)| \leq C\chi_{\geq 1}(\frac{2r}{\log^N(t)}) \begin{cases} \frac{1}{r^4 t^2 \log^{3b}(t)}, & r \leq \frac{t}{2} \\ \frac{\log(r)}{\log^{2b}(t) r^4 (t-r)^2} + \frac{1}{\log^{3b}(t) t^2 r^4}, & r \geq \frac{t}{2} \end{cases} + \frac{C|\chi'_{\geq 1}(\frac{2r}{\log^N(t)})|}{\log^{5N+2b}(t)} \frac{r}{t^2 \log^b(t)}$$

Then, we get

$$\begin{aligned} &|v_4(t,r)| \\ &\leq Cr \int_0^1 \int_t^\infty \int_{B_{s-t}(0) \cap B_{\frac{s}{2}}(-\beta x)} \frac{1}{\sqrt{(s-t)^2 - |y|^2}} |\partial_2 v_{4,c}(s, |\beta x + y|)| dA(y) ds d\beta \\ &+ Cr \int_0^1 \int_t^\infty \int_{B_{s-t}(0) \cap B_{\frac{s}{2}}(-\beta x)} \frac{1}{\sqrt{(s-t)^2 - |y|^2}} \frac{|v_{4,c}(s, |\beta x + y|)|}{|\beta x + y|} dA(y) ds d\beta \\ &+ Cr \int_0^1 \int_t^\infty \int_{B_{s-t}(0) \cap (B_{\frac{s}{2}}(-\beta x))^c} \frac{1}{\sqrt{(s-t)^2 - |y|^2}} |\partial_2 v_{4,c}(s, |\beta x + y|)| dA(y) ds d\beta \\ &+ Cr \int_0^1 \int_t^\infty \int_{B_{s-t}(0) \cap (B_{\frac{s}{2}}(-\beta x))^c} \frac{1}{\sqrt{(s-t)^2 - |y|^2}} \frac{|v_{4,c}(s, |\beta x + y|)|}{|\beta x + y|} dA(y) ds d\beta \end{aligned} \tag{3.86}$$

Each line of (3.86) is then further split into two terms, based on the decomposition

$$\frac{1}{\sqrt{(s-t)^2-|y|^2}} = \frac{1}{s-t} + \left(\frac{1}{\sqrt{(s-t)^2-|y|^2}} - \frac{1}{s-t}\right)$$

and estimated separately. For the first term of the first line, we have

$$\begin{aligned}
&r\int_0^1\int_t^\infty\int_{B_{s-t}(0)\cap B_{\frac{s}{2}}(-\beta x)} \frac{1}{(s-t)}|\partial_2 v_{4,c}(s,|\beta x+y|)|dA(y)dsd\beta\\
&\leq Cr\int_0^1\int_t^{t+\frac{1}{2}}\int_{B_{s-t}(0)} \frac{1}{(s-t)}\frac{1}{\log^{4N}(s)s^2\log^{3b}(s)}dA(y)dsd\beta\\
&+Cr\int_0^1\int_{t+\frac{1}{2}}^\infty \frac{1}{(s-t)}\int_{B_{\frac{s}{2}}(-\beta x)} \frac{\mathbb{1}_{\{|\beta x+y|\geq \log^N(s)\}}}{|\beta x+y|^4 s^2\log^{3b}(s)}dA(y)dsd\beta\\
&\leq Cr\int_0^1\int_t^{t+\frac{1}{2}} \frac{(s-t)}{s^2\log^{3b+4N}(s)}dsd\beta\\
&+Cr\int_0^1\int_{t+\frac{1}{2}}^\infty \frac{1}{(s-t)}\int_{B_{\frac{s}{2}}(0)} \frac{\mathbb{1}_{\{|z|\geq \log^N(s)\}}}{|z|^4 s^2\log^{3b}(s)}dA(z)dsd\beta\\
&\leq \frac{Cr}{t^2\log^{3b+4N}(t)} + Cr\int_0^1\int_{t+\frac{1}{2}}^\infty \frac{1}{(s-t)}\int_{\log^N(s)}^{\frac{s}{2}}\rho\int_0^{2\pi}\frac{1}{\rho^4 s^2\log^{3b}(s)}d\theta d\rho dsd\beta\\
&\leq \frac{Cr}{\log^{3b+2N-1}(t)t^2}
\end{aligned}$$

where we used

$$\frac{|\chi'_{\geq 1}(\frac{|\beta x+y|}{\log^N(s)})||\beta x+y|}{\log^{5N+3b}(s)s^2} \leq C\frac{\mathbb{1}_{\{|\beta x+y|\geq \log^N(s)\}}}{|\beta x+y|^4\log^{3b}(s)s^2}$$

Next, we estimate

(3.87)

$$
\begin{aligned}
& r\int_0^1\int_t^\infty\int_{B_{s-t}(0)\cap B_{\frac{s}{2}}(-\beta x)}\left(\frac{1}{\sqrt{(s-t)^2-|y|^2}}-\frac{1}{(s-t)}\right)|\partial_r v_{4,c}(s,|\beta x+y|)|dA(y)dsd\beta \\
& \leq Cr\int_0^1\int_t^\infty\int_{B_{s-t}(0)}\left(\frac{1}{\sqrt{(s-t)^2-|y|^2}}-\frac{1}{(s-t)}\right)\frac{1}{\log^{2N}(s)}\frac{dA(y)dsd\beta}{(\log^{2N}(t)+|\beta x+y|^2)s^2\log^{3b}(s)} \\
& \leq Cr\int_0^1\int_0^\infty\rho\int_0^{2\pi}\frac{1}{\log^{2N}(t)}\frac{1}{(\log^{2N}(t)+\beta^2r^2+\rho^2+2\beta r\rho\cos(\theta))}\frac{1}{\log^{3b}(t)}\int_{\rho+t}^\infty \\
& \qquad \frac{1}{s^2}\left(\frac{1}{\sqrt{(s-t)^2-\rho^2}}-\frac{1}{(s-t)}\right)dsd\theta d\rho d\beta \\
& \leq Cr\int_0^1\int_0^\infty\rho\int_0^{2\pi}\frac{1}{\log^{2N+3b}(t)}\frac{1}{(\log^{2N}(t)+\beta^2r^2+\rho^2+2\beta r\rho\cos(\theta))}\frac{1}{(\rho+t)^2}d\theta d\rho d\beta \\
& \leq \frac{Cr}{\log^{2N+3b}(t)}\int_0^1\int_0^\infty\frac{\rho}{(\rho+t)^2}\frac{1}{\sqrt{(\log^{2N}(t)+(\beta r+\rho)^2)(\log^{2N}(t)+(\beta r-\rho)^2)}}d\rho d\beta \\
& \leq \frac{Cr}{\log^{2N+3b}(t)}\int_0^1\int_0^t\frac{\rho}{(\rho+t)^2}\frac{1}{\sqrt{(\log^{2N}(t)+(\beta r+\rho)^2)(\log^{2N}(t)+(\beta r-\rho)^2)}}d\rho d\beta \\
& + \frac{Cr}{\log^{2N+3b}(t)}\int_0^1\int_t^\infty\frac{\rho}{\rho^4}d\rho d\beta \\
& \leq \frac{Cr(\log(2+r^2)+\log(2+t^2))}{t^2\log^{2N+3b}(t)} \\
& \leq \frac{Cr}{t^2\log^{2N+3b-1}(t)}, \quad r\leq\frac{t}{2}
\end{aligned}
$$

where, we used the fact that

$$\rho\geq t,\quad r\leq\frac{t}{2}\implies|\beta r-\rho|=\rho-\beta r\geq\frac{\rho}{2}$$

Next, we estimate

$$
\begin{aligned}
& r\int_0^1\int_t^\infty\int_{B_{s-t}(0)\cap B_{\frac{s}{2}}(-\beta x)}\frac{1}{(s-t)}\frac{|v_{4,c}(s,|\beta x+y|)|}{|\beta x+y|}dA(y)dsd\beta \\
& \leq Cr\int_0^1\int_t^\infty\int_{B_{s-t}(0)\cap B_{\frac{s}{2}}(-\beta x)}\frac{1}{(s-t)}\frac{\chi_{\geq 1}(\frac{|\beta x+y|}{\log^N(s)})}{\log^{3b}(s)s^2|\beta x+y|^4}dA(y)dsd\beta
\end{aligned}
$$

The last line in the above equation has already been estimated above, and we get

$$r\int_0^1\int_t^\infty\int_{B_{s-t}(0)\cap B_{\frac{s}{2}}(-\beta x)}\frac{1}{(s-t)}\frac{|v_{4,c}(s,|\beta x+y|)|}{|\beta x+y|}dA(y)dsd\beta\leq\frac{Cr}{t^2\log^{3b+2N-1}(t)}$$

Next, we have

$$
\begin{aligned}
& r\int_0^1\int_t^\infty\int_{B_{s-t}(0)\cap B_{\frac{s}{2}}(-\beta x)}\left(\frac{1}{\sqrt{(s-t)^2-|y|^2}}-\frac{1}{(s-t)}\right)\frac{|v_{4,c}(s,|\beta x+y|)|}{|\beta x+y|}dA(y)dsd\beta \\
& \leq Cr\int_0^1\int_t^\infty\int_{B_{s-t}(0)\cap B_{\frac{s}{2}}(-\beta x)}\left(\frac{1}{\sqrt{(s-t)^2-|y|^2}}-\frac{1}{(s-t)}\right)\frac{\chi_{\geq 1}(\frac{|\beta x+y|}{\log^N(s)})}{|\beta x+y|^4s^2\log^{3b}(s)}dA(y)dsd\beta
\end{aligned}
$$

The last line in the above equation has also been estimated above. Next, we recall that

$$r\leq\frac{t}{2},|y|\leq s-t\implies s-|\beta x+y|\geq\frac{t}{2}$$

and consider

$$r\int_0^1\int_t^\infty\int_{B_{s-t}(0)\cap(B_{\frac{s}{2}}(-\beta x))^c}\frac{1}{(s-t)}|\partial_2 v_{4,c}(s,|\beta x+y|)|dA(y)dsd\beta$$

We first treat the portion of the integral when $s-t\leq\frac{1}{2}$:

$$\begin{aligned}&r\int_0^1\int_t^{t+\frac{1}{2}}\int_{B_{s-t}(0)\cap(B_{\frac{s}{2}}(-\beta x))^c}\frac{1}{(s-t)}|\partial_2 v_{4,c}(s,|\beta x+y|)|dA(y)dsd\beta\\&\leq Cr\int_0^1\int_t^{t+\frac{1}{2}}\int_{B_{s-t}(0)}\frac{1}{(s-t)}\left(\frac{1}{\log^{2b-1}(s)s^4t^2}+\frac{1}{\log^{3b}(s)s^6}\right)dA(y)dsd\beta\\&\leq\frac{Cr}{t^6\log^{2b-1}(t)}\end{aligned}$$

We next consider the region $s-t\geq\frac{1}{2}$, and get

$$\begin{aligned}&r\int_0^1\int_{t+\frac{1}{2}}^{\infty}\int_{B_{s-t}(0)\cap(B_{\frac{s}{2}}(-\beta x))^c}\frac{1}{(s-t)}|\partial_2 v_{4,c}(s,|\beta x+y|)|dA(y)dsd\beta\\&\leq Cr\int_0^1\int_{t+\frac{1}{2}}^{\infty}\frac{1}{(s-t)}\int_{(B_{\frac{s}{2}}(0))^c}\left(\frac{\log(|z|)}{\log^{2b}(s)|z|^4t^2}+\frac{1}{\log^{3b}(s)s^2|z|^4}\right)dA(z)dsd\beta\\&\leq Cr\int_{t+\frac{1}{2}}^{\infty}\frac{1}{(s-t)}\frac{1}{s^2t^2\log^{2b-1}(s)}ds\\&\leq\frac{Cr}{t^4\log^{2b-2}(t)}\end{aligned}$$

Next, we have

$$\begin{aligned}&r\int_0^1\int_t^{\infty}\int_{B_{s-t}(0)\cap(B_{\frac{s}{2}}(-\beta x))^c}\left(\frac{1}{\sqrt{(s-t)^2-|y|^2}}-\frac{1}{(s-t)}\right)|\partial_2 v_{4,c}(s,|\beta x+y|)|dA(y)dsd\beta\\&\leq Cr\int_0^1\int_t^{\infty}\int_{B_{s-t}(0)}\left(\frac{1}{\sqrt{(s-t)^2-|y|^2}}-\frac{1}{(s-t)}\right)\frac{1}{s^2}\frac{\log(s)}{(t^2+|\beta x+y|^2)t^2\log^{2b}(s)}dA(y)dsd\beta\\&\leq Cr\int_0^1\int_0^{\infty}\rho\int_0^{2\pi}\frac{1}{(\rho+t)t^3\log^{2b-1}(t)(t^2+\beta^2r^2+\rho^2+2\beta r\rho\cos(\theta))}d\theta d\rho d\beta\end{aligned}$$

We can then treat the last line of the above equation in the same way as we treated (3.87). This results in

$$\begin{aligned}&r\int_0^1\int_t^{\infty}\int_{B_{s-t}(0)\cap(B_{\frac{s}{2}}(-\beta x))^c}\left(\frac{1}{\sqrt{(s-t)^2-|y|^2}}-\frac{1}{(s-t)}\right)|\partial_2 v_{4,c}(s,|\beta x+y|)|dA(y)dsd\beta\\&\leq\frac{Cr}{t^4\log^{2b-2}(t)}\end{aligned}$$

Next, we have

$$r\int_0^1\int_t^{\infty}\int_{B_{s-t}(0)\cap(B_{\frac{s}{2}}(-\beta x))^c}\frac{1}{(s-t)}\frac{|v_{4,c}(s,|\beta x+y|)|}{|\beta x+y|}dA(y)dsd\beta$$

The contribution to the integral from the region $s-t\leq\frac{1}{2}$ is treated in an identical manner as above, and we estimate the other contribution, using the same procedure as used above:

$$r\int_0^1\int_{t+\frac{1}{2}}^\infty\int_{B_{s-t}(0)\cap(B_{\frac{s}{2}}(-\beta x))^c}\frac{1}{(s-t)}\frac{|v_{4,c}(s,|\beta x+y|)|}{|\beta x+y|}dA(y)dsd\beta$$
$$\leq Cr\int_0^1\int_{t+\frac{1}{2}}^\infty\frac{1}{(s-t)}\int_{\frac{s}{2}}^\infty\rho\int_0^{2\pi}\left(\frac{1}{t\log^{2b-1}(s)\rho^5}+\frac{\log^{2b\alpha}(s)}{s^2\rho^4\log^{3b+1}(s)}\right)d\theta d\rho dsd\beta$$
$$\leq Cr\int_{t+\frac{1}{2}}^\infty\frac{ds}{(s-t)ts^3\log^{2b-1}(s)}+Cr\int_{t+\frac{1}{2}}^\infty\frac{\log^{2b\alpha-3b-1}(s)ds}{(s-t)s^4}$$
$$\leq\frac{Cr}{t^4\log^{2b-2}(t)}$$

The last term to estimate is

$$r\int_0^1\int_t^\infty\int_{B_{s-t}(0)\cap(B_{\frac{s}{2}}(-\beta x))^c}\left(\frac{1}{\sqrt{(s-t)^2-|y|^2}}-\frac{1}{(s-t)}\right)\frac{|v_{4,c}(s,|\beta x+y|)|}{|\beta x+y|}dA(y)dsd\beta$$
$$\leq Cr\int_0^1\int_t^\infty\int_{B_{s-t}(0)}\left(\frac{1}{\sqrt{(s-t)^2-|y|^2}}-\frac{1}{(s-t)}\right)\left(\frac{\log(s)}{\log^{2b}(s)s^5t}+\frac{\log^{2b\alpha}(s)}{s^6\log^{3b+1}(s)}\right)dA(y)dsd\beta$$
$$\leq Cr\int_0^\infty\rho\int_{\rho+t}^\infty\left(\frac{1}{\sqrt{(s-t)^2-\rho^2}}-\frac{1}{(s-t)}\right)\left(\frac{1}{(\rho+t)^4t^2\log^{2b-1}(t)}+\frac{\log^{2b\alpha-3b-1}(t)}{(\rho+t)^6}\right)dsd\rho$$
$$\leq\frac{Cr}{t^4\log^{2b-1}(t)}$$

Combining these estimates, we conclude that

$$|v_4(t,r)|\leq\frac{Cr}{t^2\log^{3b+2N-1}(t)},\quad r\leq\frac{t}{2}$$

Next, we estimate $\partial_r v_4$, starting with the region $r\leq\frac{t}{2}$. We recall the function G defined in (3.24):

$$G(s,r,\rho)=\int_0^{2\pi}\frac{v_{4,c}(s,\sqrt{r^2+2r\rho\cos(\theta)+\rho^2})}{\sqrt{r^2+2r\rho\cos(\theta)+\rho^2}}(r+\rho\cos(\theta))\,d\theta$$
$$s\geq t,\quad r\geq 0,\quad s-t\geq\rho\geq 0$$

and start with

$$v_4(t,r)=\frac{-1}{2\pi}\int_t^\infty\int_0^{s-t}\frac{\rho}{\sqrt{(s-t)^2-\rho^2}}G(s,r,\rho)d\rho ds$$

Note that, when we estimated v_4 in the region $r\leq\frac{t}{2}$, we used

$$G(s,r,\rho)=r\int_0^1\partial_2G(s,r\beta,\rho)d\beta$$

Now, we have

$$\partial_r v_4(t,r)=\frac{-1}{2\pi}\int_t^\infty\int_0^{s-t}\frac{\rho}{\sqrt{(s-t)^2-\rho^2}}\partial_2G(s,r,\rho)d\rho ds$$

Therefore, the identical procedure gives the estimate

$$|\partial_r v_4(t,r)|\leq\frac{C}{t^2\log^{3b+2N-1}(t)},\quad r\leq\frac{t}{2}$$

Now, we treat the region $r \geq \frac{t}{2}$. Here, we use a different combination of the v_2 estimates in the various regions, to obtain

(3.88)

$$|v_{4,c}(t,r)| \leq C|\chi_{\geq 1}(\frac{2r}{\log^N(t)})| \begin{cases} \frac{1}{r^3 t^2 \log^{3b}(t)}, & r \leq \frac{t}{2} \\ \frac{\log(r)}{\log^{2b}(t) r^4 |t-r|} + \frac{\log^{2b\alpha}(t)}{t^2 r^3 \log^{3b+1}(t)}, & \frac{t}{2} \leq r \leq t-\sqrt{t}, \quad r > t+\sqrt{t} \\ \frac{1}{\log^{2b}(t) r^{9/2}}, & t-\sqrt{t} \leq r \leq t+\sqrt{t} \end{cases}$$

and

(3.89)

$$|\partial_r v_{4,c}(t,r)| \leq \frac{C|\chi'_{\geq 1}(\frac{2r}{\log^N(t)})|}{\log^{2b+N}(t) r^4} \left(\frac{r}{t^2 \log^b(t)}\right) + \frac{C\chi_{\geq 1}(\frac{2r}{\log^N(t)})}{r^4 \log^{2b}(t)} \begin{cases} \frac{1}{t^2 \log^b(t)}, & r \leq \frac{t}{2} \\ \frac{1}{t^2 \log^b(t)} + \frac{\log(r)}{(t-r)^2}, & \frac{t}{2} \leq r \leq t - t^{1/4}, \text{ or } r \geq t + t^{1/4} \\ \frac{1}{\sqrt{r}}, & t - t^{1/4} \leq r \leq t + t^{1/4} \end{cases}$$

Next, we will use a different representation formula for v_4 to estimate v_4 and $\partial_r v_4$ in the region $r \geq \frac{t}{2}$. In particular, we have

$$v_4(t,r) = \int_t^\infty \int_0^\infty J_1(r\xi) \sin((t-x)\xi) \widehat{v_{4,c}}(x,\xi) d\xi dx$$

So, it suffices to estimate $\widehat{v_{4,c}}$. To do this, we consider separately the regions

$$\frac{1}{\xi} \geq t+\sqrt{t}, \quad t+\sqrt{t} \geq \frac{1}{\xi} \geq t-\sqrt{t}, \quad t-\sqrt{t} \geq \frac{1}{\xi} \geq \frac{t}{2}, \quad \frac{t}{2} \geq \frac{1}{\xi} \geq \log^N(t)$$

Then, for example, in the case $\frac{1}{\xi} \geq t+\sqrt{t}$, we have

$$\int_0^\infty J_1(r\xi) r v_{4,c}(t,r) dr = \sum_{k=1}^5 \int_{I_k} J_1(r\xi) r v_{4,c}(t,r) dr$$

where

$$I_1 = [\frac{1}{\xi}, \infty), \quad I_2 = [t+\sqrt{t}, \frac{1}{\xi}], \quad I_3 = [t-\sqrt{t}, t+\sqrt{t}], \quad I_4 = [\frac{t}{2}, t-\sqrt{t}], \quad I_5 = [0, \frac{t}{2}]$$

and we use

$$|J_1(x)| \leq \begin{cases} Cx, & 0 < x < 1 \\ \frac{C}{\sqrt{x}}, & x > 1 \end{cases}$$

and (3.88). The analogous decomposition is done for all cases of ξ mentioned above.

This procedure results in

$$|\widehat{v_{4,c}}(t,\xi)| \leq \begin{cases} \frac{C\xi \log(t)(\log(t)+|\log(\xi)|)}{t^2 \log^{2b}(t)} + \frac{C\xi^2 |\log(\xi)| \cdot |\log(1-t\xi)|}{t \log^{2b}(t)}, & \xi \leq \frac{1}{t+\sqrt{t}} \\ \frac{C\xi}{t^2 \log^{2b-2}(t)}, & \frac{1}{t+\sqrt{t}} \leq \xi \leq \frac{1}{t-\sqrt{t}} \\ \frac{C\xi \log^2(t)}{t^2 \log^{2b}(t)}, & \frac{1}{t-\sqrt{t}} \leq \xi \leq \frac{2}{t} \\ \frac{C\xi(|\log(\xi)| + \log(\log(t)))}{t^2 \log^{3b}(t)} + \frac{C}{\sqrt{\xi} t^{7/2} \log^{2b-2}(t)}, & \frac{2}{t} \leq \xi \leq \frac{1}{\log^N(t)} \end{cases}$$

Rather than recording pointwise estimates on $\widehat{v_{4,c}}(t,\xi)$ in the region $\frac{1}{\xi} \leq \log^N(t)$, we use the following argument to infer an integral estimate on $\widehat{v_{4,c}}(t,\xi)$. From (3.88) and (3.89),

$$\begin{aligned}&||\left(\partial_r + \frac{1}{r}\right) v_{4,c}(t,r)||_{L^2(rdr)} \leq ||\partial_r v_{4,c}(t,r)||_{L^2(rdr)} + ||\frac{v_{4,c}(t,r)}{r}||_{L^2(rdr)}\\&\leq \frac{C}{t^2 \log^{3b+3N}(t)}\end{aligned}$$

On the other hand,

$$\begin{aligned}&\int_0^\infty J_0(r\xi)\left(\partial_r v_{4,c}(t,r) + \frac{v_{4,c}(t,r)}{r}\right) rdr\\&= -\int_0^\infty v_{4,c}(t,r) J_0'(r\xi)\xi r dr = \xi \int_0^\infty v_{4,c}(t,r) J_1(r\xi) r dr\\&= \xi \widehat{v_{4,c}}(t,\xi)\end{aligned}$$

where the vanishing of the boundary terms arising from integration by parts is justified by (3.88).

By the L^2 isometry property of the Hankel transform of order 0, this implies

$$\frac{C}{t^2 \log^{3b+3N}(t)} \geq ||\partial_r v_{4,c}(t,r) + \frac{v_{4,c}(t,r)}{r}||_{L^2(rdr)} = ||\xi \widehat{v_{4,c}}(t,\xi)||_{L^2(\xi d\xi)} \tag{3.90}$$

Then, we use

$$|J_1(x)| + |J_1'(x)| \leq \frac{C}{\sqrt{x}}$$

in the formulae

$$v_4(t,r) = \int_t^\infty \int_0^\infty J_1(r\xi) \sin((t-x)\xi) \widehat{v_{4,c}}(x,\xi) d\xi dx$$

and

$$\partial_r v_4(t,r) = \int_t^\infty \int_0^\infty \xi J_1'(r\xi) \sin((t-x)\xi) \widehat{v_{4,c}}(x,\xi) d\xi dx$$

(Note that the differentiation under the integral sign is justified by the pointwise estimates on $\widehat{v_{4,c}}$, as well as (3.90)).

Thus, we get

$$\begin{aligned}|v_4(t,r)| &\leq C \int_t^\infty \frac{1}{\sqrt{r}} \int_0^{\frac{1}{\log^N(x)}} \frac{|\widehat{v_{4,c}}(x,\xi)|}{\sqrt{\xi}} d\xi dx\\&+ C \int_t^\infty \frac{1}{\sqrt{r}} \int_{\frac{1}{\log^N(x)}}^\infty \frac{|\widehat{v_{4,c}}(x,\xi)|}{\sqrt{\xi}} \cdot \frac{\xi^{3/2}}{\xi^{3/2}} d\xi dx\\&\leq C \int_t^\infty \frac{1}{\sqrt{r}} \int_0^{\frac{1}{\log^N(x)}} \frac{|\widehat{v_{4,c}}(x,\xi)|}{\sqrt{\xi}} d\xi dx\\&+ C \int_t^\infty \frac{1}{\sqrt{r}} \left(\int_{\frac{1}{\log^N(x)}}^\infty \frac{d\xi}{\xi^4}\right)^{1/2} ||\xi \widehat{v_{4,c}}(x,\xi)||_{L^2(\xi d\xi)} dx\end{aligned}$$

and

$$\begin{aligned}|\partial_r v_4(t,r)| &\leq C\int_t^\infty \frac{1}{\sqrt{r}}\int_0^{\frac{1}{\log^N(x)}} |\widehat{v_{4,c}}(x,\xi)|\sqrt{\xi}d\xi dx\\&+C\int_t^\infty \frac{1}{\sqrt{r}}\int_{\frac{1}{\log^N(x)}}^\infty |\widehat{v_{4,c}}(x,\xi)|\sqrt{\xi}\cdot\frac{\xi}{\xi}d\xi dx\\&\leq C\int_t^\infty \frac{1}{\sqrt{r}}\int_0^{\frac{1}{\log^N(x)}} |\widehat{v_{4,c}}(x,\xi)|\sqrt{\xi}d\xi dx\\&+C\int_t^\infty \frac{1}{\sqrt{r}}\left(\int_{\frac{1}{\log^N(x)}}^\infty \frac{d\xi}{\xi^2}\right)^{1/2} ||\xi\widehat{v_{4,c}}(x,\xi)||_{L^2(\xi d\xi)}dx\end{aligned}$$

Using our pointwise estimates on $\widehat{v_{4,c}}$, as well as (3.90), we get

$$|v_4(t,r)| \leq \frac{C}{\sqrt{r}t\log^{\frac{3N}{2}+3b-1}(t)}, \quad r\geq\frac{t}{2}$$

and

$$|\partial_r v_4(t,r)| \leq \frac{C}{\sqrt{r}t\log^{3b-1+\frac{5N}{2}}(t)}, \quad r\geq\frac{t}{2}$$

In addition, we have

$$\partial_r v_4(t,r)+\frac{v_4(t,r)}{r} = \int_t^\infty\int_0^\infty \xi J_0(r\xi)\sin((t-x)\xi)\widehat{v_{4,c}}(x,\xi)d\xi dx$$

First using Minkowski's inequality, then the L^2 isometry property of the Hankel transform of order 0, and then that of the Hankel transform of order 1, we get

$$\begin{aligned}||\partial_r v_4(t,r)+\frac{v_4(t,r)}{r}||_{L^2(rdr)} &\leq \int_t^\infty || \int_0^\infty \xi J_0(r\xi)\sin((t-x)\xi)\widehat{v_{4,c}}(x,\xi)d\xi||_{L^2(rdr)}dx\\&\leq\int_t^\infty ||\sin((t-x)\xi)\widehat{v_{4,c}}(x,\xi)||_{L^2(\xi d\xi)}dx\\&\leq\int_t^\infty ||\widehat{v_{4,c}}(x,\xi)||_{L^2(\xi d\xi)}dx\\&\leq\int_t^\infty ||v_{4,c}(x)||_{L^2(rdr)}dx \leq \int_t^\infty \frac{Cdx}{\log^{2N+3b}(x)x^2}\\&\leq\frac{C}{t\log^{2N+3b}(t)}\end{aligned}$$

where we used (3.88). Note that this is simply the energy estimate for the equation solved by v_4. Next, we note that the pointwise estimates established for v_4 imply that

$$v_4(t,0)=0, \quad \lim_{r\to\infty} v_4(t,r)=0$$

So,

$$\begin{aligned}||\left(\partial_r+\frac{1}{r}\right)v_4||^2_{L^2(rdr)} &= \int_0^\infty\left((\partial_r v_4)^2+\frac{\partial_r(v_4^2)}{r}+\frac{v_4^2}{r^2}\right)rdr\\&=||\partial_r v_4||^2_{L^2(rdr)}+||\frac{v_4}{r}||^2_{L^2(rdr)}\end{aligned}$$

We now treat $\partial_t v_4$:

$$\partial_t v_4(t,r) = \int_t^\infty \int_0^\infty J_1(r\xi)\cos((t-x)\xi)\xi\widehat{v_{4,c}}(x,\xi)d\xi dx$$

where the differentiation under the integral sign is again justified by the pointwise estimates on $\widehat{v_{4,c}}$ and (3.90).

This gives

$$|\partial_t v_4(t,r)| \leq \frac{C}{\sqrt{r}} \int_t^\infty \int_0^\infty \sqrt{\xi}|\widehat{v_{4,c}}(x,\xi)|d\xi dx$$

So, the same exact procedure used for $\partial_r v_4$ pointwise estimates in the region $r \geq \frac{t}{2}$ also applies to $\partial_t v_4$.

Finally,

$$\begin{aligned} ||\partial_t v_4(t,r)||_{L^2(rdr)} &\leq \int_t^\infty ||\int_0^\infty \xi J_1(r\xi)\cos((t-x)\xi)\widehat{v_{4,c}}(x,\xi)d\xi||_{L^2(rdr)}dx \\ &\leq \int_t^\infty ||\cos((t-x)\xi)\widehat{v_{4,c}}(x,\xi)||_{L^2(\xi d\xi)}dx \\ &\leq \int_t^\infty ||\widehat{v_{4,c}}(x,\xi)||_{L^2(\xi d\xi)}dx \\ &\leq \int_t^\infty ||v_{4,c}(x)||_{L^2(rdr)}dx \leq \int_t^\infty \frac{Cdx}{\log^{2N+3b}(x)x^2} \\ &\leq \frac{C}{t\log^{2N+3b}(t)} \end{aligned}$$

This completes the proof of the lemma. □

3.0.8.8. *Pointwise estimates on v_5.*

LEMMA 3.10. *For all λ of the form*

$$\lambda(t) = \lambda_0(t) + e(t), \quad e \in \overline{B_1(0)} \subset X$$

we have the pointwise estimates

$$|v_5(t,r)| \leq \begin{cases} \frac{Cr}{t^{7/2}\log^{3b-3+\frac{5N}{2}}(t)}, & r \leq \frac{t}{2} \\ \frac{C\log^4(t)}{\sqrt{r}t^{3/2}}, & r > \frac{t}{2} \end{cases} \tag{3.91}$$

$$|\partial_r v_5(t,r)| \leq \begin{cases} \frac{C}{t^{7/2}\log^{3b-3+\frac{5N}{2}}(t)}, & r \leq \frac{t}{2} \\ \frac{C\log^3(t)}{\sqrt{r}t^{3/2}}, & r > \frac{t}{2} \end{cases} \tag{3.92}$$

$$|\partial_t v_5(t,r)| \leq \frac{C\log^3(t)}{\sqrt{r}t^{3/2}}, \quad r > \frac{t}{2}$$

In addition, we have the L^2 estimates

$$||\partial_t v_5||_{L^2(rdr)} + ||\partial_r v_5||_{L^2(rdr)} + ||\frac{v_5}{r}||_{L^2(rdr)} \leq C\frac{\log^3(t)}{t^{7/4}} \tag{3.93}$$

PROOF. We start with the region $\frac{t}{2} > r > 0$. In order to ease notation, let $x \in \mathbb{R}^2$ be defined by $x = r\mathbf{e}_1$. Then,

$$\begin{aligned} &v_5(t,r) \\ &= \frac{-r}{2\pi}\int_0^1\int_t^\infty\int_{B_{s-t}(0)} \frac{1}{\sqrt{(s-t)^2-|y|^2}}\left(\frac{\partial_2 N_2(f_{v_5})(s,|\beta x+y|)\left((\beta x+y)\cdot\hat{x}\right)^2}{|\beta x+y|^2}\right. \\ &\quad -\frac{N_2(f_{v_5})(s,|\beta x+y|)\left((\beta x+y)\cdot\hat{x}\right)^2}{|\beta x+y|^3} \\ &\quad \left.+\frac{N_2(f_{v_5})(s,|\beta x+y|)}{|\beta x+y|}\right) dA(y)dsd\beta \end{aligned}$$

We then decompose

$$v_5(t,r) = v_{5,1}(t,r) + v_{5,2}(t,r)$$

where

$$\begin{aligned} &v_{5,1}(t,r) \\ &= \frac{-r}{2\pi}\int_0^1\int_t^\infty\int_{B_{s-t}(0)\cap B_{\frac{s}{2}}(-\beta x)} \frac{1}{\sqrt{(s-t)^2-|y|^2}}\left(\frac{\partial_2 N_2(f_{v_5})(s,|\beta x+y|)\left((\beta x+y)\cdot\hat{x}\right)^2}{|\beta x+y|^2}\right. \\ &\quad -\frac{N_2(f_{v_5})(s,|\beta x+y|)\left((\beta x+y)\cdot\hat{x}\right)^2}{|\beta x+y|^3} \\ &\quad \left.+\frac{N_2(f_{v_5})(s,|\beta x+y|)}{|\beta x+y|}\right) dA(y)dsd\beta \end{aligned}$$

and

$$v_{5,2}(t,r) = v_5(t,r) - v_{5,1}(t,r)$$

Finally, we use the pointwise estimates on v_1, v_2, v_3, v_4 to record pointwise estimates on $N_2(f_{v_5})$: We start with

$$|N_2(f)(t,r)| \leq \frac{Cr(f(t,r))^2}{\lambda(t)(1+\frac{r^2}{\lambda(t)^2})r^2} + C\frac{|f(t,r)|^3}{r^2}$$

The first set of estimates we will require on $N_2(f_{v_5})$ concern the regions $r \leq \frac{t}{2}$, and $t > r > \frac{t}{2}$. In the region $r \leq \frac{t}{2}$, we get

$$|N_2(f_{v_5})(t,r)| \leq \frac{Cr}{(\lambda(t)^2+r^2)t^4\log^{3b}(t)} + \frac{Cr}{t^6\log^{3b}(t)}, \quad r \leq \frac{t}{2}$$

In the region $t > r > \frac{t}{2}$, we use (3.51) to estimate v_2 in the region $t > r \geq \frac{t}{2}$, exactly as was done while studying v_4, and we get

$$|N_2(f_{v_5})(t,r)| \leq \frac{C\log^3(r)}{r^2|t-r|^3} + \frac{C\lambda(t)}{r^{7/2}t^{5/2}\log^{3N+6b-2}(t)}, \quad t > r \geq \frac{t}{2}$$

Next, we consider $\partial_r N_2(f_{v_5})$:

$$\begin{aligned} |\partial_r(N_2(f))(t,r)| \leq &\frac{C(f(t,r))^2}{r^2\lambda(t)\left(1+\frac{r^2}{\lambda(t)^2}\right)} + \frac{C|f(t,r)\partial_r f(t,r)|}{\lambda(t)r\left(1+\frac{r^2}{\lambda(t)^2}\right)} + \frac{C|f(t,r)|^3}{r^3} \\ &+ \frac{C|f(t,r)^2\partial_r f(t,r)|}{r^2} \end{aligned}$$

We again treat the regions $t > r > \frac{t}{2}$ and $r \leq \frac{t}{2}$. We get

$$|\partial_r N_2(f_{v_5})(t,r)| \leq \frac{C}{t^4 \log^{3b}(t)(\lambda(t)^2 + r^2)} + \frac{C}{t^6 \log^{3b}(t)}, \quad r \leq \frac{t}{2}$$

$$|\partial_r N_2(f_{v_5})(t,r)| \leq \frac{C}{r^3 t^3 \log^{4N+7b-2}(t)} + \frac{C\lambda(t)\log(r)}{r^3 t^{3/2}|t-r|\log^{3b-1+\frac{5N}{2}}(t)} + \frac{C\log^2(r)}{r^2 t^{3/2}(t-r)^2 \log^{3b-1+\frac{5N}{2}}(t)} + \frac{C\log^3(r)}{r^2(t-r)^4}, \quad t > r \geq \frac{t}{2}$$

where we used

$$\frac{1}{r} \leq \frac{1}{|t-r|}, \quad r \geq \frac{t}{2}$$

So,

(3.94)

$$|v_{5,1}(t,r)| \leq Cr \int_0^1 \int_t^\infty \int_{B_{s-t}(0) \cap B_{\frac{s}{2}}(-\beta x)} \frac{1}{\sqrt{(s-t)^2 - |y|^2}} \left(\frac{1}{s^4 \log^b(s)\left(1 + \frac{|\beta x + y|^2}{\lambda(s)^2}\right)} + \frac{1}{s^6 \log^{3b}(s)} \right) dA(y) ds d\beta$$

We treat several terms comprising (3.94) separately. First, we have

$$\begin{aligned} &r \int_0^1 \int_t^{t+\frac{1}{2}} \int_{B_{s-t}(0) \cap B_{\frac{s}{2}}(-\beta x)} \frac{1}{(s-t)} \frac{1}{s^4 \log^b(s)\left(1 + \frac{|\beta x + y|^2}{\lambda(s)^2}\right)} dA(y) ds d\beta \\ &\leq Cr \int_0^1 \int_t^{t+\frac{1}{2}} \int_0^{s-t} \frac{\rho}{(s-t)} \int_0^{2\pi} \frac{1}{s^4 \log^b(s)} d\theta d\rho ds d\beta \\ &\leq \frac{Cr}{t^4 \log^b(t)} \end{aligned}$$

Next,

$$\begin{aligned} &r \int_0^1 \int_{t+\frac{1}{2}}^\infty \int_{B_{s-t}(0) \cap B_{\frac{s}{2}}(-\beta x)} \frac{1}{(s-t)} \frac{1}{s^4 \log^b(s)\left(1 + \frac{|\beta x + y|^2}{\lambda(s)^2}\right)} dA(y) ds d\beta \\ &\leq Cr \int_0^1 \int_{t+\frac{1}{2}}^\infty \frac{1}{(s-t)} \int_0^{s-t} \rho \int_0^{2\pi} \frac{1}{s^2(\rho+t)^2 \log^b(s)(1 + \beta^2 r^2 + \rho^2 + 2\beta r \rho \cos(\theta))} d\theta d\rho ds d\beta \\ &\leq Cr \int_0^1 \int_{t+\frac{1}{2}}^\infty \frac{(\log(2 + r^2\beta^2) + \log(2 + t^2))}{(s-t)s^2 \log^b(s) t^2} ds d\beta \\ &\leq \frac{Cr}{t^4 \log^{b-2}(t)}, \quad r \leq \frac{t}{2} \end{aligned}$$

where we used the same procedure that we used in (3.87).

The third term to consider is

$$r\int_0^1\int_t^\infty\int_{B_{s-t}(0)\cap B_{\frac{s}{2}}(-\beta x)}\left(\frac{1}{\sqrt{(s-t)^2-|y|^2}}-\frac{1}{(s-t)}\right)\left(\frac{1}{s^4\log^b(s)\left(1+\frac{|\beta x+y|^2}{\lambda(s)^2}\right)}\right)dA(y)dsd\beta$$

$$\leq Cr\int_0^1\int_0^\infty\frac{\rho}{(\rho+t)^2}\int_0^{2\pi}\frac{1}{(1+\beta^2r^2+\rho^2+2\beta r\rho\cos(\theta))}\int_{\rho+t}^\infty\left(\frac{1}{\sqrt{(s-t)^2-\rho^2}}-\frac{1}{(s-t)}\right)\frac{1}{s^2\log^b(s)}dsd\theta d\rho d\beta$$

$$\leq Cr\int_0^1\left(\frac{\log(2+r^2\beta^2)+\log(2+t^2)}{t^2}\right)\frac{d\beta}{t^2\log^b(t)}$$

$$\leq\frac{Cr}{t^4\log^{b-1}(t)},\quad r\leq\frac{t}{2}$$

Finally, the last term to consider is

$$r\int_0^1\int_t^\infty\int_{B_{s-t}(0)\cap B_{\frac{s}{2}}(-\beta x)}\frac{1}{\sqrt{(s-t)^2-|y|^2}}\frac{1}{s^6\log^{3b}(s)}dA(y)dsd\beta$$

$$\leq Cr\int_0^1\int_t^\infty\frac{1}{s^6\log^{3b}(s)}\int_0^{s-t}\rho\int_0^{2\pi}\frac{1}{\sqrt{(s-t)^2-\rho^2}}d\theta d\rho dsd\beta$$

$$\leq Cr\int_t^\infty\frac{(s-t)}{s^6\log^{3b}(s)}ds$$

$$\leq\frac{Cr}{t^4\log^{3b}(t)}$$

Combining the above estimates, we conclude:

$$|v_{5,1}(t,r)|\leq C\frac{r}{t^4\log^{b-2}(t)},\quad r\leq\frac{t}{2}$$

Next, we treat $v_{5,2}$.

$$v_{5,2}(t,r)$$
$$=\frac{-r}{2\pi}\int_0^1\int_t^\infty\int_{B_{s-t}(0)\cap(B_{\frac{s}{2}}(-\beta x))^c}\frac{1}{\sqrt{(s-t)^2-|y|^2}}\left(\frac{\partial_2N_2(f)(s,|\beta x+y|)\left((\beta x+y)\cdot\hat{x}\right)^2}{|\beta x+y|^2}\right.$$
$$-\frac{N_2(f)(s,|\beta x+y|)\left((\beta x+y)\cdot\hat{x}\right)^2}{|\beta x+y|^3}$$
$$\left.+\frac{N_2(f)(s,|\beta x+y|)}{|\beta x+y|}\right)dA(y)dsd\beta$$

So,

$$|v_{5,2}(t,r)|$$
$$\leq Cr\int_0^1\int_t^\infty\int_{B_{s-t}(0)\cap(B_{\frac{s}{2}}(-\beta x))^c}\frac{1}{\sqrt{(s-t)^2-|y|^2}}\left(|\partial_2N_2(f)(s,|\beta x+y|)|\right.$$
$$\left.+\frac{|N_2(f)(s,|\beta x+y|)|}{|\beta x+y|}\right)dA(y)dsd\beta$$
$$\leq Cr\int_0^1\int_t^\infty\int_{B_{s-t}(0)\cap(B_{\frac{s}{2}}(-\beta x))^c}\frac{1}{\sqrt{(s-t)^2-|y|^2}}\text{integrand}_{v_{5,2}}(s,|\beta x+y|)dA(y)dsd\beta$$

where

$$\text{integrand}_{v_{5,2}}(s,|\beta x+y|)$$
$$=\frac{1}{|\beta x+y|^3 s^3 \log^{4N+7b-2}(s)}+\frac{\lambda(s)\log(|\beta x+y|)}{|\beta x+y|^3|s-|\beta x+y||s^{3/2}\log^{3b-1+\frac{5N}{2}}(s)}$$
$$+\frac{\log^2(|\beta x+y|)}{|\beta x+y|^2 s^{3/2}(s-|\beta x+y|)^2\log^{3b-1+\frac{5N}{2}}(s)}+\frac{\log^3(|\beta x+y|)}{|\beta x+y|^2(s-|\beta x+y|)^4}$$

Exactly as in the case of v_4, we note that, for $z\in B_{s-t}(\beta x)\cap(B_{\frac{s}{2}}(0))^c$, we have

$$|z|=|z-\beta x+\beta x|\leq|z-\beta x|+r<s-t+r\leq s-\frac{t}{2}$$

So,

$$s-|z|\geq\frac{t}{2}$$

We use this estimate for every term in $\text{integrand}_{v_{5,2}}$, except for the term

$$\frac{\log^3(|\beta x+y|)}{|\beta x+y|^2(s-|\beta x+y|)^4}\leq\frac{C\log^3(s)}{s^2t^3(s-|\beta x+y|)}$$

Then,

(3.95)

$$\begin{aligned}&|v_{5,2}(t,r)|\\&\leq Cr\int_0^1\int_t^\infty\int_0^{s-t}\frac{\rho}{\sqrt{(s-t)^2-\rho^2}}\int_0^{2\pi}\left(\frac{\log^2(s)}{s^{7/2}t^2\log^{3b-1+\frac{5N}{2}}(s)}+\frac{\log^3(s)}{s^3t^3}\right)d\theta d\rho ds d\beta\\&+Cr\int_0^1\int_t^\infty\int_0^{s-t}\frac{\rho}{\sqrt{(s-t)^2-\rho^2}}\int_0^{2\pi}\frac{\log^3(s)}{s^2t^3(s-\sqrt{\beta^2r^2+\rho^2+2\beta r\rho\cos(\theta)})}d\theta d\rho ds d\beta\\&\leq\frac{Cr}{t^{7/2}\log^{3b-3+\frac{5N}{2}}(t)}\\&+Cr\int_0^1\int_t^\infty\int_0^{s-t}\frac{\rho}{\sqrt{(s-t)^2-\rho^2}}\int_0^{2\pi}\frac{\log^3(s)}{s^2t^3(s-\sqrt{\beta^2r^2+\rho^2+2\beta r\rho\cos(\theta)})}d\theta d\rho ds d\beta\end{aligned}$$

We need only continue to estimate the last line of (3.95).

(3.96)

$$\begin{aligned}&r\int_0^1\int_t^\infty\int_0^{s-t}\frac{\rho}{\sqrt{(s-t)^2-\rho^2}}\int_0^{2\pi}\frac{\log^3(s)}{s^2t^3(s-\sqrt{\beta^2r^2+\rho^2+2\beta r\rho\cos(\theta)})}d\theta d\rho ds d\beta\\&\leq Cr\int_0^1\int_0^\infty\rho\int_0^{2\pi}\frac{1}{t^3}\int_{t+\rho}^\infty\frac{1}{\sqrt{(s-t)^2-\rho^2}}\frac{\log^3(s)}{s^2}\frac{1}{(s-\sqrt{\rho^2+2\rho\beta r\cos(\theta)+\beta^2r^2})}ds d\theta d\rho d\beta\\&\leq Cr\int_0^1\int_0^\infty\rho\int_0^{2\pi}\frac{1}{t^3}\int_{t+\rho}^\infty\frac{1}{\sqrt{s-t+\rho}}\frac{1}{\sqrt{s-t-\rho}}\frac{\log^3(s)}{s^2}\frac{1}{(s-\sqrt{\rho^2+2\rho\beta r\cos(\theta)+\beta^2r^2})}ds d\theta d\rho d\beta\\&\leq Cr\int_0^1\int_0^\infty\rho\int_0^{2\pi}\frac{1}{t^3}\frac{1}{\sqrt{\rho}}\frac{\log^3(t+\rho)}{(t+\rho)^2}\int_{t+\rho}^\infty\frac{1}{\sqrt{s-t-\rho}}\frac{1}{(s-\sqrt{\rho^2+2\rho\beta r\cos(\theta)+\beta^2r^2})}ds d\theta d\rho d\beta\\&\leq Cr\int_0^1\int_0^\infty\frac{\sqrt{\rho}\log^3(t+\rho)}{(t+\rho)^2}\int_0^{2\pi}\frac{1}{t^3}\frac{1}{\sqrt{t}}d\theta d\rho d\beta\\&\leq C\frac{r\log^3(t)}{t^4}\end{aligned}$$

So, we get

$$|v_{5,2}(t,r)|\leq\frac{Cr}{t^{7/2}\log^{3b-3+\frac{5N}{2}}(t)},\quad r\leq\frac{t}{2}$$

In total, we have

$$|v_5(t,r)| \leq \frac{Cr}{t^{7/2}\log^{3b-3+\frac{5N}{2}}(t)}, \quad r \leq \frac{t}{2}$$

Exactly as for $\partial_r v_4$, we have

(3.97)

$$\begin{aligned}&|\partial_r v_5(t,r)|\\ &\leq C\int_t^{\infty}\int_0^{s-t}\frac{\rho}{\sqrt{(s-t)^2-\rho^2}}\int_0^{2\pi}\Big(|\partial_2 N_2(f_{v_5})(s,\sqrt{r^2+\rho^2+2r\rho\cos(\theta)})|\\ &+\frac{|N_2(f_{v_5})(s,\sqrt{r^2+\rho^2+2r\rho\cos(\theta)})|}{\sqrt{r^2+\rho^2+2r\rho\cos(\theta)}}\Big)d\theta d\rho ds\end{aligned}$$

As was also the case for v_4, the integrals appearing in (3.97) were already estimated above. So, we get

$$|\partial_r v_5(t,r)| \leq \frac{C}{t^{7/2}\log^{3b-3+\frac{5N}{2}}(t)}, \quad r \leq \frac{t}{2}$$

We will now prove estimates on $v_5(t,r)$ and $\partial_r v_5$ in the region $r \geq \frac{t}{2}$. For the next steps, we will use slightly different combinations of the estimates (3.51) and (3.50) for v_2 in various subsets of the region $r \geq \frac{t}{2}$, in order to estimate $N_2(f_{v_5})$, resulting in

(3.98) $|N_2(f_{v_5})(t,r)|$

$$\leq \begin{cases}\frac{Cr}{(\lambda(t)^2+r^2)t^4\log^{3b}(t)}+\frac{Cr}{t^6\log^{3b}(t)}, & r\leq\frac{t}{2}\\ \frac{C\log^3(r)}{r^2|t-r|^3}+\frac{C}{r^{7/2}t^{5/2}\log^{3N+7b-2}(t)}, & \frac{t}{2}\leq r\leq t-\sqrt{t} \text{ or } r\geq t+\sqrt{t}\\ \frac{C}{r^{7/2}}, & t-\sqrt{t}\leq r\leq t+\sqrt{t}\end{cases}$$

We will also need estimates on $\partial_r N_2(f_{v_5})$. To obtain these, we use (3.49) in the region $r \leq \frac{t}{2}$. In the region $r > \frac{t}{2}$, we use (3.50) in the region $t-\sqrt{t} \leq r \leq t+\sqrt{t}$, and (3.51) in the regions $\frac{t}{2} < r < t-\sqrt{t}$ and $r > t+\sqrt{t}$ for v_2. For $\partial_r v_2$, we use (3.50) in the region $t-t^{1/4} \leq r \leq t+t^{1/4}$ and (3.51) in the regions $\frac{t}{2} \leq r \leq t-t^{1/4}$ and $t+t^{1/4} \leq r$. Then, we get

$$|\partial_r N_2(f_{v_5})(t,r)| \leq \frac{C}{t^4\log^{3b}(t)(\lambda(t)^2+r^2)}+\frac{C}{t^6\log^{3b}(t)}, \quad r\leq\frac{t}{2}$$

$$\begin{aligned}|\partial_r N_2(f_{v_5})(t,r)| &\leq \frac{C}{r^3t^3\log^{4N+7b-2}(t)}+\frac{C\log(r)}{r^3|t-r|t^{3/2}\log^{4b-1+\frac{5N}{2}}(t)}\\ &+\frac{C\log^2(r)}{r^2t^{3/2}(t-r)^2\log^{3b-1+\frac{5N}{2}}(t)}\\ &+\frac{C\log^3(r)}{r^2(t-r)^4}, \quad t-\sqrt{t}>r>\frac{t}{2} \text{ or } r>t+\sqrt{t}\end{aligned}$$

$$|\partial_r N_2(f_{v_5})(t,r)| \leq \begin{cases}\frac{C\log(r)}{r^3(t-r)^2}, & t-\sqrt{t}\leq r\leq t-t^{1/4}, \text{ or } t+t^{1/4}\leq r\leq t+\sqrt{t}\\ \frac{C}{r^{7/2}}, & t-t^{1/4}\leq r\leq t+t^{1/4}\end{cases}$$

Then, we have

$$v_5(t,r) = \int_t^\infty \int_0^\infty J_1(r\xi)\sin((t-s)\xi)\widehat{N_2(f_{v_5})}(s,\xi)d\xi ds$$

and we will prove estimates on $\widehat{N_2(f_{v_5})}(t,\xi)$, for $\frac{1}{\xi} \geq 1$. We consider separately the cases

$$\frac{1}{\xi} \geq t+\sqrt{t}, \quad t-\sqrt{t} \leq \frac{1}{\xi} \leq t+\sqrt{t}, \quad \frac{t}{2} \leq \frac{1}{\xi} \leq t-\sqrt{t}, \quad 1 \leq \frac{1}{\xi} \leq \frac{t}{2}$$

and proceed in exactly the same manner as was done for $\widehat{v_{4,c}}$.

This results in

$$|\widehat{N_2(f_{v_5})}(t,\xi)| \leq \frac{C\xi}{t}\left(\log^3(t)+|\log(\xi)|^3\right), \quad \frac{1}{\xi} \geq t+\sqrt{t}$$

$$|\widehat{N_2(f_{v_5})}(t,\xi)| \leq \frac{C\log^3(t)}{t^2}, \quad t-\sqrt{t} \leq \frac{1}{\xi} \leq t+\sqrt{t}$$

$$|\widehat{N_2(f_{v_5})}(t,\xi)| \leq \frac{C\log^3(t)}{t^2}, \quad \frac{t}{2} \leq \frac{1}{\xi} \leq t-\sqrt{t}$$

$$|\widehat{N_2(f_{v_5})}(t,\xi)| \leq \frac{C}{\xi t^4\log^{3b}(t)} + \frac{C}{\xi^3 t^6 \log^{3b}(t)} + \frac{C\log^3(t)}{\sqrt{\xi}t^{5/2}}, \quad 1 \leq \frac{1}{\xi} \leq \frac{t}{2}$$

For the region $\frac{1}{\xi} < 1$, we again use the following argument:
From (3.98) and the $\partial_r N_2(f_{v_5})$ estimates which follow it,

$$\begin{aligned}
&||\left(\partial_r + \frac{1}{r}\right)N_2(f_{v_5})(t,r)||_{L^2(rdr)}\\
&\leq ||\partial_r N_2(f_{v_5})(t,r)||_{L^2(rdr)} + ||\frac{N_2(f_{v_5})(t,r)}{r}||_{L^2(rdr)}\\
&\leq \frac{C\log(t)}{t^{23/8}}
\end{aligned}$$

Then, the same procedure used for $v_{4,c}$ gives

$$\frac{C\log(t)}{t^{23/8}} \geq ||\partial_r N_2(f_{v_5})(t,r) + \frac{N_2(f_{v_5})(t,r)}{r}||_{L^2(rdr)} = ||\xi\widehat{N_2(f_{v_5})}(t,\xi)||_{L^2(\xi d\xi)} \tag{3.99}$$

Now, we return to

$$v_5(t,r) = \int_t^\infty \int_0^\infty J_1(r\xi)\sin((t-s)\xi)\widehat{N_2(f_{v_5})}(s,\xi)d\xi ds$$

use

$$|J_1(x)| \leq \frac{C}{\sqrt{x}}$$

and the same procedure used for v_4:

$$\begin{aligned}
|v_5(t,r)| &\leq \frac{C}{\sqrt{r}}\int_t^\infty \int_0^1 \frac{1}{\sqrt{\xi}}|\widehat{N_2(f_{v_5})}(s,\xi)|d\xi ds\\
&+ \frac{C}{\sqrt{r}}\int_t^\infty \left(\int_1^\infty \frac{d\xi}{\xi^4}\right)^{1/2} ||\xi\widehat{N_2(f_{v_5})}(s,\xi)||_{L^2(\xi d\xi)}ds\\
&\leq \frac{C}{\sqrt{r}}\frac{\log^4(t)}{t^{3/2}}
\end{aligned}$$

Next, the pointwise estimates on $\widehat{N_2(f_{v_5})}$, as well as (3.99) justify the following differentiation under the integral sign:

$$\partial_r v_5(t,r) = \int_t^\infty \int_0^\infty \xi J_1'(r\xi) \sin((t-s)\xi) \widehat{N_2(f_{v_5})}(s,\xi) d\xi ds$$

and we also have

$$|J_1'(x)| \leq \frac{C}{\sqrt{x}}$$

So,

$$|\partial_r v_5(t,r)| \leq \frac{C}{\sqrt{r}} \int_t^\infty \int_0^1 \sqrt{\xi} |\widehat{N_2(f_{v_5})}(s,\xi)| d\xi ds + \frac{C}{\sqrt{r}} \int_t^\infty ||\xi \widehat{N_2(f_{v_5})}(s)||_{L^2(\xi d\xi)} ds$$
$$\leq \frac{C \log^3(t)}{\sqrt{r} t^{3/2}}$$

For $\partial_t v_5$, we can similarly differentiate under the integral sign to get

$$\partial_t v_5(t,r) = \int_t^\infty \int_0^\infty J_1(r\xi) \cos((t-x)\xi) \xi \widehat{N_2(f_{v_5})}(x,\xi) d\xi dx$$

so, the same argument used for $\partial_r v_5$ also applies to $\partial_t v_5$.

Finally, the same procedure used for the energy estimates for v_4 also applies here, to give

$$||\partial_t v_5||_{L^2(rdr)} + ||\left(\partial_r + \frac{1}{r}\right) v_5||_{L^2(rdr)} \leq \int_t^\infty ||N_2(f_{v_5})(x,r)||_{L^2(rdr)} dx$$
$$\leq C \frac{\log^3(t)}{t^{7/4}}$$

(We again have

$$||\left(\partial_r + \frac{1}{r}\right) v_5||^2_{L^2(rdr)} = ||\partial_r v_5||^2_{L^2(rdr)} + ||\frac{v_5}{r}||^2_{L^2(rdr)}$$

by the same argument used while studying v_4). □

3.0.8.9. *Solving the modulation equation.* The main result of this section is to prove Proposition 3.1, which we recall is the following statement:

There exists $T_3 > 0$ such that, for all $T_0 \geq T_3$, there exists $\lambda \in C^2([T_0,\infty))$ which solves (3.63). Moreover,

$$\lambda(t) = \lambda_0(t) + e_0(t), \quad ||e_0||_X \leq 1$$

($||\cdot||_X$ was defined in (3.30)).

We use our pointwise estimates on v_k, E_5 to get

$$|-\lambda(t)\langle \left(\frac{\cos(2Q_{\frac{1}{\lambda(t)}}) - 1}{r^2}\right) v_4 \left(1 - \chi_{\geq 1}(\frac{4r}{t})\right)|_{r=R\lambda(t)}, \phi_0\rangle|$$
$$\leq \frac{C}{\lambda(t)} \int_0^{\frac{t}{2\lambda(t)}} \frac{R^2}{(1+R^2)^3} |v_4(t, R\lambda(t))| dR$$
$$\leq \frac{C}{t^2 \log^{3b+2N-1}(t)}$$

$$|-\lambda(t)\langle \left(\frac{\cos(2Q_{\frac{1}{\lambda(t)}}) - 1}{r^2}\right) v_5 \left(1 - \chi_{\geq 1}(\frac{4r}{t})\right)|_{r=R\lambda(t)}, \phi_0\rangle| \leq \frac{C}{t^{7/2} \log^{3b-3+\frac{5N}{2}}(t)}$$

$$|-\lambda(t)\langle\left(\frac{\cos(2Q_{\frac{1}{\lambda(t)}})-1}{r^2}\right)E_5|_{r=R\lambda(t)},\phi_0\rangle|\leq\frac{C}{t^2\log^{b+1}(t)}$$

$$|-\lambda(t)\langle\left(\left(\frac{\cos(2Q_{\frac{1}{\lambda(t)}})-1}{r^2}\right)(v_1+v_2+v_3)+F_{0,2}\right)\chi_{\geq 1}(\frac{2r}{\log^N(t)})|_{r=R\lambda(t)},\phi_0\rangle|$$

$$\leq C\lambda(t)\int_0^\infty |v_{4,c}(t,R\lambda(t))|\phi_0(R)RdR\leq\frac{C}{t^2\log^{3b+2N}(t)}$$

Combining the above with our previous estimates, there exists a constant $D_1>0$ such that

$$|G(t,\lambda(t))|\leq\frac{D_1}{t^2\log^{b+1}(t)},\quad \lambda=\lambda_0+e,\quad e\in\overline{B}_1(0)\subset X$$

Then, combining this with the estimate of all the terms comprising RHS except for G, we get, for a constant D_2 independent of e, T_0:

$$|RHS(e,t)|\leq\frac{D_2}{\log(\log(t))t^2\log^{b+1}(t)},\quad e\in\overline{B}_1(0)\subset X$$

Let $T_{0,2}>e^{e^{900}}$ be such that the following three inequalities are true:

$$\frac{1}{\sqrt{\log(\log(T_{0,2}))}}\leq\frac{1}{100}\cdot\frac{\alpha}{3D_2}$$

$$\frac{(b+1)}{\log(T_{0,2})}+\frac{1}{2\log(T_{0,2})\log(\log(T_{0,2}))}\leq\frac{1}{100}$$

$$\frac{1}{2b\log(\log(T_{0,2}))}+\frac{b}{\log(T_{0,2})}+\frac{b}{2(b+1)\log(T_{0,2})\log(\log(T_{0,2}))}\leq\frac{1}{100}$$

Finally, there exists $T_{0,3}>700+700b$, such that

$$\frac{\log^{700+700b}(T_{0,3})}{T_{0.3}}<\frac{1}{2}$$

From now on, we will work under the additional restriction that

$$T_0\geq 2e^{e^{1000(b+1)}}+T_{0,1}+T_{0,2}+T_{0,3}$$

(Recall that (3.1) was, up to now, the only constraint on T_0). From the discussion preceding (3.75), and our estimates on $RHS(e,t)$ above, for $e\in\overline{B_1}(0)\subset X$, the map

$$x\mapsto\frac{RHS(e,x)}{4\alpha}-\int_x^\infty\frac{RHS(e,z)}{4\alpha}r(-x,-z)dz,\ \text{a.e. } x\in[T_0,\infty)$$

a priori defined only almost everywhere, admits a continuous extension to $[T_0,\infty)$. Then, we define T on $\overline{B_1}(0)\subset X$ by

$$T(f)(t)=\int_t^\infty\int_{q_1}^\infty S_f(q_2)dq_2dq_1$$

where we define S_f by (the above mentioned continuous extension of)

$$S_f(x)=\frac{RHS(f,x)}{4\alpha}-\int_x^\infty\frac{RHS(f,z)}{4\alpha}r(-x,-z)dz$$

By definition of T, we have

$$T(e)''(t)=S_e(t)=\frac{RHS(e,t)}{4\alpha}-\int_t^\infty\frac{RHS(e,z)}{4\alpha}r(-t,-z)dz$$

Using (3.76), we get

$$\begin{aligned}|T(e)''(t)| &\leq \frac{3D_2}{\alpha \log(\log(t))t^2 \log^{b+1}(t)} \\ &\leq \frac{1}{100\sqrt{\log(\log(t))}t^2 \log^{b+1}(t)}, \quad e \in \overline{B}_1(0) \subset X\end{aligned}$$

Next, we have

$$\begin{aligned}T(e)'(t) &= -\int_t^\infty T(e)''(x)dx \\ |T(e)'(t)| &\leq \frac{1}{100}\int_t^\infty \frac{dx}{\sqrt{\log(\log(x))}x^2 \log^{b+1}(x)}\end{aligned}$$

Then, we note that

$$\begin{aligned}\int_t^\infty \frac{dx}{\sqrt{\log(\log(x))}x^2 \log^{b+1}(x)} &= \frac{1}{t\log^{b+1}(t)\sqrt{\log(\log(t))}} \\ &\quad -(b+1)\int_t^\infty \frac{dx}{x^2\log^{b+2}(x)\sqrt{\log(\log(x))}} \\ &\quad -\frac{1}{2}\int_t^\infty \frac{1}{x^2}\frac{dx}{\log^{b+2}(x)(\log(\log(x)))^{3/2}} \\ &= \frac{1}{t\log^{b+1}(t)\sqrt{\log(\log(t))}} + E_{int,1}\end{aligned}$$

where

$$\begin{aligned}|E_{int,1}| &\leq \frac{(b+1)}{t\log^{b+2}(t)\sqrt{\log(\log(t))}} + \frac{1}{2t\log^{b+2}(t)(\log(\log(t)))^{3/2}} \\ &\leq \frac{1}{100}\frac{1}{t\log^{b+1}(t)\sqrt{\log(\log(t))}}, \quad t \geq T_0\end{aligned}$$

which gives

$$|T(e)'(t)| \leq \frac{1}{100}\left(1+\frac{1}{100}\right)\left(\frac{1}{t\log^{b+1}(t)\sqrt{\log(\log(t))}}\right), \quad t \geq T_0$$

Similarly,

$$T(e)(t) = \int_t^\infty \int_{x_1}^\infty T(e)''(x_2)dx_2 dx_1$$

Then we use

$$\int_t^\infty \int_{x_1}^\infty \frac{dx_2 dx_1}{\sqrt{\log(\log(x_2))}x_2^2\log^{b+1}(x_2)} = \frac{1}{b\log^b(t)\sqrt{\log(\log(t))}} + E_{int,2}$$

where

$$\begin{aligned}|E_{int,2}| &\leq \frac{1}{2b^2(\log(\log(t)))^{3/2}\log^b(t)} + \frac{1}{\sqrt{\log(\log(t))}\log^{b+1}(t)} \\ &\quad + \frac{1}{2(b+1)(\log(\log(t)))^{3/2}\log^{b+1}(t)} \\ &\leq \frac{1}{100}\frac{1}{b\log^b(t)\sqrt{\log(\log(t))}}, \quad t \geq T_0\end{aligned}$$

to get

$$|T(e)(t)| \leq \frac{1}{100}\left(1+\frac{1}{100}\right)\frac{1}{b\log^b(t)\sqrt{\log(\log(t))}}, \quad t \geq T_0$$

Then, we conclude that $T(e) \in \overline{B_1(0)}$ for $e \in \overline{B_1(0)}$.

Now, we will show that T is a contraction on the space $\overline{B_1(0)} \subset X$. We start by estimating, for $e_1, e_2 \in \overline{B_1(0)} \subset X$, the expression

$$RHS(e_1(t),t) - RHS(e_2(t),t)$$

In particular, we will prove the following proposition.

PROPOSITION 3.2. *There exists* $C_{lip} > 0$ independent *of* T_0, *such that for* $e_1, e_2 \in \overline{B_1(0)} \subset X$,

$$|RHS(e_1,t) - RHS(e_2,t)| \leq \frac{C_{lip}||e_1-e_2||_X}{t^2\log^{b+1}(t)(\log(\log(t)))^{3/2}}$$

To prove the proposition, it will be useful to define the functions

$$\lambda_i(t) = \lambda_0(t) + e_i(t), \quad i = 1,2$$

We start with the terms of RHS which don't involve G:

$$\begin{aligned}|\frac{-4\alpha\lambda_0''(t)\left(\log(\lambda_1(t))-\log(\lambda_2(t))\right)}{\log(\lambda_0(t))}| &\leq \frac{C}{t^2\log^{b+1}(t)\log(\log(t))}\frac{|\lambda_1(t)-\lambda_2(t)|}{\lambda_{0,0}(t)}\\ &\leq \frac{C||e_1-e_2||_X}{t^2\log^{b+1}(t)(\log(\log(t)))^{3/2}}\end{aligned}$$

$$\begin{aligned}&|\frac{-4\alpha}{\log(\lambda_0(t))}\left(e_1''(t)\left(\log(\lambda_1(t))-\log(\lambda_0(t))\right)-e_2''(t)\left(\log(\lambda_2(t))-\log(\lambda_0(t))\right)\right)|\\ &\leq \frac{C}{\log(\log(t))}\left(|e_1''(t)-e_2''(t)|\log\left(\frac{\lambda_0(t)+e_1(t)}{\lambda_0(t)}\right)+|e_2''(t)|\left(\log(\lambda_1(t))-\log(\lambda_2(t))\right)\right)\\ &\leq C\frac{||e_1-e_2||_X}{t^2\log^{b+1}(t)(\log(\log(t)))^2}\end{aligned}$$

$$\begin{aligned}&|\frac{4}{\log(\lambda_0(t))}\left(\int_t^\infty\frac{(e_1''(s)-e_2''(s))}{(1+s-t)^3}\left(\frac{1}{\lambda_1(t)^{1-\alpha}+s-t}-\frac{1}{\lambda_0(t)^{1-\alpha}+s-t}\right)ds\right.\\ &\left.+\int_t^\infty\frac{e_2''(s)}{(1+s-t)^3}\left(\frac{1}{\lambda_1(t)^{1-\alpha}+s-t}-\frac{1}{\lambda_2(t)^{1-\alpha}+s-t}\right)ds\right)|\\ &\leq C\frac{||e_1-e_2||_X}{(\log(\log(t)))^2t^2\log^{b+1}(t)}\end{aligned}$$

$$\begin{aligned}&\frac{1}{\log(\log(t))}\int_t^\infty\frac{1}{s^2\log^{b+1}(s)(1+s-t)^3}|\frac{1}{\lambda_1(t)^{1-\alpha}+s-t}-\frac{1}{\lambda_2(t)^{1-\alpha}+s-t}|ds\\ &\leq \frac{C||e_1-e_2||_X}{(\log(\log(t)))^{3/2}t^2\log^{b+1}(t)}\end{aligned}$$

$$\int_t^\infty|e_1''(s)-e_2''(s)||\frac{1}{\log(\lambda_0(t))}-\frac{1}{\log(\lambda_0(s))}|\frac{ds}{1+s-t} \leq \frac{C||e_1-e_2||_X}{t^2\log^{b+2}(t)(\log(\log(t)))^{5/2}}$$

$$\begin{aligned}&\int_t^\infty|e_1''(s)-e_2''(s)||\frac{1}{\log(\lambda_0(t))}-\frac{1}{\log(\lambda_0(s))}|\frac{ds}{(\lambda_0(t)^{1-\alpha}+s-t)(1+s-t)^3}\\ &\leq \frac{C||e_1-e_2||_X}{t^3\log^{b+2}(t)(\log(\log(t)))^{5/2}}\end{aligned}$$

We now proceed to study each term appearing in the expression

$$G(t,\lambda_1(t))-G(t,\lambda_2(t))$$

By writing

$$\begin{aligned}&\lambda_1(t)E_{0,1}(\lambda_1(t),\lambda_1'(t),\lambda_1''(t))-\lambda_2(t)E_{0,1}(\lambda_2(t),\lambda_2'(t),\lambda_2''(t))\\&=\int_0^1 DF_{0,0,1}(\lambda_\sigma)\cdot(\lambda_1(t)-\lambda_2(t),\lambda_1'(t)-\lambda_2'(t),\lambda_1''(t)-\lambda_2''(t))d\sigma\end{aligned}$$

where

$$F_{0,0,1}(x,y,z)=xE_{0,1}(x,y,z)$$

$$\lambda_\sigma=\sigma(\lambda_1(t),\lambda_1'(t),\lambda_1''(t))+(1-\sigma)(\lambda_2(t),\lambda_2'(t),\lambda_2''(t))$$

we get

$$\begin{aligned}&\lambda_1(t)E_{0,1}(\lambda_1(t),\lambda_1'(t),\lambda_1''(t))-\lambda_2(t)E_{0,1}(\lambda_2(t),\lambda_2'(t),\lambda_2''(t))\\&\leq\frac{C||e_1-e_2||_X}{t^2\log^{b+1}(t)\sqrt{\log(\log(t))}}\end{aligned}$$

Next, we consider

$$\begin{aligned}&K_3(w,\lambda_1(t))-K_{3,0}(w,\lambda_1(t))-(K_3(w,\lambda_2(t))-K_{3,0}(w,\lambda_2(t)))\\&=\frac{w^5}{4(1+w^2)}\left(\frac{1}{(w^2+36\lambda_1(t)^2)^2}-\frac{1}{(w^2+36\lambda_2(t)^2)^2}\right)\\&-\frac{1}{4}\left(\frac{w^5}{(\lambda_1(t)^{2-2\alpha}+w^2)(w^2+36\lambda_1(t)^2)^2}-\frac{w^5}{(\lambda_2(t)^{2-2\alpha}+w^2)(w^2+36\lambda_1(t)^2)^2}\right.\\&+\frac{w^5}{(\lambda_2(t)^{2-2\alpha}+w^2)(w^2+36\lambda_1(t)^2)^2}-\frac{w^5}{(\lambda_2(t)^{2-2\alpha}+w^2)(w^2+36\lambda_2(t)^2)^2}\\&\left.-\frac{1}{(\lambda_1(t)^{1-\alpha}+w)(1+w)^3}+\frac{1}{(\lambda_2(t)^{1-\alpha}+w)(1+w)^3}\right)\end{aligned}$$

So,

$$\begin{aligned}&\int_t^\infty|K_3(s-t,\lambda_1(t))-K_{3,0}(s-t,\lambda_1(t))-(K_3(s-t,\lambda_2(t))-K_{3,0}(s-t,\lambda_2(t)))|ds\\&\leq\frac{C||e_1-e_2||_X}{\log^b(t)\sqrt{\log(\log(t))}}+C\frac{||e_1-e_2||_X}{\sqrt{\log(\log(t))}}+\frac{C||e_1-e_2||_X}{\log^{b\alpha}(t)\sqrt{\log(\log(t))}}+\frac{C||e_1-e_2||_X}{\sqrt{\log(\log(t))}}\\&\leq\frac{C||e_1-e_2||_X}{\sqrt{\log(\log(t))}}\end{aligned}$$

From this, we get

$$
\begin{aligned}
&|\int_t^\infty \lambda_1''(s)\left(K_3(s-t,\lambda_1(t))-K_{3,0}(s-t,\lambda_1(t))\right)ds\\
&\quad-\int_t^\infty \lambda_2''(s)\left(K_3(s-t,\lambda_2(t))-K_{3,0}(s-t,\lambda_2(t))\right)ds|\\
&\quad\leq C\int_t^\infty |e_1''(s)-e_2''(s)||K_3(s-t,\lambda_1(t))-K_{3,0}(s-t,\lambda_1(t))|ds\\
&+C\int_t^\infty |\lambda_2''(s)||K_3(s-t,\lambda_1(t))-K_{3,0}(s-t,\lambda_1(t))\\
&\quad-\left(K_3(s-t,\lambda_2(t))-K_{3,0}(s-t,\lambda_2(t))\right)|ds\\
&\quad\leq C\frac{||e_1-e_2||_X}{t^2\log^{b+1}(t)\sqrt{\log(\log(t))}}\int_t^\infty |K_3(s-t,\lambda_1(t))-K_{3,0}(s-t,\lambda_1(t))|ds\\
&\quad+\frac{C}{t^2\log^{b+1}(t)}\int_t^\infty |K_3(s-t,\lambda_1(t))-K_{3,0}(s-t,\lambda_1(t))\\
&\quad-\left(K_3(s-t,\lambda_2(t))-K_{3,0}(s-t,\lambda_2(t))\right)|ds\\
&\quad\leq C\frac{||e_1-e_2||_X}{t^2\log^{b+1}(t)\sqrt{\log(\log(t))}}+\frac{C||e_1-e_2||_X}{t^2\log^{b+1}(t)\sqrt{\log(\log(t))}}\\
&\quad\leq \frac{C||e_1-e_2||_X}{t^2\log^{b+1}(t)\sqrt{\log(\log(t))}}
\end{aligned}
$$

(where we recall that $\int_t^\infty |K_3(s-t,\lambda_1(t))-K_{3,0}(s-t,\lambda_1(t))|ds$ was previously estimated).

Next, let us consider the following term which appears in the expression $G(t,\lambda_0(t)+e_1(t))-G(t,\lambda_0(t)+e_2(t))$:

$$
\begin{aligned}
&\frac{16}{(\lambda_0(t)+e_1(t))^2}\int_t^\infty (\lambda_0''(x)+e_1''(x))\left(K_1(x-t,\lambda_0(t)+e_1(t))-\frac{(\lambda_0(t)+e_1(t))^2}{4(1+x-t)}\right)dx\\
&-\frac{16}{(\lambda_0(t)+e_2(t))^2}\int_t^\infty (\lambda_0''(x)+e_2''(x))\left(K_1(x-t,\lambda_0(t)+e_2(t))-\frac{(\lambda_0(t)+e_2(t))^2}{4(1+x-t)}\right)dx
\end{aligned}
$$

which we re-write as

(3.100)

$$
\begin{aligned}
&\left(\frac{16}{(\lambda_0(t)+e_1(t))^2}-\frac{16}{(\lambda_0(t)+e_2(t))^2}\right)\int_t^\infty (\lambda_0''(x)+e_1''(x))\left(K_1(x-t,\lambda_0(t)+e_1(t))-\frac{(\lambda_0(t)+e_1(t))^2}{4(1+x-t)}\right)dx\\
&+\frac{16}{(\lambda_0(t)+e_2(t))^2}\int_t^\infty (e_1''(x)-e_2''(x))\left(K_1(x-t,\lambda_0(t)+e_1(t))-\frac{(\lambda_0(t)+e_1(t))^2}{4(1+x-t)}\right)dx\\
&+\frac{16}{(\lambda_0(t)+e_2(t))^2}\int_t^\infty (\lambda_0''(x)+e_2''(x))\left(L_1(x-t,\lambda_0(t)+e_1(t))-L_1(x-t,\lambda_0(t)+e_2(t))\right)dx
\end{aligned}
$$

where

$$L_1(w,z)=K_1(w,z)-\frac{z^2}{4(1+w)}$$

From (3.37) and (3.39) there exists $C>0$, independent of z such that

$$|K_1(w,z)-\frac{z^2}{4(1+w)}|\leq\frac{Cz^2}{1+w^2},\quad |z|\leq\frac{1}{2}$$

(Recall that, by the choice of T_0, $|\lambda_0(t)+f_i(t)| \leq \frac{1}{2}, \quad i=1,2$). Finally, let us note that

$$|\partial_z \left(K_1(w,z) - \frac{z^2}{4(1+w)}\right)| \leq \frac{Cz}{1+w^2}, \quad w \geq 0$$

Using the facts that

$$|\frac{16}{(\lambda_0(t)+e_1(t))^2} - \frac{16}{(\lambda_0(t)+e_2(t))^2}| \leq C\frac{\log^{3b}(t)||e_1-e_2||_X}{\sqrt{\log(\log(t))}\log^b(t)}$$

and

$$|L_1(w,z_1) - L_1(w,z_2)| \leq ||\partial_2 L_1(w, sz_1+(1-s)z_2)||_{L^\infty_s([0,1])}|z_1-z_2|$$

we get that the absolute value of (3.100) is bounded above by

$$\frac{C||e_1-e_2||_X}{t^2\log^{b+1}(t)\sqrt{\log(\log(t))}}$$

Another term arising in
$G(t,\lambda_0(t)+e_1(t)) - G(t,\lambda_0(t)+e_2(t))$ is

$$\frac{16}{(\lambda_0(t)+e_1(t))^2}\int_t^\infty (\lambda_0''(x)+e_1''(x))K(x-t,\lambda_0(t)+e_1(t))dx$$
$$-\frac{16}{(\lambda_0(t)+e_2(t))^2}\int_t^\infty (\lambda_0''(x)+e_2''(x))K(x-t,\lambda_0(t)+e_2(t))dx$$

This can be split analogously to the previous term:

(3.101)
$$\begin{aligned}&\left(\frac{16}{(\lambda_0(t)+e_1(t))^2} - \frac{16}{(\lambda_0(t)+e_2(t))^2}\right)\int_t^\infty (\lambda_0''(x)+e_1''(x))K(x-t,\lambda_0(t)+e_1(t))dx\\ &+\frac{16}{(\lambda_0(t)+e_2(t))^2}\int_t^\infty (e_1''(x)-e_2''(x))K(x-t,\lambda_0(t)+e_1(t))dx\\ &+\frac{16}{(\lambda_0(t)+e_2(t))^2}\int_t^\infty (\lambda_0''(x)+e_2''(x))\left(K(x-t,\lambda_0+e_1(t)) - K(x-t,\lambda_0(t)+e_2(t))\right)dx\end{aligned}$$

Note that

$$K(x-t,z_1) - K(x-t,z_2) = (z_1-z_2)\int_0^1 \partial_2 K(x-t, z_2+q(z_1-z_2))dq$$

So,

$$\begin{aligned}&\int_t^\infty |K(x-t,z_1) - K(x-t,z_2)|dx\\ &\leq |z_1-z_2|\int_0^1\int_0^\infty |\partial_2 K(w, z_2+(z_1-z_2)q)|dwdq\end{aligned} \tag{3.102}$$

But, using the formula for K from the previous sections, we see that $\partial_z K(w,z)$ has a sign:

$$\begin{aligned}&\partial_z K(w,z)\\ &=\int_0^\infty\int_0^w \left(\frac{4\rho R^3 z\left(\rho^2+R^2z^2+1\right)}{(R^2+1)^3\left((\rho^2-R^2z^2+1)^2+4R^2z^2\right)^{3/2}}\right)\left(\frac{1}{\sqrt{w^2-\rho^2}}-\frac{1}{w}\right)d\rho dR\\ &\geq 0\end{aligned}$$

So, to control the integral in (3.102), we can note the following: If

$$z_q = z_2 + (z_1 - z_2)q$$

then

(3.103)

$$\begin{aligned}
&\int_0^1 \int_0^\infty |\partial_2 K(w, z_q)| dw dq \\
&= \int_0^1 \int_0^\infty \int_0^\infty \int_0^w \left(\frac{4\rho R^3 z_q \left(\rho^2 + R^2 z_q^2 + 1\right)}{(R^2+1)^3 \left(\left(\rho^2 - R^2 z_q^2 + 1\right)^2 + 4R^2 z_q^2\right)^{3/2}} \right) \left(\frac{1}{\sqrt{w^2 - \rho^2}} - \frac{1}{w} \right) d\rho dR dw dq \\
&= \int_0^1 \int_0^\infty \int_0^\infty \int_\rho^\infty \left(\frac{4\rho R^3 z_q \left(\rho^2 + R^2 z_q^2 + 1\right)}{(R^2+1)^3 \left(\left(\rho^2 - R^2 z_q^2 + 1\right)^2 + 4R^2 z_q^2\right)^{3/2}} \right) \left(\frac{1}{\sqrt{w^2 - \rho^2}} - \frac{1}{w} \right) dw d\rho dR dq \\
&= \int_0^1 \int_0^\infty \int_0^\infty \frac{4\rho R^3 z_q \log(2) \left(\rho^2 + R^2 z_q^2 + 1\right)}{(R^2+1)^3 \left(\left(\rho^2 - R^2 z_q^2 + 1\right)^2 + 4R^2 z_q^2\right)^{3/2}} d\rho dR dq \\
&= \int_0^1 \int_0^\infty \frac{R^3 z_q \log(4)}{(R^2+1)^3} dR dq \\
&= \int_0^1 \frac{1}{2} z_q \log(2) dq \\
&= \frac{1}{4} \log(2)(z_1 + z_2)
\end{aligned}$$

Using (3.34), we see that the absolute value of (3.101) is bounded above by

$$\frac{C||e_1 - e_2||_X}{t^2 \log^{b+1}(t) \sqrt{\log(\log(t))}}$$

Next, we will consider the terms involving $E_{v_2,ip}$, starting with the case $b \neq 1$. We recall

(3.104)

$$\begin{aligned}
\lambda(t) E_{v_2,ip}(t, \lambda(t)) &= 2c_b \lambda(t) \int_0^\infty \frac{\sin(t\xi)}{t^2} \psi_{v_2}(\xi, \lambda(t)) d\xi \\
&+ 2c_b \lambda(t) \int_0^\infty \chi_{\leq \frac{1}{4}}(\xi) \frac{\sin(t\xi)}{t^2} F_{v_2}(\xi, \lambda(t)) d\xi \\
&+ 2c_b \int_0^{\frac{1}{2}} \left(\chi_{\leq \frac{1}{4}}(\xi) - 1\right) \frac{\sin(t\xi)}{t^2} \left(\frac{(b-1)}{\xi \log^b(\frac{1}{\xi})} + \frac{b(b-1)}{\xi \log^{b+1}(\frac{1}{\xi})} \right) d\xi \\
&+ 2c_b \left(\int_0^{\frac{t}{2}} \frac{\sin(u)(b-1) du}{t^2 u \log^b(\frac{t}{u})} - \frac{(b-1)\pi}{2t^2 \log^b(t)} + \int_0^{\frac{t}{2}} \frac{\sin(u) b(b-1) du}{t^2 u \log^{b+1}(\frac{t}{u})} \right)
\end{aligned}$$

where F_{v_2} and ψ_{v_2} were defined in (3.42) and (3.41), respectively. Notice that the last two lines of (3.104) are *independent* of $\lambda(t)$. So,

(3.105)
$$\begin{aligned}&\lambda_1(t)E_{v_2,ip}(t,\lambda_1(t)) - \lambda_2(t)E_{v_2,ip}(t,\lambda_2(t))\\
&\quad = 2c_b\lambda_1(t)\int_0^\infty \frac{\sin(t\xi)}{t^2}\psi_{v_2}(\xi,\lambda_1(t))d\xi - 2c_b\lambda_2(t)\int_0^\infty \frac{\sin(t\xi)}{t^2}\psi_{v_2}(\xi,\lambda_2(t))d\xi\\
&\quad + 2c_b\lambda_1(t)\int_0^\infty \chi_{\leq\frac{1}{4}}(\xi)\frac{\sin(t\xi)}{t^2}F_{v_2}(\xi,\lambda_1(t))d\xi\\
&\quad - 2c_b\lambda_2(t)\int_0^\infty \chi_{\leq\frac{1}{4}}(\xi)\frac{\sin(t\xi)}{t^2}F_{v_2}(\xi,\lambda_2(t))d\xi\end{aligned}$$

First, we consider the second line of (3.105):

(3.106)
$$\begin{aligned}&|2c_b\lambda_1(t)\int_0^\infty \frac{\sin(t\xi)}{t^2}\psi_{v_2}(\xi,\lambda_1(t))d\xi - 2c_b\lambda_2(t)\int_0^\infty \frac{\sin(t\xi)}{t^2}\psi_{v_2}(\xi,\lambda_2(t))d\xi|\\
&\leq |2c_b(e_1(t)-e_2(t))\int_0^\infty \frac{\sin(t\xi)}{t^2}\psi_{v_2}(\xi,\lambda_1(t))d\xi|\\
&+ |2c_b\lambda_2(t)\int_0^\infty \frac{\sin(t\xi)}{t^2}\left(\psi_{v_2}(\xi,\lambda_1(t)) - \psi_{v_2}(\xi,\lambda_2(t))\right)d\xi|\end{aligned}$$

The second line of (3.106) is estimated as follows:

$$\begin{aligned}&|2c_b(e_1(t)-e_2(t))\int_0^\infty \frac{\sin(t\xi)}{t^2}\psi_{v_2}(\xi,\lambda_1(t))d\xi|\\
&\leq C|e_1(t)-e_2(t)||\int_0^\infty \frac{\sin(t\xi)}{t^2}\psi_{v_2}(\xi,\lambda_1(t))d\xi|\\
&\leq C\frac{|e_1(t)-e_2(t)|}{t^3\lambda_1(t)}\end{aligned}$$

where we use the same procedure as used in the previous subsections to estimate the integral. For the third line of (3.106), we first note that

$$\psi_{v_2}(\xi,\lambda(t)) = 2\chi'_{\leq\frac{1}{4}}(\xi)\partial_\xi\left(\frac{\xi^2K_1(\xi\lambda(t))}{\log^{b-1}(\frac{1}{\xi})}\right) + \frac{\chi''_{\leq\frac{1}{4}}(\xi)\xi^2K_1(\xi\lambda(t))}{\log^{b+1}(\frac{1}{\xi})}$$

$$\begin{aligned}\partial_\xi\psi_{v_2}(\xi,\lambda(t)) &= 3\chi''_{\leq\frac{1}{4}}(\xi)\partial_\xi\left(\frac{\xi^2K_1(\xi\lambda(t))}{\log^{b-1}(\frac{1}{\xi})}\right) + 2\chi'_{\leq\frac{1}{4}}(\xi)\partial_\xi^2\left(\frac{\xi^2K_1(\xi\lambda(t))}{\log^{b-1}(\frac{1}{\xi})}\right)\\
&+ \chi'''_{\leq\frac{1}{4}}(\xi)\frac{\xi^2K_1(\xi\lambda(t))}{\log^{b-1}(\frac{1}{\xi})}\end{aligned}$$

Then, we use

$$\begin{aligned}K_1'(y) &= -\left(\frac{yK_0(y)+K_1(y)}{y}\right)\\
K_1''(y) &= K_1(y) + \frac{K_2(y)}{y}\\
K_1'''(y) &= \frac{-(3+y^2)}{y^2}K_2(y)\end{aligned}$$

and the estimates, for $0 < x < \frac{1}{2}$ and $\frac{1}{8} \leq \xi \leq \frac{1}{4}$,

$$|K_1(\xi x)| \leq \frac{C}{\xi x}, \quad |K_0(\xi x)| \leq C|\log(\xi x)|, \quad |K_2(\xi x)| \leq \frac{C}{\xi^2 x^2} \tag{3.107}$$

Since $\text{supp}\left(\chi^{(j)}_{\leq \frac{1}{4}}\right) \subset [\frac{1}{8}, \frac{1}{4}]$, we have

$$|\partial_{12}\psi_{v_2}(\xi, y)| \leq \frac{C\mathbb{1}_{[\frac{1}{8},\frac{1}{4}]}(\xi)}{\xi y^2 \log^{b-1}(\frac{1}{\xi})}, \quad 0 < y < \frac{1}{2} \tag{3.108}$$

Note that

$$\begin{aligned}&\int_0^\infty \frac{\sin(t\xi)}{t^2}\left(\psi_{v_2}(\xi, \lambda_1(t)) - \psi_{v_2}(\xi, \lambda_2(t))\right) d\xi \\ &= \int_0^\infty \frac{\cos(t\xi)}{t^3}\partial_\xi \left(\psi_{v_2}(\xi, \lambda_1(t)) - \psi_{v_2}(\xi, \lambda_2(t))\right) d\xi\end{aligned}$$

and

$$|\partial_\xi \left(\psi_{v_2}(\xi, \lambda_1(t)) - \psi_{v_2}(\xi, \lambda_2(t))\right)| \leq \sup_{x \in [\frac{\lambda_0(t)}{2}, \frac{1}{2}]} |\partial_{12}\psi_{v_2}(\xi, x)||\lambda_1(t) - \lambda_2(t)|$$

where we used the fact that, for $e_i \in \overline{B_1(0)} \subset X, \quad i = 1, 2$, we have

$$\frac{1}{2}\lambda_0(t) < \lambda_i(t) < 2\lambda_0(t) < \frac{1}{2}, \quad i = 1, 2$$

Now, we use (3.108) to get

$$\begin{aligned}&|\int_0^\infty \frac{\cos(t\xi)}{t^3}\partial_\xi \left(\psi_{v_2}(\xi, \lambda_1(t)) - \psi_{v_2}(\xi, \lambda_2(t))\right) d\xi| \\ &\leq \frac{C}{t^3}\int_0^\infty \sup_{x \in [\frac{\lambda_0(t)}{2}, \frac{1}{2}]} |\partial_{12}\psi_{v_2}(\xi, x)||\lambda_1(t) - \lambda_2(t)| d\xi \\ &\leq C\frac{|e_1(t) - e_2(t)|}{t^3}\int_0^\infty \frac{\mathbb{1}_{[\frac{1}{8},\frac{1}{4}]}(\xi) d\xi}{\lambda_0(t)^2 \xi \log^{b-1}(\frac{1}{\xi})} \\ &\leq C\frac{|e_1(t) - e_2(t)| \log^{2b}(t)}{t^3}\end{aligned}$$

We conclude that

$$\begin{aligned}&|2c_b(\lambda_0(t) + e_2(t))\int_0^\infty \frac{\cos(t\xi)}{t^3}\left(\partial_\xi\psi_{v_2}(\xi, \lambda_1(t)) - \partial_\xi\psi_{v_2}(\xi, \lambda_2(t))\right) d\xi| \\ &\leq \frac{C||e_1 - e_2||_X}{t^3\sqrt{\log(\log(t))}}\end{aligned}$$

We now need to consider the third line of (3.105). Here, we use a similar procedure as above:

(3.109)

$$\begin{aligned}&|2c_b\lambda_1(t)\int_0^\infty \chi_{\leq \frac{1}{4}}(\xi)\frac{\sin(t\xi)}{t^2}F_{v_2}(\xi, \lambda_1(t)) d\xi - 2c_b\lambda_2(t)\int_0^\infty \chi_{\leq \frac{1}{4}}(\xi)\frac{\sin(t\xi)}{t^2}F_{v_2}(\xi, \lambda_2(t)) d\xi| \\ &\leq C|\left(e_1(t) - e_2(t)\right)\int_0^\infty \chi_{\leq \frac{1}{4}}(\xi)\frac{\sin(t\xi)}{t^2}F_{v_2}(\xi, \lambda_1(t)) d\xi| \\ &+ C\lambda_2(t)|\int_0^\infty \frac{\chi_{\leq \frac{1}{4}}(\xi)\sin(t\xi)}{t^2}\left(F_{v_2}(\xi, \lambda_1(t)) - F_{v_2}(\xi, \lambda_2(t))\right) d\xi|\end{aligned}$$

For the second line of (3.109), we use the same procedure to estimate the integral as that used in previous subsections, to get

$$\begin{aligned}&|\left(e_1(t)-e_2(t)\right)\int_0^\infty \chi_{\leq \frac{1}{4}}(\xi)\frac{\sin(t\xi)}{t^2}F_{v_2}(\xi,\lambda_1(t))d\xi|\\&\leq C|e_1(t)-e_2(t)|\frac{\lambda_1(t)|\log(\lambda_1(t))|}{t^3}\end{aligned}$$

For the third line of (3.109), we first integrate by parts once to get

$$\begin{aligned}&\int_0^\infty \frac{\chi_{\leq \frac{1}{4}}(\xi)\sin(t\xi)}{t^2}\left(F_{v_2}(\xi,\lambda_1(t))-F_{v_2}(\xi,\lambda_2(t))\right)d\xi\\&=\int_0^\infty \frac{\cos(t\xi)}{t^3}\Big(\chi'_{\leq \frac{1}{4}}(\xi)\left(F_{v_2}(\xi,\lambda_1(t))-F_{v_2}(\xi,\lambda_2(t))\right)\\&\qquad\qquad+\chi_{\leq \frac{1}{4}}(\xi)\partial_\xi\left(F_{v_2}(\xi,\lambda_1(t))-F_{v_2}(\xi,\lambda_2(t))\right)\Big)d\xi\end{aligned}$$

Now, we recall that

$$F_{v_2}(\xi,y)=\partial_\xi^2\left(\frac{\xi^2}{\log^{b-1}(\frac{1}{\xi})}\left(K_1(\xi y)-\frac{1}{\xi y}\right)\right)$$

Using the same estimates on K_j as used to obtain (3.108), we get

$$|\partial_2 F_{v_2}(\xi,y)|\leq \frac{C\xi(|\log(\xi)|+|\log(y)|)}{\log^{b-1}(\frac{1}{\xi})},\quad 0<\xi<\frac{1}{4},\quad 0<y<\frac{1}{2}$$

and

$$|\partial_{12} F_{v_2}(\xi,y)|\leq \frac{C(|\log(\xi)|+|\log(y)|)}{\log^{b-1}(\frac{1}{\xi})},\quad 0<y<\frac{1}{2},\quad 0<\xi<\frac{1}{4}$$

We then get

$$|F_{v_2}(\xi,\lambda_2(t))-F_{v_2}(\xi,\lambda_1(t))|\leq\left(\sup_{x\in[\frac{\lambda_0(t)}{2},\frac{1}{2}]}|\partial_2 F_{v_2}(\xi,x)|\right)|\lambda_1(t)-\lambda_2(t)|$$

and

$$|\partial_1 F_{v_2}(\xi,\lambda_2(t))-\partial_1 F_{v_2}(\xi,\lambda_1(t))|\leq\left(\sup_{x\in[\frac{\lambda_0(t)}{2},\frac{1}{2}]}|\partial_{12} F_{v_2}(\xi,x)|\right)|\lambda_1(t)-\lambda_2(t)|$$

Using the above estimates, we obtain,

$$\begin{aligned}&|2c_b\lambda_2(t)\int_0^\infty \chi_{\leq \frac{1}{4}}(\xi)\frac{\sin(t\xi)}{t^2}\left(F_{v_2}(\xi,\lambda_1(t))-F_{v_2}(\xi,\lambda_2(t))\right)d\xi|\\&\leq C\frac{|\lambda_1(t)-\lambda_2(t)|\lambda_0(t)|\log(\lambda_0(t))|}{t^3}\leq C\frac{||e_1-e_2||_X\sqrt{\log(\log(t))}}{t^3\log^{2b}(t)}\end{aligned}$$

The same procedure for the case $b=1$ yields the same estimates, since ψ_{v_2} and F_{v_2} have the same form for all b.

Combining the above estimates then gives, for all $b>0$:

$$|-\lambda_1(t)E_{v_2,ip}(t,\lambda_1(t))+\lambda_2(t)E_{v_2,ip}(t,\lambda_2(t))|\leq\frac{C||e_1-e_2||_X}{t^3\sqrt{\log(\log(t))}}$$

If

$$G_2(t,\lambda(t)) = -\lambda(t)\langle\left(\frac{\cos(2Q_{\frac{1}{\lambda(t)}})-1}{r^2}\right)\left((v_4+v_5)\left(1-\chi_{\geq 1}(\frac{4r}{t})\right)\right)|_{r=R\lambda(t)},\phi_0\rangle$$
$$-\lambda(t)\langle\left(\frac{\cos(2Q_{\frac{1}{\lambda(t)}})-1}{r^2}\right)E_5|_{r=R\lambda(t)},\phi_0\rangle$$
$$+\lambda(t)\langle\left(\frac{\cos(2Q_{\frac{1}{\lambda(t)}})-1}{r^2}\right)\chi_{\geq 1}(\frac{2r}{\log^N(t)})\,(v_1+v_2+v_3)\,|_{r=R\lambda(t)},\phi_0\rangle$$

Then, we need to estimate

$$G_2(t,\lambda_1(t))-G_2(t,\lambda_2(t))$$

From now on, the fact that $v_k, \quad k\neq 2$ depends on $\lambda(t)$ is important, so we will denote these functions by v_k^λ to emphasize the dependence of these functions on λ.

We first note that

$$\lambda(t)\langle\left(\frac{\cos(2Q_{\frac{1}{\lambda(t)}})-1}{r^2}\right)v_4^\lambda\left(1-\chi_{\geq 1}(\frac{4r}{t})\right)|_{r=R\lambda(t)},\phi_0\rangle$$
$$=\frac{1}{\lambda(t)}\int_0^\infty\left(\frac{\cos(2Q_1(\frac{r}{\lambda(t)}))-1}{r^2}\right)v_4^\lambda(t,r)\left(1-\chi_{\geq 1}(\frac{4r}{t})\right)\phi_0(\frac{r}{\lambda(t)})rdr$$

so,

$$\lambda_1(t)\langle\left(\frac{\cos(2Q_{\frac{1}{\lambda_1(t)}})-1}{r^2}\right)v_4^{\lambda_1}\left(1-\chi_{\geq 1}(\frac{4r}{t})\right)|_{r=R\lambda_1(t)},\phi_0\rangle$$
$$-\left(\lambda_2(t)\langle\left(\frac{\cos(2Q_{\frac{1}{\lambda_2(t)}})-1}{r^2}\right)v_4^{\lambda_2}\left(1-\chi_{\geq 1}(\frac{4r}{t})\right)|_{r=R\lambda_2(t)},\phi_0\rangle\right)$$
$$=-16\int_0^\infty\left(\frac{\lambda_1(t)^2}{(r^2+\lambda_1(t)^2)^3}-\frac{\lambda_2(t)^2}{(r^2+\lambda_2(t)^2)^3}\right)v_4^{\lambda_1}(t,r)\left(1-\chi_{\geq 1}(\frac{4r}{t})\right)r^2dr$$
$$+\int_0^\infty\left(\frac{\cos(2Q_1(\frac{r}{\lambda_2(t)})-1}{r^2\lambda_2(t)}\right)\phi_0(\frac{r}{\lambda_2(t)})\left(v_4^{\lambda_1}-v_4^{\lambda_2}\right)(t,r)\left(1-\chi_{\geq 1}(\frac{4r}{t})\right)rdr$$

and we have the analogous formulae for all the other terms in G_2. Using the pointwise estimates for $v_i, E_5, \quad 1\leq i\leq 5$, we get

$$|-16\int_0^\infty\left(\frac{\lambda_1(t)^2}{(r^2+\lambda_1(t)^2)^3}-\frac{\lambda_2(t)^2}{(r^2+\lambda_2(t)^2)^3}\right)\left((v_4^{\lambda_1}+v_5^{\lambda_1})\left(1-\chi_{\geq 1}(\frac{4r}{t})\right)\right.$$
$$\left.+E_5^{\lambda_1}-\chi_{\geq 1}(\frac{2r}{\log^N(t)})\left(v_1^{\lambda_1}+v_2+v_3^{\lambda_1}\right)\right)r^2dr|$$
$$\leq\frac{C||e_1-e_2||_X}{t^2\log^{3b+2N-1}(t)\sqrt{\log(\log(t))}}+\frac{C||e_1-e_2||_X}{t^{7/2}\log^{3b-3+\frac{5N}{2}}(t)\sqrt{\log(\log(t))}}$$
$$+\frac{C||e_1-e_2||_X}{t^2\log^{b+1}(t)\sqrt{\log(\log(t))}}+\frac{C||e_1-e_2||_X}{t^2\log^{3b+2N}(t)\sqrt{\log(\log(t))}}$$
$$\leq\frac{C||e_1-e_2||_X}{t^2\log^{b+1}(t)\sqrt{\log(\log(t))}}$$

In order to estimate the rest of the terms arising in $G_2(t,\lambda_1)-G_2(t,\lambda_2)$, we must first prove estimates on $v_{3,1,2}$ defined by

$$v_{3,1,2}:=v_3^{\lambda_1}-v_3^{\lambda_2}$$

LEMMA 3.11.

$$|v_{3,1,2}(t,r)|\leq\begin{cases}\frac{Cr||e_1-e_2||_X}{t^2\sqrt{\log(\log(t))}\log^b(t)}, & r\leq\frac{t}{2}\\ \frac{C||e_1-e_2||_X}{r\log^b(t)\sqrt{\log(\log(t))}}, & r>\frac{t}{2}\end{cases}\tag{3.110}$$

PROOF. Note that $v_{3,1,2}$ solves the following equation with 0 Cauchy data at infinity:

$$-\partial_{tt}v_{3,1,2}+\partial_{rr}v_{3,1,2}+\frac{1}{r}\partial_r v_{3,1,2}-\frac{v_{3,1,2}}{r^2}=F_{0,1}^{\lambda_1}(t,r)-F_{0,1}^{\lambda_2}(t,r)$$

With the definitions from the v_3 subsection:

$$\begin{aligned}&v_{3,1}^{\lambda}(t,r)\\&=\frac{-1}{r}\int_t^\infty\int_0^{s-t}\frac{\rho}{(s-t)}\lambda''(s)\left(\frac{-1-\rho^2+r^2}{\sqrt{(1+\rho^2-r^2)^2+4r^2}}+F_3(r,\rho,\lambda(s))\right)d\rho ds\end{aligned}$$

and

$$v_{3,2}^{\lambda}=v_3^{\lambda}-v_{3,1}^{\lambda}$$

we then consider

$$\begin{aligned}&v_{3,2,a}^{\lambda}(t,r)\\&:=\frac{-1}{r}\int_t^\infty\int_0^{s-t}\rho\left(\frac{1}{\sqrt{(s-t)^2-\rho^2}}-\frac{1}{(s-t)}\right)\lambda''(s)\left(\frac{-1-\rho^2+r^2}{\sqrt{(1+\rho^2-r^2)^2+4r^2}}+1\right)d\rho ds\\&v_{3,2,b}^{\lambda}(t,r)\\&:=\frac{-1}{r}\int_t^\infty\int_0^{s-t}\rho\left(\frac{1}{\sqrt{(s-t)^2-\rho^2}}-\frac{1}{(s-t)}\right)\lambda''(s)\left(-1+F_3(r,\rho,\lambda(s))\right)d\rho ds\end{aligned}$$

and get

$$\begin{aligned}&|v_{3,2,a}^{\lambda_1}(t,r)-v_{3,2,a}^{\lambda_2}(t,r)|\\&\leq\frac{C}{r}\int_t^\infty\int_0^{s-t}\rho\left(\frac{1}{\sqrt{(s-t)^2-\rho^2}}-\frac{1}{(s-t)}\right)|\lambda_1''(s)-\lambda_2''(s)|\left(\frac{-1-\rho^2+r^2}{\sqrt{(1+\rho^2-r^2)^2+4r^2}}+1\right)d\rho ds\\&\leq\frac{C}{r}\int_0^\infty\rho\left(1+\frac{-1-\rho^2+r^2}{\sqrt{(1+\rho^2-r^2)^2+4r^2}}\right)\int_{\rho+t}^\infty\left(\frac{1}{\sqrt{(s-t)^2-\rho^2}}-\frac{1}{(s-t)}\right)\frac{||e_1-e_2||_X\,dsd\rho}{s^2\log^{b+1}(s)\sqrt{\log(\log(s))}}\end{aligned}$$

which gives

$$|v_{3,2,a}^{\lambda_1}(t,r)-v_{3,2,a}^{\lambda_2}(t,r)|\leq\frac{Cr||e_1-e_2||_X}{t^2\log^{b+1}(t)\sqrt{\log(\log(t))}}$$

For $v_{3,2,b}$, we have

$$\begin{aligned}&v_{3,2,b}^{\lambda_1}(t,r)-v_{3,2,b}^{\lambda_2}(t,r)\\&=\frac{-1}{r}\int_t^\infty\int_0^{s-t}\rho\left(\frac{1}{\sqrt{(s-t)^2-\rho^2}}-\frac{1}{(s-t)}\right)\left(\lambda_1''(s)-\lambda_2''(s)\right)\left(-1+F_3(r,\rho,\lambda_1(s)\right)d\rho ds\\&+\frac{-1}{r}\int_t^\infty\int_0^{s-t}\rho\left(\frac{1}{\sqrt{(s-t)^2-\rho^2}}-\frac{1}{(s-t)}\right)\lambda_2''(s)\left(F_3(r,\rho,\lambda_1(s))-F_3(r,\rho,\lambda_2(s))\right)d\rho ds\end{aligned}\tag{3.111}$$

The first line on the right-hand side of (3.111) is treated identically to the analogous term arising in the pointwise estimates for v_3, and it is bounded above in absolute value by

$$\frac{Cr||e_1-e_2||_X}{t^2\log^{b+1}(t)\sqrt{\log(\log(t))}}$$

For the second line on the right-hand side of (3.111), we start with

$$|F_3(r,\rho,z_1)-F_3(r,\rho,z_2)|\leq \max_{\sigma\in[0,1]}|\partial_3 F_3(r,\rho,\sigma z_1+(1-\sigma)z_2)||z_1-z_2|$$

we then note that

$$\partial_3 F_3(r,\rho,z)=\frac{-4(-1+\alpha)r^2z^{-3+2\alpha}(1+(r^2-\rho^2)z^{2\alpha-2})}{((1+(r^2-\rho^2)z^{2\alpha-2})^2+4\rho^2z^{2\alpha-2})^{3/2}}$$

so, for $z=\sigma\lambda_1(s)+(1-\sigma)\lambda_2(s)$,

$$\begin{aligned}|\partial_3 F_3(r,\rho,z)|&\leq \frac{Cr^2\lambda_{0,0}(s)^{2\alpha-3}}{(1+(-\rho^2+r^2)\lambda_{0,0}(s)^{2\alpha-2})^2+4\rho^2\lambda_{0,0}(s)^{2\alpha-2}}\\&\leq\frac{Cr^2\lambda_{0,0}(s)^{2\alpha-3}}{1+2(\rho^2+r^2)\lambda_{0,0}(t)^{2\alpha-2}+(\rho^2-r^2)^2\lambda_{0,0}(t)^{4\alpha-4}}\end{aligned}$$

Then, we get

$$\begin{aligned}&|\frac{-1}{r}\int_t^\infty\int_0^{s-t}\rho\left(\frac{1}{\sqrt{(s-t)^2-\rho^2}}-\frac{1}{(s-t)}\right)\lambda_2''(s)\left(F_3(r,\rho,\lambda_1(s))-F_3(r,\rho,\lambda_2(s))\right)d\rho ds|\\&\leq\frac{C}{r}\int_0^\infty\frac{\rho}{t^2\log^{b+1}(t)}\frac{r^2||e_1-e_2||_X}{\log^{b(2\alpha-3)}(t)\log^b(t)\sqrt{\log(\log(t))}}\frac{d\rho}{(1+2(\rho^2+r^2)\lambda_{0,0}(t)^{2\alpha-2}+(\rho^2-r^2)^2\lambda_{0,0}(t)^{4\alpha-4})}\end{aligned}$$

The ρ integral appearing on the last line of the above estimate has been treated in the v_3 pointwise estimates, and, in total, we get

$$|v_{3,2,b}^{\lambda_1}(t,r)-v_{3,2,b}^{\lambda_2}(t,r)|\leq\frac{Cr||e_1-e_2||_X}{t^2\log^{b+1}(t)\sqrt{\log(\log(t))}}$$

We similarly divide $v_{3,1}^{\lambda_1}-v_{3,1}^{\lambda_2}$ into two parts. First, we consider

$$\begin{aligned}&|\frac{-1}{r}\int_t^\infty\int_0^{s-t}\frac{\rho}{(s-t)}\left(\lambda_1''(s)-\lambda_2''(s)\right)\left(\frac{-1-\rho^2+r^2}{\sqrt{(1+\rho^2-r^2)^2+4r^2}}+F_3(r,\rho,\lambda_1(s))\right)d\rho ds|\\&\leq\frac{C}{r}\int_t^{t+6r}\int_0^{s-t}\frac{\rho}{(s-t)}|\lambda_1''(s)-\lambda_2''(s)|d\rho ds\\&+|\int_{t+6r}^\infty\frac{1}{r}\int_0^{s-t}\frac{\rho}{(s-t)}\left(\lambda_1''(s)-\lambda_2''(s)\right)\left(\frac{-1-\rho^2+r^2}{\sqrt{(-1-\rho^2+r^2)^2+4r^2}}+F_3(r,\rho,\lambda_1(s))\right)d\rho ds|\\&\leq\frac{Cr||e_1-e_2||_X}{t^2\log^{b+1}(t)\sqrt{\log(\log(t))}}+II\end{aligned}$$

where, using a procedure similar to the analogous term treated in the v_3 pointwise estimates, we have

$$\begin{aligned}II=&|-2r\int_{6r}^\infty\left(\lambda_1''(t+w)-\lambda_2''(t+w)\right)\left(\frac{w}{2(1+w^2)}-\frac{w}{2(\lambda_1(t+w)^{2-2\alpha}+w^2)}\right)dw|\\&+E_{II}\end{aligned}$$

with

$$|E_{II}|\leq\frac{Cr||e_1-e_2||_X}{t^2\log^{b+1}(t)\sqrt{\log(\log(t))}}$$

We already estimated the integral appearing in the above expression in (3.80) (except with the replacement of λ'' with $\lambda_1''-\lambda_2''$). Therefore, the same procedure used there shows

$$\begin{aligned}|II| &\leq Cr\log(\log(t))\sup_{x\geq t}|\lambda_1''(x)-\lambda_2''(x)|+\frac{Cr||e_1-e_2||_X}{t^2\log^{b+1}(t)\sqrt{\log(\log(t))}}\\ &\leq \frac{Cr\log(\log(t))||e_1-e_2||_X}{t^2\log^{b+1}(t)\sqrt{\log(\log(t))}}+\frac{Cr||e_1-e_2||_X}{t^2\log^{b+1}(t)\sqrt{\log(\log(t))}}\end{aligned}$$

Next, we consider the second part of $v_{3,1}^{\lambda_1}-v_{3,1}^{\lambda_2}$:

$$\begin{aligned}&|\frac{-1}{r}\int_t^\infty\int_0^{s-t}\frac{\rho}{(s-t)}\lambda_2''(s)\left(F_3(r,\rho,\lambda_1(s))-F_3(r,\rho,\lambda_2(s))\right)d\rho ds|\\ &\leq \frac{C}{r}\int_t^\infty\int_0^{s-t}\frac{\rho}{(s-t)}|\lambda_2''(s)|\frac{r^2(\lambda_{0,0}(s)^{2\alpha-3})|\lambda_1(s)-\lambda_2(s)|d\rho ds}{(1+2(\rho^2+r^2)\lambda_{0,0}(t)^{2\alpha-2}+(\rho^2-r^2)^2\lambda_{0,0}(t)^{4\alpha-4})}\\ &\leq Cr\int_t^{t+\lambda_{0,0}(t)^{1-\alpha}}\int_0^{s-t}\frac{\rho}{(s-t)}|\lambda_2''(s)||e_1(s)-e_2(s)|\lambda_{0,0}(s)^{2\alpha-3}d\rho ds\\ &+Cr\int_{t+\lambda_{0,0}(t)^{1-\alpha}}^\infty\frac{1}{(s-t)}|\lambda_2''(s)|\frac{||e_1-e_2||_X\lambda_{0,0}(s)^{2\alpha-3}ds}{\log^b(s)\sqrt{\log(\log(s))}}\int_0^\infty\\ &\qquad\frac{\rho d\rho}{(1+2(\rho^2+r^2)\lambda_{0,0}(t)^{2\alpha-2}+(\rho^2-r^2)^2\lambda_{0,0}(t)^{4\alpha-4})}\\ &\leq Cr\int_t^{t+\lambda_{0,0}(t)^{1-\alpha}}(s-t)|\lambda_2''(s)||e_1(s)-e_2(s)|\lambda_{0,0}(s)^{2\alpha-3}ds\\ &+\frac{Cr||e_1-e_2||_X}{\log^b(t)\sqrt{\log(\log(t))}}\int_{t+\lambda_{0,0}(t)^{1-\alpha}}^\infty\frac{\lambda_{0,0}(s)^{2\alpha-3}\lambda_{0,0}(t)^{2-2\alpha}ds}{(s-t)s^2\log^{b+1}(s)}\\ &\leq \frac{Cr||e_1-e_2||_X}{t^2\log^{b+1}(t)\sqrt{\log(\log(t))}}+\frac{Cr||e_1-e_2||_X}{t^2\sqrt{\log(\log(t))}\log^b(t)}\\ &\leq \frac{Cr||e_1-e_2||_X}{t^2\sqrt{\log(\log(t))}\log^b(t)}\end{aligned}$$

Combining the above estimates, we get

$$|(v_3^{\lambda_1}-v_3^{\lambda_2})(t,r)|\leq \frac{Cr||e_1-e_2||_X}{t^2\sqrt{\log(\log(t))}\log^b(t)}$$

Next, we prove an estimate on $v_{3,1,2}$ which is more useful for large r.

$$\begin{aligned}&v_3^\lambda(t,r)\\ &=\frac{-1}{r}\int_t^\infty\int_0^{s-t}\frac{\rho}{\sqrt{(s-t)^2-\rho^2}}\lambda''(s)\left(\frac{-1-\rho^2+r^2}{\sqrt{(-1-\rho^2+r^2)^2+4r^2}}+F_3(r,\rho,\lambda(s))\right)d\rho ds\end{aligned}$$

and get

(3.112)

$$\begin{aligned}&\left(v_3^{\lambda_1}-v_3^{\lambda_2}\right)(t,r)\\ &=\frac{-1}{r}\int_t^\infty\int_0^{s-t}\frac{\rho}{\sqrt{(s-t)^2-\rho^2}}\left(\lambda_1''(s)-\lambda_2''(s)\right)\left(\frac{-1-\rho^2+r^2}{\sqrt{(-1-\rho^2+r^2)^2+4r^2}}+F_3(r,\rho,\lambda_1(s))\right)d\rho ds\\ &-\frac{1}{r}\int_t^\infty\int_0^{s-t}\frac{\rho}{\sqrt{(s-t)^2-\rho^2}}\lambda_2''(s)\left(F_3(r,\rho,\lambda_1(s))-F_3(r,\rho,\lambda_2(s))\right)d\rho ds\end{aligned}$$

For the first line of the right-hand side of (3.112), we have

$$\begin{aligned}
&|\frac{-1}{r}\int_t^\infty\int_0^{s-t}\frac{\rho}{\sqrt{(s-t)^2-\rho^2}}\left(\lambda_1''(s)-\lambda_2''(s)\right)\left(\frac{-1-\rho^2+r^2}{\sqrt{(-1-\rho^2+r^2)^2+4r^2}}+F_3(r,\rho,\lambda_1(s))\right)d\rho ds| \\
&\leq \frac{C}{r}\int_t^\infty (s-t)\frac{||e_1-e_2||_X}{s^2\log^{b+1}(s)\sqrt{\log(\log(s))}}ds \\
&\leq \frac{C||e_1-e_2||_X}{r\log^b(t)\sqrt{\log(\log(t))}}
\end{aligned}$$

For the second line on the right-hand side of (3.112), we use the same method used previously to estimate

$$F_3(r,\rho,\lambda_1)-F_3(r,\rho,\lambda_2)$$

and we get

$$\begin{aligned}
&|-\frac{1}{r}\int_t^\infty\int_0^{s-t}\frac{\rho}{\sqrt{(s-t)^2-\rho^2}}\lambda_2''(s)\left(F_3(r,\rho,\lambda_1(s))-F_3(r,\rho,\lambda_2(s))\right)d\rho ds| \\
&\leq \frac{C}{r}\int_t^\infty\int_0^{s-t}\frac{\rho}{\sqrt{(s-t)^2-\rho^2}}|\lambda_2''(s)|\frac{\lambda_{0,0}(s)^{-3+2\alpha}}{\log^b(s)\sqrt{\log(\log(s))}}\frac{r^2||e_1-e_2||_X\,d\rho ds}{(1+\lambda_{0,0}(s)^{4\alpha-4}(\rho^2-r^2)^2+2\lambda_{0,0}(s)^{2\alpha-2}(\rho^2+r^2))} \\
&\leq \frac{C}{r}\int_t^\infty\int_0^{s-t}\frac{\rho}{\sqrt{(s-t)^2-\rho^2}}\frac{|\lambda_2''(s)|\lambda_{0,0}(s)^{-3+2\alpha}}{\log^b(s)\sqrt{\log(\log(s))}}\frac{r^2||e_1-e_2||_X}{r^2\lambda_{0,0}(s)^{2\alpha-2}}d\rho ds \\
&\leq \frac{C||e_1-e_2||_X}{r\log^b(t)\sqrt{\log(\log(t))}}
\end{aligned}$$

whence, we conclude the final estimate

$$|v_3^{\lambda_1}-v_3^{\lambda_2}|(t,r)\leq\frac{C||e_1-e_2||_X}{r\log^b(t)\sqrt{\log(\log(t))}},\quad r>\frac{t}{2}$$ □

Next, we recall the function E_5 defined in the v_3 subsection:

$$v_3^\lambda(t,r)=-r\int_{t+6r}^\infty\lambda''(s)(s-t)\left(\frac{1}{1+(s-t)^2}-\frac{1}{\lambda(t)^{2-2\alpha}+(s-t)^2}\right)ds+E_5^\lambda(t,r)$$

We will need to estimate $E_5^{\lambda_1}-E_5^{\lambda_2}$. For this purpose, some of the estimates already proven for $v_3^{\lambda_1}-v_3^{\lambda_2}$ will suffice, but we will need to use a slightly more complicated argument for some terms. In particular, we have

$$E_5^\lambda(t,r)=v_{3,1,a}^\lambda(t,r)+v_{3,1,b,i,1}^\lambda(t,r)+v_{3,1,b,ii}^\lambda(t,r)+v_{3,2}^\lambda(t,r)$$

with

$$\begin{aligned}
&v_{3,1,b,i,1}^\lambda(t,r) \\
&\quad=-2r\int_{6r}^\infty\lambda''(t+w)w\left(\frac{-1}{2(\lambda(t+w)^{2-2\alpha}+w^2)}+\frac{1}{2(\lambda(t)^{2-2\alpha}+w^2)}\right)dw \\
&v_{3,1,b,ii}^\lambda(t,r) \\
&\quad=\frac{-1}{r}\int_{t+6r}^\infty\int_0^{s-t}\frac{\rho}{(s-t)}\lambda''(s)\left(\left(\frac{-1-\rho^2+r^2}{\sqrt{(-1-\rho^2+r^2)^2+4r^2}}+F_3(r,\rho,\lambda(s))\right)\right. \\
&\qquad\left.-\left(\frac{2r^2}{(1+\rho^2)^2}-\frac{2r^2\lambda(s)^{2-2\alpha}}{(\lambda(s)^{2-2\alpha}+\rho^2)^2}\right)\right)d\rho ds
\end{aligned}$$

and

$$v_{3,1,a}^{\lambda}(t,r)$$
$$=\frac{-1}{r}\int_t^{t+6r}\int_0^{s-t}\frac{\rho}{(s-t)}\lambda''(s)\left(\frac{-1-\rho^2+r^2}{\sqrt{(-1-\rho^2+r^2)^2+4r^2}}+F_3(r,\rho,\lambda(s))\right)d\rho ds$$

(Note that $v_{3,2}$ was already defined when proving estimates on $v_3^{\lambda_1}-v_3^{\lambda_2}$).

$$v_{3,1,a}^{\lambda_1}-v_{3,1,a}^{\lambda_2}$$
$$=-\frac{1}{r}\int_t^{t+6r}\int_0^{s-t}\frac{\rho}{(s-t)}\left(\lambda_1''(s)-\lambda_2''(s)\right)\left(\frac{-1-\rho^2+r^2}{\sqrt{(-1-\rho^2+r^2)^2+4r^2}}+F_3(r,\rho,\lambda_1(s))\right)d\rho ds$$
$$-\frac{1}{r}\int_t^{t+6r}\int_0^{s-t}\frac{\rho}{(s-t)}\lambda_2''(s)\left(F_3(r,\rho,\lambda_1(s))-F_3(r,\rho,\lambda_2(s))\right)d\rho ds$$

This gives

$$|v_{3,1,a}^{\lambda_1}-v_{3,1,a}^{\lambda_2}|$$
$$\leq\frac{C}{r}\int_t^{t+6r}\int_0^{s-t}\frac{\rho}{(s-t)}|e_1''(s)-e_2''(s)|d\rho ds$$
$$+\frac{C}{r}\int_t^{t+6r}\int_0^{s-t}\frac{\rho}{(s-t)}\frac{|\lambda_2''(s)|(r^2\lambda_{0,0}(t)^{2\alpha-2})\frac{\lambda_{0,0}(s)^{2\alpha-3}}{\lambda_{0,0}(t)^{2\alpha-2}}|e_1(s)-e_2(s)|d\rho ds}{(1+2(\rho^2+r^2)\lambda_{0,0}(t)^{2\alpha-2}+(\rho^2-r^2)^2\lambda_{0,0}(t)^{4\alpha-4})}$$
$$\leq\frac{Cr||e_1-e_2||_X}{t^2\log^{b+1}(t)\sqrt{\log(\log(t))}}$$
$$+\frac{C}{r}\int_t^{t+6r}\int_0^{s-t}\frac{\rho}{(s-t)}|\lambda_2''(s)|\frac{\lambda_{0,0}(s)^{2\alpha-3}}{\lambda_{0,0}(t)^{2\alpha-2}}|e_1(s)-e_2(s)|d\rho ds$$
$$\leq\frac{Cr||e_1-e_2||_X}{t^2\log^{b+1}(t)\sqrt{\log(\log(t))}}$$

For the next term, we have

(3.113)
$$v_{3,1,b,i,1}^{\lambda_1}-v_{3,1,b,i,1}^{\lambda_2}$$
$$=-2r\int_{6r}^{\infty}\left(\lambda_1''(t+w)-\lambda_2''(t+w)\right)\left(\frac{-w}{2(\lambda_1(t+w)^{2-2\alpha}+w^2)}+\frac{w}{2(\lambda_1(t)^{2-2\alpha}+w^2)}\right)dw$$
$$-2r\int_{6r}^{\infty}\lambda_2''(t+w)\left(\frac{-w}{2(\lambda_1(t+w)^{2-2\alpha}+w^2)}+\frac{w}{2(\lambda_1(t)^{2-2\alpha}+w^2)}\right)dw$$
$$+2r\int_{6r}^{\infty}\lambda_2''(t+w)\left(\frac{-w}{2(\lambda_2(t+w)^{2-2\alpha}+w^2)}+\frac{w}{2(\lambda_2(t)^{2-2\alpha}+w^2)}\right)dw$$

For the first line on the right-hand side of (3.113), we have

$$|-2r\int_{6r}^{\infty}\left(\lambda_1''(t+w)-\lambda_2''(t+w)\right)\left(\frac{-w}{2(\lambda_1(t+w)^{2-2\alpha}+w^2)}+\frac{w}{2(\lambda_1(t)^{2-2\alpha}+w^2)}\right)dw|$$
$$\leq Cr\int_{6r}^{\infty}\frac{|e_1''(t+w)-e_2''(t+w)|w\left(w\lambda_{0,0}(t)^{1-2\alpha}|\lambda_{0,0}'(t)|\right)}{w^2\lambda_{0,0}(t+w)^{2-2\alpha}(1+w^2\lambda_{0,0}(t)^{2\alpha-2})}dw$$
$$\leq\frac{Cr||e_1-e_2||_X}{t^3\log^{b+2}(t)\sqrt{\log(\log(t))}}\int_0^{\infty}\frac{dw}{(1+w^2\lambda_{0,0}(t)^{2\alpha-2})}$$
$$\leq\frac{Cr||e_1-e_2||_X}{t^3\log^{b+2}(t)\sqrt{\log(\log(t))}}$$

The second and third lines of (3.113) can be treated together:

$$\begin{aligned}
&-2r\int_{6r}^{\infty}\lambda_2''(t+w)\left(\frac{-w}{2(\lambda_1(t+w)^{2-2\alpha}+w^2)}+\frac{w}{2(\lambda_1(t)^{2-2\alpha}+w^2)}\right)dw\\
&+2r\int_{6r}^{\infty}\lambda_2''(t+w)\left(\frac{-w}{2(\lambda_2(t+w)^{2-2\alpha}+w^2)}+\frac{w}{2(\lambda_2(t)^{2-2\alpha}+w^2)}\right)dw\\
&=-r\int_{6r}^{\infty}\lambda_2''(t+w)w\left(F(t+w)-F(t)\right)dw
\end{aligned}$$

where

$$F(x)=\frac{\lambda_1(x)^{2-2\alpha}-\lambda_2(x)^{2-2\alpha}}{(\lambda_1(x)^{2-2\alpha}+w^2)(\lambda_2(x)^{2-2\alpha}+w^2)}$$

We first note that

$$|F'(\sigma s+(1-\sigma)t)|\leq\frac{C\lambda_{0,0}(t)^{1-2\alpha}||e_1-e_2||_X}{t\log^b(t)\log^{-b}(s)\log^{b+1}(t)\sqrt{\log(\log(t))}(w^2+\lambda_{0,0}(s)^{2-2\alpha})^2}\quad,\quad 0\leq\sigma\leq 1$$

Then, we get

$$\begin{aligned}
&|-r\int_{6r}^{\infty}\lambda_2''(t+w)w\left(F(t+w)-F(t)\right)dw|\\
&\leq\frac{Cr||e_1-e_2||_X\log^{-b(1-2\alpha)}(t)}{t\log^{b+1}(t)\sqrt{\log(\log(t))}}\int_{6r}^{\infty}\frac{|\lambda_2''(t+w)|dw}{(w^2+\lambda_{0,0}(s)^{2-2\alpha})\log^b(t)\log^{-b}(t+w)}\\
&\leq\frac{Cr\log^{-b(1-2\alpha)}(t)||e_1-e_2||_X}{t\log^{b+1}(t)\sqrt{\log(\log(t))}\log^b(t)}\int_{t+6r}^{\infty}\frac{ds}{s^2\log(s)\lambda_{0,0}(s)^{2-2\alpha}(\lambda_{0,0}(s)^{2\alpha-2}(s-t)^2+1)}\\
&\leq\frac{Cr||e_1-e_2||_X}{t^3\log^{b+2}(t)\sqrt{\log(\log(t))}}
\end{aligned}$$

It then remains to consider

$$\begin{aligned}
&v_{3,1,b,ii}^{\lambda_1}-v_{3,1,b,ii}^{\lambda_2}\\
&=\frac{-1}{r}\int_{t+6r}^{\infty}\int_0^{s-t}\frac{\rho}{(s-t)}\left(\lambda_1''(s)-\lambda_2''(s)\right)\left(\frac{-1-\rho^2+r^2}{\sqrt{(-1-\rho^2+r^2)^2+4r^2}}+F_3(r,\rho,\lambda_1(s))\right.\\
&\left.-\left(\frac{2r^2}{(1+\rho^2)^2}-\frac{2r^2\lambda_1(s)^{2-2\alpha}}{(\lambda_1(s)^{2-2\alpha}+\rho^2)^2}\right)\right)d\rho ds\\
&-\frac{1}{r}\int_{t+6r}^{\infty}\int_0^{s-t}\frac{\rho}{(s-t)}\lambda_2''(s)\left(F_3(r,\rho,\lambda_1(s))+\frac{2r^2\lambda_1(s)^{2-2\alpha}}{(\lambda_1(s)^{2-2\alpha}+\rho^2)^2}\right.\\
&\left.-\left(F_3(r,\rho,\lambda_2(s))+\frac{2r^2\lambda_2(s)^{2-2\alpha}}{(\lambda_2(s)^{2-2\alpha}+\rho^2)^2}\right)\right)d\rho ds
\end{aligned}\tag{3.114}$$

The first two lines on the right-hand side of (3.114) are estimated in the same way as the analogous term appearing in the v_3 pointwise estimates, and we get

$$\begin{aligned}
&|\frac{-1}{r}\int_{t+6r}^{\infty}\int_0^{s-t}\frac{\rho}{(s-t)}\left(\lambda_1''(s)-\lambda_2''(s)\right)\left(\frac{-1-\rho^2+r^2}{\sqrt{(-1-\rho^2+r^2)^2+4r^2}}+F_3(r,\rho,\lambda_1(s))\right.\\
&\left.-\left(\frac{2r^2}{(1+\rho^2)^2}-\frac{2r^2\lambda_1(s)^{2-2\alpha}}{(\lambda_1(s)^{2-2\alpha}+\rho^2)^2}\right)\right)d\rho ds|\\
&\leq\frac{Cr||e_1-e_2||_X}{t^2\log^{b+1}(t)\sqrt{\log(\log(t))}}
\end{aligned}$$

To treat the third line on the right-hand side of (3.114), we start with defining $G_{3,1}$ by

$$G_{3,1}(w,r,\lambda(s)) := \int_0^w \rho\left(F_3(r,\rho,\lambda(s)) + \frac{2r^2\lambda(s)^{2-2\alpha}}{(\lambda(s)^{2-2\alpha}+\rho^2)^2}\right)d\rho$$

Then, we get

$$|G_{3,1}(w,r,\lambda_1(s)) - G_{3,1}(w,r,\lambda_2(s))| \leq \frac{|e_1(s)-e_2(s)|\lambda_{0,0}(s)^{-3+2\alpha}r^4}{(1+\lambda_{0,0}(s)^{2\alpha-2}w^2)^2}$$

which gives

$$\begin{aligned}&|-\frac{1}{r}\int_{t+6r}^{\infty}\int_0^{s-t}\frac{\rho}{(s-t)}\lambda_2''(s)\left(F_3(r,\rho,\lambda_1(s)) + \frac{2r^2\lambda_1(s)^{2-2\alpha}}{(\lambda_1(s)^{2-2\alpha}+\rho^2)^2}\right.\\ &\qquad\left.-\left(F_3(r,\rho,\lambda_2(s)) + \frac{2r^2\lambda_2(s)^{2-2\alpha}}{(\lambda_2(s)^{2-2\alpha}+\rho^2)^2}\right)\right)d\rho ds|\\ &\leq \frac{C}{r}\int_{t+6r}^{\infty}\frac{|\lambda_{0,0}''(s)|}{(s-t)}\frac{|e_1(s)-e_2(s)|\lambda_{0,0}(s)^{1-2\alpha}r^4}{(\lambda_{0,0}(s)^{2-2\alpha})((s-t)^2)}ds\\ &\leq \frac{Cr||e_1-e_2||_X}{t^2\log^{b+1}(t)\sqrt{\log(\log(t))}}\end{aligned}$$

Combining the above, we get

$$|\left(E_5^{\lambda_1}-E_5^{\lambda_2}\right)(t,r)| \leq \frac{Cr||e_1-e_2||_X}{t^2\log^{b+1}(t)\sqrt{\log(\log(t))}}$$

LEMMA 3.12. *We have the following estimates*

$$|v_4^{\lambda_1}-v_4^{\lambda_2}|(t,r) \leq \begin{cases}\frac{C||e_1-e_2||_X}{t^2\sqrt{\log(\log(t))}\log^{N+3b-1}(t)}, & r\leq\frac{t}{2}\\ \frac{C||e_1-e_2||_X}{t\log^{3b+2N}(t)\sqrt{\log(\log(t))}}, & r\geq\frac{t}{2}\end{cases}$$

PROOF. We proceed to estimate $v_{4,1,2} := v_4^{\lambda_1} - v_4^{\lambda_2}$, by noting that $v_{4,1,2}$ solves the following equation with 0 Cauchy data at infinity:

$$-\partial_{tt}v_{4,1,2} + \partial_{rr}v_{4,1,2} + \frac{1}{r}\partial_r v_{4,1,2} - \frac{v_{4,1,2}}{r^2} = v_{4,c}^{\lambda_1} - v_{4,c}^{\lambda_2}$$

So, we start by estimating $v_{4,c}^{\lambda_1} - v_{4,c}^{\lambda_2}$:

$$\begin{aligned}|v_{4,c}^{\lambda_1}-v_{4,c}^{\lambda_2}| &\leq C\frac{\chi_{\geq 1}(\frac{2r}{\log^N(t)})||e_1-e_2||_X}{\log^{2b}(t)\sqrt{\log(\log(t))}}\frac{|v_1^{\lambda_1}+v_2+v_3^{\lambda_1}|}{(\lambda_{0,0}(t)^2+r^2)^2}\\ &+C\chi_{\geq 1}(\frac{2r}{\log^N(t)})\frac{||e_1-e_2||_X}{\log^b(t)\sqrt{\log(\log(t))}}\frac{1}{t^2\log^{2b+1-2\alpha b}(t)r^3}\\ &+C\frac{\chi_{\geq 1}(\frac{2r}{\log^N(t)})\lambda_{0,0}(t)^2}{(r^2+\lambda_{0,0}(t)^2)^2}\left(|v_1^{\lambda_1}-v_1^{\lambda_2}|+|v_3^{\lambda_1}-v_3^{\lambda_2}|\right)\end{aligned}$$

where we used the explicit formula for $F_{0,2}$.

We note that the right-hand side of the equation for v_1^{λ} depends linearly on λ''. Therefore $v_1^{\lambda_1} - v_1^{\lambda_2} = v_1^{\lambda_1-\lambda_2}$, so we can use our estimates for v_1^{λ} which were previously recorded.

We thus have, in addition to (3.110), the following estimates, valid for any λ of the form

$$\lambda(t) = \lambda_0(t) + f(t), \quad f\in\overline{B}_1(0)\subset X$$

$$|v_1^\lambda + v_2 + v_3^\lambda| \leq \begin{cases} \frac{Cr}{t^2 \log^b(t)}, & r \leq \frac{t}{2} \\ \frac{C \log(r)}{|t-r|}, & t > r \geq \frac{t}{2} \end{cases}$$

$$|v_1^{\lambda_1} - v_1^{\lambda_2}| \leq \begin{cases} \frac{Cr||e_1-e_2||_X}{t^2 \log^b(t)\sqrt{\log(\log(t))}}, & r \leq \frac{t}{2} \\ \frac{C||e_1-e_2||_X}{r\sqrt{\log(\log(t))}\log^b(t)}, & r > \frac{t}{2} \end{cases}$$

This gives

(3.115)

$$|v_{4,c}^{\lambda_1} - v_{4,c}^{\lambda_2}| \leq C\chi_{\geq 1}(\frac{2r}{\log^N(t)})||e_1 - e_2||_X$$
$$\cdot \begin{cases} \frac{1}{r^3 t^2 \log^{3b}(t)\sqrt{\log(\log(t))}}, & r \leq \frac{t}{2} \\ \frac{\log(r)}{r^4 \log^{2b}(t)\sqrt{\log(\log(t))}|t-r|} + \frac{1}{t^2 \log^{3b+1-2\alpha b}(t)\sqrt{\log(\log(t))}r^3}, & t > r > \frac{t}{2} \end{cases}$$

Now, we start with

$$\left(v_4^{\lambda_1} - v_4^{\lambda_2}\right)(t,r)$$
$$= \frac{-1}{2\pi}\int_t^\infty \int_0^{s-t} \frac{\rho}{\sqrt{(s-t)^2-\rho^2}} \int_0^{2\pi} \frac{(r+\rho\cos(\theta))(v_{4,c}^{\lambda_1} - v_{4,c}^{\lambda_2})(s, \sqrt{r^2+\rho^2+2r\rho\cos(\theta)})}{\sqrt{r^2+\rho^2+2r\rho\cos(\theta)}} d\theta d\rho ds$$

which gives

$$|v_4^{\lambda_1} - v_4^{\lambda_2}|$$
$$\leq C\int_t^\infty \int_0^{s-t} \frac{\rho}{\sqrt{(s-t)^2-\rho^2}} \int_0^{2\pi} |v_{4,c}^{\lambda_1} - v_{4,c}^{\lambda_2}|(s, \sqrt{r^2+\rho^2+2r\rho\cos(\theta)}) d\theta d\rho ds$$

Note that (3.115) has the same right-hand side as (3.85), except for an extra factor of $\frac{||e_1-e_2||_X}{\sqrt{\log(\log(t))}}$. So, we can inspect the intermediate steps of the procedure used when obtaining pointwise estimates on v_4, thereby getting

$$|v_4^{\lambda_1} - v_4^{\lambda_2}|(t,r) \leq \frac{C||e_1-e_2||_X}{t^2\sqrt{\log(\log(t))}\log^{N+3b-1}(t)}, \quad r \leq \frac{t}{2}$$

In the region $r \geq \frac{t}{2}$, we use the following, slightly different estimate for $v_1^\lambda + v_2 + v_3^\lambda$, again valid for all $\lambda(t) = \lambda_0(t) + f(t), \quad f \in \overline{B}_1(0) \subset X$:

$$|v_1^\lambda + v_2 + v_3^\lambda| \leq \begin{cases} \frac{Cr}{t^2 \log^b(t)}, & r \leq \frac{t}{2} \\ \frac{C}{\sqrt{r}}, & t - \sqrt{t} \leq r \leq t + \sqrt{t} \\ \frac{C \log(r)}{|t-r|}, & \frac{t}{2} \leq r \leq t - \sqrt{t}, \text{ or } r \geq t + \sqrt{t} \end{cases}$$

We then get

$$|v_{4,c}^{\lambda_1} - v_{4,c}^{\lambda_2}| \leq C\chi_{\geq 1}(\frac{2r}{\log^N(t)})||e_1 - e_2||_X$$

(3.116)
$$\cdot \begin{cases} \frac{1}{r^3 t^2 \log^{3b}(t)\sqrt{\log(\log(t))}}, & r \leq \frac{t}{2} \\ \frac{1}{\log^{2b}(t)\sqrt{\log(\log(t))}r^{9/2}}, & t - \sqrt{t} \leq r \leq t + \sqrt{t} \\ \frac{\log(r)}{r^4 \log^{2b}(t)\sqrt{\log(\log(t))}|t-r|} + \frac{1}{t^2 \log^{3b+1-2\alpha b}(t)\sqrt{\log(\log(t))}r^3} \\ , \quad t - \sqrt{t} > r > \frac{t}{2} \text{ or } r > t + \sqrt{t} \end{cases}$$

Note that (3.116) again has the same right-hand side as (3.88), except for the extra factor of $\frac{||e_1-e_2||_X}{\sqrt{\log(\log(t))}}$, so, we can infer the following estimates from our study of v_4:

$$||\widehat{v_{4,c}^{\lambda_1} - v_{4,c}^{\lambda_2}}(x,\xi)||_{L^2(\xi d\xi)} \leq \frac{C||e_1-e_2||_X}{x^2 \log^{3b+2N}(x)\sqrt{\log(\log(x))}}$$

Then, we simply note:

$$v_{4,1,2}(t,r) = \int_t^\infty \int_0^\infty J_1(r\xi)\sin((t-s)\xi)\left(\widehat{v_{4,c}^{\lambda_1} - v_{4,c}^{\lambda_2}}(s,\xi)\right) d\xi ds$$

So,

$$\begin{aligned}
|v_{4,1,2}(t,r)| &\leq C\int_t^\infty \int_0^{\frac{1}{r}} r\xi|\widehat{v_{4,c}^{\lambda_1} - v_{4,c}^{\lambda_2}}|(s,\xi) d\xi ds\\
&+ C\int_t^\infty \int_{\frac{1}{r}}^\infty \frac{|\widehat{v_{4,c}^{\lambda_1} - v_{4,c}^{\lambda_2}}|(s,\xi)}{\sqrt{r\xi}} d\xi ds\\
&\leq Cr\int_t^\infty \left(\int_0^{\frac{1}{r}} \xi d\xi\right)^{1/2} \left(\int_0^{\frac{1}{r}} \xi|\widehat{v_{4,c}^{\lambda_1} - v_{4,c}^{\lambda_2}}(s,\xi)|^2 d\xi\right)^{1/2} ds\\
&+ \frac{C}{\sqrt{r}}\int_t^\infty \left(\int_{\frac{1}{r}}^\infty \frac{d\xi}{\xi^2}\right)^{1/2} \left(\int_{\frac{1}{r}}^\infty \xi|\widehat{v_{4,c}^{\lambda_1} - v_{4,c}^{\lambda_2}}(s,\xi)|^2 d\xi\right)^{1/2} ds\\
&\leq C\int_t^\infty ||\widehat{v_{4,c}^{\lambda_1} - v_{4,c}^{\lambda_2}}(s,\xi)||_{L^2(\xi d\xi)} ds\\
&\leq \frac{C||e_1-e_2||_X}{t\sqrt{\log(\log(t))}\log^{3b+2N}(t)}
\end{aligned}$$

which finishes the proof of the lemma. □

LEMMA 3.13. *We have the following pointwise estimate*

$$|v_5^{\lambda_1} - v_5^{\lambda_2}|(t,r) \leq C\frac{||e_1-e_2||_X \log^2(t)}{t^3\sqrt{\log(\log(t))}\log^b(t)}, \quad r \leq \frac{t}{2}$$

PROOF. As was the case for v_4, we start with noting that

$$v_{5,1,2} := v_5^{\lambda_1} - v_5^{\lambda_2}$$

solves the equation (with 0 Cauchy data at infinity)

$$-\partial_{tt}v_{5,1,2} + \partial_{rr}v_{5,1,2} + \frac{1}{r}\partial_r v_{5,1,2} - \frac{v_{5,1,2}}{r^2} = N_2(f_{v_5}^{\lambda_1}) - N_2(f_{v_5}^{\lambda_2})$$

where

$$f_{v_5}^\lambda = v_1^\lambda + v_2 + v_3^\lambda + v_4^\lambda$$

Collecting the estimates from previous subsections, one estimate on $f(v_5^\lambda)$, valid for all $\lambda = \lambda_0 + f, \quad f \in \overline{B}_1(0) \subset X$, is

$$|f_{v_5}^\lambda| \leq \begin{cases} \frac{Cr}{t^2\log^b(t)}, & r \leq \frac{t}{2}\\ \frac{C\log(r)}{|t-r|}, & t > r > \frac{t}{2}\end{cases}$$

Now, we start to estimate the right-hand side of the $v_{5,1,2}$ equation.

(3.117)

$$\begin{aligned}N_2(f_{v_5}^{\lambda_1})-N_2(f_{v_5}^{\lambda_2})&=\left(\frac{\sin(2Q_{\frac{1}{\lambda_1(t)}}(r))}{2r^2}-\frac{\sin(2Q_{\frac{1}{\lambda_2(t)}}(r))}{2r^2}\right)(\cos(2f_{v_5}^{\lambda_1})-1)\\&+\frac{\sin(2Q_{\frac{1}{\lambda_2(t)}}(r))}{2r^2}(\cos(2f_{v_5}^{\lambda_1})-\cos(2f_{v_5}^{\lambda_2}))\\&+\left(\frac{\cos(2Q_{\frac{1}{\lambda_1(t)}}(r))-\cos(2Q_{\frac{1}{\lambda_2(t)}}(r))}{2r^2}\right)(\sin(2f_{v_5}^{\lambda_1})-2f_{v_5}^{\lambda_1})\\&+\frac{\cos(2Q_{\frac{1}{\lambda_2(t)}}(r))}{2r^2}\left(\sin(2f_{v_5}^{\lambda_1})-2f_{v_5}^{\lambda_1}-\left(\sin(2f_{v_5}^{\lambda_2})-2f_{v_5}^{\lambda_2}\right)\right)\end{aligned}$$

For the first line of (3.117), we have

$$\begin{aligned}&|\left(\frac{\sin(2Q_{\frac{1}{\lambda_1(t)}}(r))}{2r^2}-\frac{\sin(2Q_{\frac{1}{\lambda_2(t)}}(r))}{2r^2}\right)(\cos(2f_{v_5}^{\lambda_1})-1)|\\&\leq\frac{C||e_1-e_2||_X}{r\log^b(t)\sqrt{\log(\log(t))}}\frac{\left(f_{v_5}^{\lambda_1}\right)^2}{(\lambda_{0,0}(t)^2+r^2)}\\&\leq\frac{C||e_1-e_2||_X}{r\log^b(t)\sqrt{\log(\log(t))}}\begin{cases}\frac{r^2}{t^4\log^{2b}(t)(\lambda_{0,0}(t)^2+r^2)}, & r\leq\frac{t}{2}\\ \frac{\log^2(r)}{r^2(t-r)^2}, \quad t>r\geq\frac{t}{2}\end{cases}\end{aligned}$$

For the second line of (3.117), we have

$$\begin{aligned}&|\frac{\sin(2Q_{\frac{1}{\lambda_2(t)}}(r))}{2r^2}\left(\cos(2f_{v_5}^{\lambda_1})-\cos(2f_{v_5}^{\lambda_2})\right)|\\&\leq\frac{C\lambda_2(t)}{(r^2+\lambda_2(t)^2)r}\left(|f_{v_5}^{\lambda_1}|+|f_{v_5}^{\lambda_2}|\right)\left(|v_1^{\lambda_1}-v_1^{\lambda_2}|+|v_3^{\lambda_1}-v_3^{\lambda_2}|+|v_4^{\lambda_1}-v_4^{\lambda_2}|\right)\\&\leq\begin{cases}\frac{Cr\lambda_{0,0}(t)||e_1-e_2||_X}{(r^2+\lambda_{0,0}(t)^2)t^4\log^{2b}(t)\sqrt{\log(\log(t))}}+\frac{C||e_1-e_2||_X\lambda_{0,0}(t)}{(r^2+\lambda_{0,0}(t)^2)t^4\sqrt{\log(\log(t))}\log^{N+4b-1}(t)}, & r\leq\frac{t}{2}\\ \frac{C\log(t)||e_1-e_2||_X}{r^3|t-r|t\log^{2b}(t)\sqrt{\log(\log(t))}}, \quad t>r>\frac{t}{2}\end{cases}\end{aligned}$$

We consider the third line of (3.117), and get

$$\begin{aligned}&|\left(\frac{\cos(2Q_{\frac{1}{\lambda_1(t)}}(r))-\cos(2Q_{\frac{1}{\lambda_2(t)}}(r))}{2r^2}\right)(\sin(2f_{v_5}^{\lambda_1})-2f_{v_5}^{\lambda_1})|\\&\leq\frac{C||e_1-e_2||_X}{\log^{2b}(t)\sqrt{\log(\log(t))}(r^2+\lambda_{0,0}(t)^2)^2}\begin{cases}\frac{r^3}{t^6\log^{3b}(t)}, & r\leq\frac{t}{2}\\ \frac{\log^3(r)}{|t-r|^3}, & t>r>\frac{t}{2}\end{cases}\end{aligned}$$

For the fourth line of (3.117), we have

$$\begin{aligned}&|\frac{\cos(2Q_{\frac{1}{\lambda_2(t)}}(r))}{2r^2}\left(\sin(2f_{v_5}^{\lambda_1})-2f_{v_5}^{\lambda_1}-\left(\sin(2f_{v_5}^{\lambda_2})-2f_{v_5}^{\lambda_2}\right)\right)|\\&\leq\frac{C}{r^2}\left(|f_{v_5}^{\lambda_1}|^2+|f_{v_5}^{\lambda_2}|^2\right)|f_{v_5}^{\lambda_1}-f_{v_5}^{\lambda_2}|\\&\leq\begin{cases}\frac{Cr||e_1-e_2||_X}{t^6\sqrt{\log(\log(t))}\log^{3b}(t)}+\frac{C||e_1-e_2||_X}{t^6\sqrt{\log(\log(t))}\log^{N+5b-1}(t)}, & r\leq\frac{t}{2}\\ \frac{C||e_1-e_2||_X\log^2(t)}{r^3(t-r)^2\log^b(t)\sqrt{\log(\log(t))}}, \quad t>r>\frac{t}{2}\end{cases}\end{aligned}$$

Combining all of the above estimates, we obtain

$$|N_2(f_{v_5}^{\lambda_1})(t,r) - N_2(f_{v_5}^{\lambda_2})(t,r)|$$
$$\leq \begin{cases} \frac{Cr\lambda_{0,0}(t)||e_1-e_2||_X}{(r^2+\lambda_{0,0}(t)^2)t^4\log^{2b}(t)\sqrt{\log(\log(t))}} + \frac{C||e_1-e_2||_X\lambda_{0,0}(t)}{(r^2+\lambda_{0,0}(t)^2)t^4\sqrt{\log(\log(t))}\log^{N+4b-1}(t)}, & r \leq \frac{t}{2} \\ \frac{C\log(t)||e_1-e_2||_X}{r^3\log^b(t)\sqrt{\log(\log(t))}}\left(\frac{\log(t)}{(t-r)^2} + \frac{1}{t|t-r|\log^b(t)} + \frac{\log^2(t)}{\log^b(t)r|t-r|^3}\right), & t > r > \frac{t}{2} \end{cases}$$

We now proceed to estimate $v_{5,1,2}$ in the region $r \leq \frac{t}{2}$. We start with

$$v_{5,1,2}(t,r) = \frac{-1}{2\pi}\int_t^\infty \int_0^{s-t} \frac{\rho}{\sqrt{(s-t)^2-\rho^2}}$$
$$\left(\int_0^{2\pi} \left(N_2(f_{v_5}^{\lambda_1}) - N_2(f_{v_5}^{\lambda_2})\right)(s, \sqrt{r^2+\rho^2+2\rho r\cos(\theta)})\frac{(r+\rho\cos(\theta))}{\sqrt{r^2+2r\rho\cos(\theta)+\rho^2}}\right) d\theta d\rho ds$$

We then estimate as follows

$$\begin{aligned} |v_{5,1,2}(t,r)| &\leq C\int_t^\infty \int_{B_{s-t}(0)} \frac{|N_2(f_{v_5}^{\lambda_1}) - N_2(f_{v_5}^{\lambda_2})|(s,|x+y|)}{\sqrt{(s-t)^2-|y|^2}} dA(y)ds \\ &\leq C\int_t^\infty \int_{B_{s-t}(0)\cap B_{\frac{s}{2}}(-x)} \frac{|N_2(f_{v_5}^{\lambda_1}) - N_2(f_{v_5}^{\lambda_2})|(s,|x+y|)}{\sqrt{(s-t)^2-|y|^2}} dA(y)ds \\ &+ C\int_t^\infty \int_{B_{s-t}(0)\cap(B_{\frac{s}{2}}(-x))^c} \frac{|N_2(f_{v_5}^{\lambda_1}) - N_2(f_{v_5}^{\lambda_2})|(s,|x+y|)}{\sqrt{(s-t)^2-|y|^2}} dA(y)ds \end{aligned}$$

Now, we carry out the identical steps which were done to obtain the v_5^λ estimates, and get

$$|v_{5,1,2}(t,r)| \leq C\frac{||e_1-e_2||_X\log^2(t)}{t^3\sqrt{\log(\log(t))}\log^b(t)}, \quad r \leq \frac{t}{2}$$

which completes the proof of the lemma. □

We can now use all of the above estimates to get

$$|\int_0^\infty \left(\frac{\cos(2Q_1(\frac{r}{\lambda_2(t)}))-1}{r^2\lambda_2(t)}\right)\phi_0(\frac{r}{\lambda_2(t)})r\left(v_4^{\lambda_1}-v_4^{\lambda_2}\right)(t,r)\left(1-\chi_{\geq 1}(\frac{4r}{t})\right)dr|$$
$$\leq C\lambda_0(t)^2\int_0^{\frac{t}{2}}\frac{r^2}{(r^2+\lambda_0(t)^2)^3}\frac{||e_1-e_2||_X}{t^2\sqrt{\log(\log(t))}\log^{N+3b-1}(t)}dr$$
$$\leq \frac{C||e_1-e_2||_X}{t^2\sqrt{\log(\log(t))}\log^{N+2b-1}(t)}$$
$$|\int_0^\infty \left(\frac{\cos(2Q_1(\frac{r}{\lambda_2(t)}))-1}{r^2\lambda_2(t)}\right)\phi_0(\frac{r}{\lambda_2(t)})r\left(v_5^{\lambda_1}-v_5^{\lambda_2}\right)(t,r)\left(1-\chi_{\geq 1}(\frac{4r}{t})\right)dr|$$
$$\leq C\lambda_0(t)^2\int_0^{\frac{t}{2}}\frac{r^2}{(r^2+\lambda_0(t)^2)^3}\frac{||e_1-e_2||_X\log^2(t)}{t^3\sqrt{\log(\log(t))}\log^b(t)}dr\leq\frac{C||e_1-e_2||_X\log^2(t)}{t^3\sqrt{\log(\log(t))}}$$
$$|\int_0^\infty \left(\frac{\cos(2Q_1(\frac{r}{\lambda_2(t)}))-1}{r^2\lambda_2(t)}\right)\phi_0(\frac{r}{\lambda_2(t)})r\left(E_5^{\lambda_1}-E_5^{\lambda_2}\right)(t,r)dr|$$
$$\leq C\int_0^\infty\frac{\lambda_2(t)^2r^2}{(\lambda_2(t)^2+r^2)^3}\left(\frac{r||e_1-e_2||_X}{t^2\log^{b+1}(t)\sqrt{\log(\log(t))}}\right)dr$$
$$\leq\frac{C||e_1-e_2||_X}{t^2\log^{b+1}(t)\sqrt{\log(\log(t))}}\int_0^\infty\frac{u^3du}{(1+u^2)^3}\leq\frac{C||e_1-e_2||_X}{t^2\log^{b+1}(t)\sqrt{\log(\log(t))}}$$
$$|\int_0^\infty \left(\frac{\cos(2Q_1(\frac{r}{\lambda_2(t)}))-1}{r^2\lambda_2(t)}\right)\phi_0(\frac{r}{\lambda_2(t)})r\chi_{\geq 1}(\frac{2r}{\log^N(t)})\left(-v_1^{\lambda_1}+v_1^{\lambda_2}-v_3^{\lambda_1}+v_3^{\lambda_2}\right)dr|$$
$$\leq C\int_{\frac{\log^N(t)}{2}}^{\frac{t}{2}}\frac{\lambda_2(t)^2r^2}{r^6}\left(\frac{||e_1-e_2||_Xr}{t^2\log^b(t)\sqrt{\log(\log(t))}}\right)dr$$
$$+C\int_{\frac{t}{2}}^\infty\frac{\lambda_2(t)^2r^2}{r^6}\frac{||e_1-e_2||_X}{r\sqrt{\log(\log(t))}\log^b(t)}dr$$
$$\leq\frac{C||e_1-e_2||_X}{t^2\log^{3b+2N}(t)\sqrt{\log(\log(t))}}$$
$$|\int_0^\infty\chi_{\geq 1}(\frac{2r}{\log^N(t)})(F_{0,2}^{\lambda_1}-F_{0,2}^{\lambda_2})(t,r)\frac{\phi_0(\frac{r}{\lambda_1(t)})}{\lambda_1(t)}rdr|$$
$$+|\int_0^\infty\chi_{\geq 1}(\frac{2r}{\log^N(t)})F_{0,2}^{\lambda_2}(t,r)\left(\frac{\phi_0(\frac{r}{\lambda_1(t)})}{\lambda_1(t)}-\frac{\phi_0(\frac{r}{\lambda_2(t)})}{\lambda_2(t)}\right)rdr|$$
$$\leq\frac{C||e_1-e_2||_X}{t^2\log^{3b+1-2\alpha b}(t)\sqrt{\log(\log(t))}\log^{2N}(t)}$$

Combining the above estimates, we firstly have

$$|G(t,\lambda_1(t))-G(t,\lambda_2(t))|\leq\frac{C||e_1-e_2||_X}{t^2\sqrt{\log(\log(t))}\log^{b+1}(t)}$$

Then, combining this with the earlier estimates of the terms in $RHS(e_1,t)-RHS(e_2,t)$ which don't involve G, we finally get: there exists $C_{lip}>0$ *independent* of T_0, such that for $e_1,e_2\in\overline{B_1(0)}\subset X$

$$|RHS(e_1,t)-RHS(e_2,t)|\leq\frac{C_{lip}||e_1-e_2||_X}{t^2\log^{b+1}(t)(\log(\log(t)))^{3/2}}$$

(Recall that G appears in the expression of RHS in the term $\frac{G(t,\lambda(t))}{\log(\lambda_0(t))}$). This concludes the proof of Proposition 3.2.

Using the L^1 estimate on the resolvent kernel, r, we get

$$\begin{aligned}&|(T(e_1)-T(e_2))''(t)|\\&\leq \frac{|RHS(e_1,t)-RHS(e_2,t)|}{\alpha}+\frac{2}{\alpha}\sup_{z\geq t}\left(|RHS(e_1,z)-RHS(e_2,z)|\right)\\&\leq \frac{3C_{lip}}{\alpha}\frac{||e_1-e_2||_X}{t^2\log^{b+1}(t)(\log(\log(t)))^{3/2}}\end{aligned}$$

Then, with the same procedure used when estimating $|T(e)'(t)|$ earlier, we get

$$|(T(e_1)-T(e_2))'(t)|\leq \frac{3C_{lip}}{\alpha}\left(1+\frac{1}{100}\right)\frac{||e_1-e_2||_X}{t\log^{b+1}(t)(\log(\log(t)))^{3/2}}$$

$$|(T(e_1)-T(e_2))(t)|\leq \frac{3C_{lip}}{\alpha}\left(1+\frac{1}{100}\right)\frac{||e_1-e_2||_X}{b\log^{b}(t)(\log(\log(t)))^{3/2}}$$

whence,

$$||T(e_1)-T(e_2)||_X\leq \frac{10C_{lip}}{\alpha\log(\log(T_0))}||e_1-e_2||_X,\quad e_1,e_2\in\overline{B_1(0)}$$

Since C_{lip} is independent of T_0, and the above estimates are valid for all T_0 satisfying

$$T_0>2e^{e^{1000(b+1)}}+T_{0,1}+T_{0,2}+T_{0,3}$$

if we further restrict T_0 to satisfy

$$T_0>2e^{e^{1000(b+1)}}+T_{0,1}+T_{0,2}+T_{0,3}+e^{e^{\frac{1500(C_{lip}+1000(1+\frac{1}{b}))}{\alpha}}}$$

then, T is a strict contraction on the complete metric space $\overline{B_1(0)}\subset X$. By Banach's fixed point theorem, T has a fixed point, say e_0, in $\overline{B_1(0)}\subset X$. Then, if

$$\lambda(t)=\lambda_0(t)+e_0(t)$$

we have

$$\lambda\in C^2([T_0,\infty)),\quad ||e_0||_X\leq 1$$

and λ solves (3.63).

3.0.8.10. *Estimating* $\lambda'''(t)$. The main proposition of this section is:

PROPOSITION 3.3. *$\lambda\in C^3([T_0,\infty))$, with*

$$|\lambda'''(t)|\leq \frac{C}{t^3\log^{b+1}(t)},\quad t\geq T_0$$

In addition, we have

$$|\partial_t v_1(t,r)|\leq\begin{cases}\frac{Cr}{t^3\log^b(t)}, & r\leq\frac{t}{2}\\ \frac{C}{rt\log^{b+1}(t)}, & r>\frac{t}{2}\end{cases}\tag{3.118}$$

$$|\partial_t v_3(t,r)|\leq\begin{cases}\frac{Cr\log(\log(t))}{t^3\log^{b+1}(t)}, & r\leq\frac{t}{2}\\ \frac{C}{rt\log^{b+1}(t)}, & r>\frac{t}{2}\end{cases}\tag{3.119}$$

$$|\partial_{tr}v_3(t,r)|\leq\frac{C}{t^3\log^b(t)}\tag{3.120}$$

$$|\partial_{tr}v_1(t,r)| \leq \begin{cases} \frac{C}{t^3\log^b(t)}, & r \leq \frac{t}{2} \\ \frac{C}{r^2 t\log^{b+1}(t)}, & r > \frac{t}{2} \end{cases} \tag{3.121}$$

$$|\partial_t v_4(t,r)| \leq \begin{cases} \frac{Cr}{t^3\log^{3b+2N-2}(t)}, & r \leq \frac{t}{2} \\ \frac{C}{t^2\log^{3b+2N-1}(t)}, & r > \frac{t}{2} \end{cases} \tag{3.122}$$

PROOF. Since $\lambda(t) = \lambda_0(t) + e_0(t)$, it suffices to show that $e_0 \in C^3([T_0,\infty))$, and estimate $e_0'''(t)$. Recall that λ solves (3.63), which can be re-written as

$$\lambda''(t) = \frac{RHS_2(t)}{g_2(t)} \tag{3.123}$$

where

$$g_2(t) = \frac{-2}{\lambda(t)} + \frac{4\alpha\log(\lambda(t))}{\lambda(t)}\left(\frac{1}{-1+\lambda(t)^{2\alpha}}\right) - \frac{2(\lambda(t)^{-2\alpha}-1)}{\lambda(t)}g_5(t)$$

$$g_5(t) = \int_0^\infty \frac{\chi_{\geq 1}(\frac{2R\lambda(t)}{\log^N(t)})\phi_0(R)R^2dR}{(R^2+1)(\lambda(t)^{-2\alpha}+R^2)}$$

and

$$\begin{aligned} RHS_2(t) = &-\frac{16}{\lambda(t)^3}\int_t^\infty \lambda''(s)\left(K_1(s-t,\lambda(t)) + K(s-t,\lambda(t))\right)ds + 2\frac{(\lambda'(t))^2}{\lambda(t)^2} \\ &+ \frac{4b}{\lambda(t)t^2\log^b(t)} + E_{v_2,ip}(t,\lambda(t)) \\ &+ \langle\left(\frac{\cos(2Q_{\frac{1}{\lambda(t)}})-1}{r^2}\right)\left(v_3 + (v_4+v_5)\left(1-\chi_{\geq 1}(\frac{4r}{t})\right)\right)|_{r=R\lambda(t)}, \phi_0\rangle \\ &- \langle\left(\frac{\cos(2Q_{\frac{1}{\lambda(t)}})-1}{r^2}\right)\chi_{\geq 1}\left(\frac{2r}{\log^N(t)}\right)(v_1+v_2+v_3)|_{r=R\lambda(t)}, \phi_0\rangle \\ &- 4\int_0^\infty \chi_{\geq 1}(\frac{2R\lambda(t)}{\log^N(t)})\frac{(\lambda'(t))^2R^2\phi_0(R)dR}{\lambda(t)^2(R^2+1)^2} \end{aligned}$$

The point of this re-writing is that the right-hand side of (3.123) is in $C^1([T_0,\infty))$, since $\lambda \in C^2([T_0,\infty))$. So, $\lambda \in C^3([T_0,\infty))$, and we have

$$\lambda'''(t) = \partial_t\left(\frac{RHS_2(t)}{g_2(t)}\right)$$

We will first prove a preliminary estimate on $\lambda'''(t)$, which will then allow us to obtain a more useful formula for e_0'''. Using the formula for g_2, and estimates on e_0 from the fact that $e_0 \in \overline{B_1(0)}$, we get

$$|\frac{1}{g_2(t)}| \leq \frac{C}{\log^b(t)\log(\log(t))}, \quad |\left(\frac{1}{g_2}\right)'(t)| \leq \frac{C}{t\log^{b+1}(t)\log(\log(t))}$$

and

$$|\lambda'''(t)| \leq \frac{C|RHS_2'(t)|}{\log^b(t)\log(\log(t))} + \frac{C|RHS_2(t)|}{t\log^{b+1}(t)\log(\log(t))}$$

For the preliminary estimate, it will suffice to estimate each term in RHS_2 separately, despite the fact that there is some cancellation between some terms in

RHS_2. When we prove the final estimate on $\lambda'''(t)$, we will take this cancellation into account.

Using the same procedures and estimates used to estimate $G(t,\lambda_0(t)+f(t))$ for arbitrary $f\in\overline{B_1(0)}\subset X$, we get

$$|RHS_2(t)|\leq\frac{C}{t^2}$$

Now, we estimate $|RHS_2'(t)|$. Using estimates on $\lambda^{(j)}(t),\quad j=0,1,2$, we get

$$\begin{aligned}(3.124)\quad &|RHS_2'(t)|\\
&\leq|\partial_t\left(\frac{-16}{\lambda(t)^3}\int_t^\infty\lambda''(s)\left(K_1(s-t,\lambda(t))+K(s-t,\lambda(t))\right)ds\right)|\\
&+\frac{C}{t^3\log^2(t)}+\frac{C}{t^3}+|\partial_t E_{v_2,ip}(t,\lambda(t))|\\
&+|\partial_t\left(\langle\left(\frac{\cos(2Q_{\frac{1}{\lambda(t)}})-1}{r^2}\right)\left(v_3+(v_4+v_5)\left(1-\chi_{\geq1}(\frac{4r}{t})\right)\right)|_{r=R\lambda(t)},\phi_0\rangle\right)|\\
&+|\partial_t\left(\langle\left(\frac{\cos(2Q_{\frac{1}{\lambda(t)}})-1}{r^2}\right)\chi_{\geq1}(\frac{2r}{\log^N(t)})\left(v_1+v_2+v_3\right)|_{r=R\lambda(t)},\phi_0\rangle\right)|\end{aligned}$$

For the $E_{v_2,ip}(t)$ term, we recall the definition in (3.104), and start with

$$\begin{aligned}&\partial_t\left(\int_0^{\frac{1}{2}}\left(\chi_{\leq\frac{1}{4}}(\xi)-1\right)\frac{\sin(t\xi)}{t^2}\left(\frac{(b-1)}{\xi\log^b(\frac{1}{\xi})}+\frac{b(b-1)}{\xi\log^{b+1}(\frac{1}{\xi})}\right)d\xi\right)\\
&=\partial_t\left(\int_0^{\frac{t}{2}}\left(\chi_{\leq\frac{1}{4}}(\frac{u}{t})-1\right)\frac{\sin(u)}{t^2}\left(\frac{b-1}{u\log^b(\frac{t}{u})}+\frac{b(b-1)}{u\log^{b+1}(\frac{t}{u})}\right)du\right)\end{aligned}$$

After taking the derivative, and integrating by parts once in the remaining integrals, we get

$$\begin{aligned}&\partial_t\left(\int_0^{\frac{1}{2}}\left(\chi_{\leq\frac{1}{4}}(\xi)-1\right)\frac{\sin(t\xi)}{t^2}\left(\frac{(b-1)}{\xi\log^b(\frac{1}{\xi})}+\frac{b(b-1)}{\xi\log^{b+1}(\frac{1}{\xi})}\right)d\xi\right)\\
&=\frac{-\sin(\frac{t}{2})}{t^3}\left(\frac{b-1}{\log^b(2)}+\frac{b(b-1)}{\log^{b+1}(2)}\right)+\mathrm{Err}\end{aligned}$$

where

$$|\mathrm{Err}|\leq\frac{C}{t^4}$$

On the other hand,

$$\begin{aligned}&\partial_t\left(\int_0^{\frac{t}{2}}\frac{\sin(u)b(b-1)du}{t^2u\log^{b+1}(\frac{t}{u})}\right)\\
&=\frac{\sin(\frac{t}{2})b(b-1)}{t^3\log^{b+1}(2)}+\int_0^{\frac{t}{2}}\frac{\sin(u)b(b-1)}{u}\left(\frac{-2}{t^3\log^{b+1}(\frac{t}{u})}-\frac{(b+1)}{\log^{b+2}(\frac{t}{u})t^3}\right)du\end{aligned}$$

So,

$$\begin{aligned}E_{t,v_2,ip,1} &:= \partial_t \left(2c_b \int_0^{\frac{1}{2}} \left(\chi_{\leq \frac{1}{4}}(\xi)-1\right) \frac{\sin(t\xi)}{t^2} \left(\frac{b-1}{\xi \log^b(\frac{1}{\xi})} + \frac{b(b-1)}{\xi \log^{b+1}(\frac{1}{\xi})}\right) d\xi\right.\\
&\left.+ 2c_b \left(\int_0^{\frac{t}{2}} \frac{\sin(u)(b-1)du}{t^2 u \log^b(\frac{t}{u})} - \frac{(b-1)\pi}{2t^2 \log^b(t)} + \int_0^{\frac{t}{2}} \frac{\sin(u)b(b-1)du}{t^2 u \log^{b+1}(\frac{t}{u})}\right)\right)\\
&= 2c_b \mathrm{Err} + 2c_b \left(\int_0^{\frac{t}{2}} \sin(u)(b-1)\left(\frac{-2}{t^3 u \log^b(\frac{t}{u})} - \frac{b}{\log^{b+1}(\frac{t}{u}) t^3 u}\right) du\right.\\
&\left.+ \int_0^{\frac{t}{2}} \frac{1}{u}\sin(u) b(b-1)\left(\frac{-2}{t^3 \log^{b+1}(t)} - \frac{(b+1)}{t^3 \log^{b+2}(\frac{t}{u})}\right) du\right)\\
&+ 2c_b \left(\frac{(b-1)\pi}{t^3 \log^b(t)} + \frac{b(b-1)\pi}{2t^3 \log^{b+1}(t)}\right)\end{aligned}$$

The asymptotics of the integrals in lines 3 and 4 of the above expression were computed previously, in the section which constructed v_2. The following is the most important asymptotic, which shows that there is some cancellation between the integral in line 3 and the terms on line 5.

$$-2\int_0^{\frac{t}{2}} \frac{\sin(u)(b-1)du}{t^3 u \log^b(\frac{t}{u})} = \frac{-2(b-1)}{t^3} \frac{\pi}{2} \left(\frac{1}{\log^b(t)} + O\left(\frac{1}{\log^{b+1}(t)}\right)\right)$$

In total, we get

$$|E_{t,v_2,ip,1}(t)| \leq \frac{C}{t^3 \log^{b+1}(t)}$$

Next, we estimate

$$\partial_t \left(2c_b \lambda(t) \int_0^\infty \frac{\sin(t\xi)}{t^2} \psi_{v_2}(\xi, \lambda(t)) d\xi\right)$$

Here, we use estimates on the derivatives of ψ_{v_2}, proven earlier, when showing that the map T was a contraction, as well as (3.107), which shows (among other things) that

$$|\partial_\xi^2 \psi_{v_2}(\xi, \lambda(t))| \leq C\frac{p(\xi)}{\lambda(t)}, \quad p \in C_c^\infty([\frac{1}{8}, \frac{1}{4}])$$

This gives us

$$|\partial_t \left(2c_b \lambda(t) \int_0^\infty \frac{\sin(t\xi)}{t^2} \psi_{v_2}(\xi, \lambda(t)) d\xi\right)| \leq \frac{C}{t^4}$$

Similarly, we use estimates on derivatives of F_{v_2} which were proven while showing that T is a contraction, as well as (3.107), and asymptotics of K_1, to get

$$|\partial_t \left(2c_b \lambda(t) \int_0^\infty \chi_{\leq \frac{1}{4}}(\xi) \frac{\sin(t\xi)}{t^2} F_{v_2}(\xi, \lambda(t)) d\xi\right)| \leq \frac{C\log(\log(t))}{t^4 \log^{2b}(t)}$$

In total, we have

$$|\partial_t \left(\lambda(t) E_{v_2,ip}(t, \lambda(t))\right)| \leq \frac{C}{t^3 \log^{b+1}(t)}$$

By the same procedure, this estimate is also true for the case $b = 1$.

Some of the terms in (3.124) which remain to be estimated will be estimated in two different ways. One way of estimating these terms will give rise to preliminary estimates on e''', which will then allow us to obtain stronger estimates. For this preliminary estimate, we will start with the first line of (3.124):

$$\partial_t \left(\frac{16}{\lambda(t)^3} \int_t^\infty \lambda''(s) \left(K(s-t, \lambda(t)) + K_1(s-t, \lambda(t)) \right) ds \right) = I + II + III + IV$$

where

$$I = \frac{16}{\lambda(t)^3} \int_t^\infty \lambda''(s) \partial_2 K(s-t, \lambda(t)) \lambda'(t) ds$$

$$II = \frac{16}{\lambda(t)^3} \int_t^\infty \lambda''(s) \partial_2 K_1(s-t, \lambda(t)) \lambda'(t) ds$$

$$III = -\frac{16}{\lambda(t)^3} \int_t^\infty \lambda''(s) \left(\partial_1 K(s-t, \lambda(t)) + \partial_1 K_1(s-t, \lambda(t)) \right) ds$$

$$IV = \frac{-48\lambda'(t)}{\lambda(t)^4} \int_t^\infty \lambda''(s) \left(K(s-t, \lambda(t)) + K_1(s-t, \lambda(t)) \right) ds$$
$$|IV| \leq \frac{C}{t^3 \log(t)}$$

In order to estimate I, let us start with

$$\partial_2 K(x, \lambda(t))$$
$$= \int_0^\infty \frac{R}{(1+R^2)^3} \int_0^x \rho \left(\frac{1}{\sqrt{x^2-\rho^2}} - \frac{1}{x} \right) \left(\frac{4\lambda(t) R^2 (1+\rho^2+\lambda(t)^2 R^2)}{(4\lambda(t)^2 R^2 + (1+\rho^2 - R^2 \lambda(t)^2)^2)^{3/2}} \right) d\rho dR$$

Then, the same procedure used to obtain (3.103) gives

$$| \int_t^\infty \lambda''(s) \partial_2 K(s-t, \lambda(t)) \lambda'(t) ds | \leq \frac{C}{t^3 \log^{2b+2}(t)} \int_t^\infty |\partial_2 K(s-t, \lambda(t))| ds$$
$$\leq \frac{C}{t^3 \log^{3b+2}(t)}$$

So,

$$|I| \leq \frac{C}{t^3 \log^2(t)}$$

Next, we recall the definition of K_1:

$$K_1(x, \lambda(t)) = \int_0^\infty \frac{r}{\lambda(t)^2 (1 + \frac{r^2}{\lambda(t)^2})^3} \int_0^x \frac{\rho}{x} \left(1 + \frac{r^2 - 1 - \rho^2}{\sqrt{(r^2-1-\rho^2)^2 + 4r^2}} \right) d\rho dr$$

So,

$$|\partial_2 K_1(x, \lambda(t)) \lambda'(t)|$$
$$\leq C \lambda(t)^3 |\lambda'(t)| \int_0^\infty \frac{r}{(r^2+\lambda(t)^2)^3} \int_0^x \frac{\rho}{x} \left(1 + \frac{r^2-1-\rho^2}{\sqrt{(r^2-1-\rho^2)^2+4r^2}} \right) d\rho dr$$

and we get

$$\begin{aligned}
|II| &\leq |\frac{16}{\lambda(t)^3}\int_t^\infty \lambda''(x)\partial_2 K_1(x-t,\lambda(t))\lambda'(t)dx| \\
&\leq C|\lambda'(t)|\int_t^\infty |\lambda''(x)|\int_0^\infty \frac{r}{(r^2+\lambda(t)^2)^3} \\
&\qquad \begin{cases} \frac{1}{(x-t)}\int_0^{x-t} 2\rho d\rho, & x-t\leq 1 \\ \frac{1}{(x-t)}\int_0^\infty \rho d\rho\left(1+\frac{r^2-1-\rho^2}{\sqrt{(r^2-1-\rho^2)^2+4r^2}}\right), & x-t\geq 1 \end{cases} drdx \\
&\leq C|\lambda'(t)|\int_t^\infty |\lambda''(x)|\int_0^\infty \frac{r}{(r^2+\lambda(t)^2)^3}\begin{cases}(x-t), & x-t\leq 1 \\ \frac{r^2}{(x-t)}, & x-t\geq 1\end{cases} drdx \\
&\leq C|\lambda'(t)|\left(\int_t^{t+1}\frac{1}{x^2\log^{b+1}(x)}\frac{(x-t)dx}{\lambda(t)^4}\right. \\
&\qquad \left.+\int_{t+1}^\infty \frac{dx}{x^2\log^{b+1}(x)(x-t)}\int_0^\infty \frac{r^3dr}{(r^2+\lambda(t)^2)^3}\right) \\
&\leq \frac{C}{t^3\log^{1-2b}(t)}
\end{aligned}$$

We will now treat the term III. Let $L=K+K_1$. Then, we have

$$\partial_1 L(x,\lambda(t)) = L_a(x,\lambda(t)) + L_b(x,\lambda(t))$$

with

$$\begin{aligned}
L_a(x,\lambda(t)) &= \int_0^\infty \frac{R}{(1+R^2)^3}\int_0^1 \frac{u}{\sqrt{1-u^2}}\left(1+\frac{R^2\lambda(t)^2-1-u^2x^2}{\sqrt{(R^2\lambda(t)^2-1-u^2x^2)^2+4R^2\lambda(t)^2}}\right)dudR \\
&= \int_0^\infty \frac{R}{(1+R^2)^3}\int_0^x \frac{\rho}{x\sqrt{x^2-\rho^2}}\left(1+\frac{R^2\lambda(t)^2-1-\rho^2}{\sqrt{(R^2\lambda(t)^2-1-\rho^2)^2+4R^2\lambda(t)^2}}\right)d\rho dR
\end{aligned}$$

$$\begin{aligned}
L_b(x,\lambda(t)) &= \int_0^\infty \frac{R}{(1+R^2)^3}\int_0^1 \frac{ux}{\sqrt{1-u^2}}\left(\frac{-8R^2\lambda(t)^2u^2x}{(4R^2\lambda(t)^2+(1-R^2\lambda(t)^2+u^2x^2)^2)^{3/2}}\right)dudR \\
&= \int_0^\infty \frac{R}{(1+R^2)^3}\int_0^x \frac{\rho}{\sqrt{x^2-\rho^2}}\left(\frac{-8R^2\lambda(t)^2\rho^2}{x(4R^2\lambda(t)^2+(1-R^2\lambda(t)^2+\rho^2)^2)^{3/2}}\right)d\rho dR
\end{aligned}$$

Now,

(3.125)

$$\begin{aligned}
&\int_t^\infty |L_a(s-t,\lambda(t))|ds \\
&\leq \int_0^\infty \frac{R}{(1+R^2)^3}\int_0^\infty \rho\left(1+\frac{R^2\lambda(t)^2-1-\rho^2}{\sqrt{(R^2\lambda(t)^2-1-\rho^2)^2+4R^2\lambda(t)^2}}\right)\int_{\rho+t}^\infty \frac{dsd\rho dR}{(s-t)\sqrt{(s-t)^2-\rho^2}} \\
&\leq C\int_0^\infty \frac{R}{(1+R^2)^3}\int_0^\infty \left(1+\frac{R^2\lambda(t)^2-1-\rho^2}{\sqrt{(R^2\lambda(t)^2-1-\rho^2)^2+4R^2\lambda(t)^2}}\right)d\rho dR \\
&\leq C\int_0^\infty \frac{R}{(1+R^2)^3}\int_{2(R\lambda(t)+1)}^\infty \left(1+\frac{R^2\lambda(t)^2-1-\rho^2}{\sqrt{(R^2\lambda(t)^2-1-\rho^2)^2+4R^2\lambda(t)^2}}\right)d\rho dR \\
&+C\int_0^\infty \frac{R}{(1+R^2)^3}\int_0^{2(R\lambda(t)+1)} \left(1+\frac{R^2\lambda(t)^2-1-\rho^2}{\sqrt{(R^2\lambda(t)^2-1-\rho^2)^2+4R^2\lambda(t)^2}}\right)d\rho dR
\end{aligned}$$

When $\rho > 2(R\lambda(t)+1)$, we have

$$|1+\frac{R^2\lambda(t)^2-1-\rho^2}{\sqrt{(R^2\lambda(t)^2-1-\rho^2)^2+4R^2\lambda(t)^2}}|$$
$$=|1+\frac{(-1-\frac{1}{\rho^2}+\frac{R^2\lambda(t)^2}{\rho^2})}{\sqrt{1-2(\frac{R^2\lambda(t)^2}{\rho^2}-\frac{1}{\rho^2})+\frac{(R^2\lambda(t)^2+1)^2}{\rho^4}}}|\leq C\frac{(R\lambda(t)+1)^2}{\rho^2}$$

If $\rho < 2(R\lambda(t)+1)$, then, we use

$$|1+\frac{R^2\lambda(t)^2-1-\rho^2}{\sqrt{(R^2\lambda(t)^2-1-\rho^2)^2+4R^2\lambda(t)^2}}|\leq 2$$

to continue the estimate in (3.125)

$$\begin{aligned}
&\int_t^\infty |L_a(s-t,\lambda(t))|ds\\
&\leq C\int_0^\infty \frac{R}{(1+R^2)^3}\int_{2(R\lambda(t)+1)}^\infty \left(1+\frac{R^2\lambda(t)^2-1-\rho^2}{\sqrt{(R^2\lambda(t)^2-1-\rho^2)^2+4R^2\lambda(t)^2}}\right)d\rho dR\\
&+C\int_0^\infty \frac{R}{(1+R^2)^3}\int_0^{2(R\lambda(t)+1)} \left(1+\frac{R^2\lambda(t)^2-1-\rho^2}{\sqrt{(R^2\lambda(t)^2-1-\rho^2)^2+4R^2\lambda(t)^2}}\right)d\rho dR\\
&\leq C\int_0^\infty \frac{R}{(1+R^2)^3}\int_{2(R\lambda(t)+1)}^\infty \frac{(R\lambda(t)+1)^2}{\rho^2}d\rho dR + C\int_0^\infty \frac{R}{(1+R^2)^3}(R\lambda(t)+1)dR\\
&\leq C
\end{aligned}$$

Similarly, we have

(3.126)
$$\begin{aligned}
&|\int_t^\infty L_b(s-t,\lambda(t))ds|\\
&\leq \int_0^\infty \frac{R}{(1+R^2)^3}\int_0^\infty \rho\left(\frac{8R^2\lambda(t)^2\rho^2}{(4R^2\lambda(t)^2+(1-R^2\lambda(t)^2+\rho^2)^2)^{3/2}}\right)\int_{\rho+t}^\infty \frac{1}{\sqrt{(s-t)^2-\rho^2}}\frac{dsd\rho dR}{(s-t)}\\
&\leq C\int_0^\infty \frac{R}{(1+R^2)^3}\int_0^\infty \frac{R^2\lambda(t)^2\rho^2}{(4R^2\lambda(t)^2+(1-R^2\lambda(t)^2+\rho^2)^2)^{3/2}}d\rho dR\\
&\leq C\int_0^\infty \frac{R}{(1+R^2)^3}\int_{2(R\lambda(t)+1)}^\infty \frac{R^2\lambda(t)^2\rho^2}{(4R^2\lambda(t)^2+(1-R^2\lambda(t)^2+\rho^2)^2)^{3/2}}d\rho dR\\
&+C\int_0^\infty \frac{R}{(1+R^2)^3}\int_0^{2(R\lambda(t)+1)} \frac{R^2\lambda(t)^2\rho^2}{(4R^2\lambda(t)^2+(1-R^2\lambda(t)^2+\rho^2)^2)^{3/2}}d\rho dR
\end{aligned}$$

When $\rho > 2(R\lambda(t)+1)$, we have

$$\frac{1}{(4R^2\lambda(t)^2+(1-R^2\lambda(t)^2+\rho^2)^2)^{3/2}}\leq\frac{C}{\rho^6}$$

When $\rho < 2(R\lambda(t)+1)$, we use

$$\frac{1}{(4R^2\lambda(t)^2+(1-R^2\lambda(t)^2+\rho^2)^2)^{3/2}}\leq\frac{C}{R^3\lambda(t)^3}$$

and then continue the estimate from (3.126):

$$
\begin{aligned}
&|\int_t^\infty L_b(s-t,\lambda(t))ds| \\
&\leq C\int_0^\infty \frac{R}{(1+R^2)^3}\int_{2(R\lambda(t)+1)}^\infty \frac{R^2\lambda(t)^2\rho^2}{(4R^2\lambda(t)^2+(1-R^2\lambda(t)^2+\rho^2)^2)^{3/2}}d\rho dR \\
&+C\int_0^\infty \frac{R}{(1+R^2)^3}\int_0^{2(R\lambda(t)+1)} \frac{R^2\lambda(t)^2\rho^2}{(4R^2\lambda(t)^2+(1-R^2\lambda(t)^2+\rho^2)^2)^{3/2}}d\rho dR \\
&\leq C\int_0^\infty \frac{R}{(1+R^2)^3}R^2\lambda(t)^2\int_{2(R\lambda(t)+1)}^\infty \frac{\rho^2}{\rho^6}d\rho dR \\
&+C\int_0^\infty \frac{R}{(1+R^2)^3}\int_0^{2(R\lambda(t)+1)} \frac{R^2\lambda(t)^2\rho^2}{R^3\lambda(t)^3}d\rho dR \\
&\leq \frac{C}{\lambda(t)}
\end{aligned}
$$

Then, we combine the above estimates to get

$$\int_t^\infty |\partial_1 L(s-t,\lambda(t))|ds \leq \frac{C}{\lambda(t)}$$

So,

$$|III| \leq C\frac{\log^{3b-1}(t)}{t^2}$$

We finally conclude that

$$|I|+|II|+|III|+|IV| \leq \frac{C\log^{3b-1}(t)}{t^2}$$

Next, we estimate the last two lines of (3.124). First, we note that

$$
\begin{aligned}
&\partial_t\langle\left(\frac{\cos(2Q_{\frac{1}{\lambda(t)}})-1}{r^2}\right)f|_{r=R\lambda(t)},\phi_0\rangle \\
&=\int_0^\infty \partial_t\left(\left(\frac{\cos(2Q_1(\frac{r}{\lambda(t)})-1}{r^2}\right)\frac{\phi_0(\frac{r}{\lambda(t)})}{\lambda(t)^2}\right)f(t,r)rdr \\
&+\int_0^\infty \left(\frac{\cos(2Q_1(\frac{r}{\lambda(t)}))-1}{r^2}\right)\frac{\phi_0(\frac{r}{\lambda(t)})}{\lambda(t)^2}\partial_t f(t,r)rdr
\end{aligned}
$$

which gives

$$
\begin{aligned}
|\partial_t\langle\left(\frac{\cos(2Q_{\frac{1}{\lambda(t)}})-1}{r^2}\right)f|_{r=R\lambda(t)},\phi_0\rangle| &\leq C\int_0^\infty \frac{r|\lambda'(t)|}{(r^2+\lambda(t)^2)^3}|f(t,r)|rdr \\
&+C\int_0^\infty \frac{r\lambda(t)}{(r^2+\lambda(t)^2)^3}|\partial_t f(t,r)|rdr
\end{aligned}
\tag{3.127}
$$

We now proceed to estimate $\partial_t v_k, \quad k=1,3,4,5$. We first prove a preliminary estimate on these quantities in order to obtain the preliminary estimate on λ'''. Once this is done, we can improve our estimates for $\partial_t v_k$.

LEMMA 3.14 (Preliminary Estimates on $\partial_t v_k, \quad k=1,3,4,5$). *We have the following* preliminary *estimates on* $\partial_t v_k, \quad k=1,3,4,5$.

$$|\partial_t v_1(t,r)| \leq \frac{Cr}{t^2 \log^{b+1}(t)}$$

(3.128)
$$|\partial_t v_3(t,r)| \leq \frac{Cr}{t^2 \log^{1+b\alpha}(t)}$$

(3.129)
$$|\partial_t v_4(t,r)| \leq \frac{C}{t^2 \log^{3b+2N-1}(t)}, \quad r \leq \frac{t}{2}$$

(3.130)
$$|\partial_t v_5(t,r)| \leq \frac{C}{t^{7/2} \log^{3b-3+\frac{5N}{2}}(t)} + \frac{Cr\log(t)}{t^4 \log^b(t)}, \quad r \leq \frac{t}{2}$$

PROOF. For $\partial_t v_3$, we have

(3.131)
$$\begin{aligned}
&\partial_t v_3(t,r)\\
&= \frac{1}{r}\int_t^\infty \int_0^{s-t} \frac{\rho}{(s-t)\sqrt{(s-t)^2-\rho^2}} \lambda''(s) \left(\frac{-1-\rho^2+r^2}{\sqrt{(1+\rho^2-r^2)^2+4r^2}} + F_3(r,\rho,\lambda(s))\right) d\rho ds\\
&- \frac{1}{r}\int_t^\infty \int_0^{s-t} \frac{\rho}{(s-t)\sqrt{(s-t)^2-\rho^2}} \lambda''(s) \left(\frac{8\rho^2 r^2}{(4r^2+(1+\rho^2-r^2)^2)^{3/2}}\right.\\
&+ \left.\frac{-8\lambda(s)^{4\alpha-4}\rho^2 r^2}{(4\lambda(s)^{2\alpha-2}r^2+(1+\lambda(s)^{2\alpha-2}(\rho^2-r^2))^2)^{3/2}}\right) d\rho ds
\end{aligned}$$

For the first line on the right-hand side of (3.131), we first note that

$$|1-F_3(r,\rho,\lambda(s))| = 1-F_3(r,\rho,\lambda(s)) = F_3(0,\rho,\lambda(s)) - F_3(r,\rho,\lambda(s))$$

So,

$$\begin{aligned}
&|1-F_3(r,\rho,\lambda(s))|\\
&\leq Cr\int_0^1 \frac{r_\sigma \lambda(s)^{2\alpha-2}(1+\lambda(s)^{2\alpha-2}(\rho^2+r_\sigma^2))}{(1+2(\rho^2+r_\sigma^2)\lambda(s)^{2\alpha-2}+(\rho^2-r_\sigma^2)^2\lambda(s)^{4\alpha-4})^{3/2}} d\sigma
\end{aligned}$$

where

$$r_\sigma = (1-\sigma)r$$

Then, because $\lambda'(x) \leq 0, \quad x \geq T_0$, we have $\lambda(s)^{2\alpha-2} \geq \lambda(t)^{2\alpha-2}, \quad s \geq t$. So,

$$\begin{aligned}
&\left|\frac{1}{r}\int_t^\infty \int_0^{s-t} \frac{\rho\lambda''(s)}{(s-t)\sqrt{(s-t)^2-\rho^2}} \left(\frac{-1-\rho^2+r^2}{\sqrt{(1+\rho^2-r^2)^2+4r^2}} + 1 + F_3(r,\rho,\lambda(s)) - 1\right) d\rho ds\right|\\
&\leq Cr \sup_{x\geq t}|\lambda''(x)| \int_0^\infty \int_0^1 \frac{d\sigma d\rho}{(1+2(\rho^2+r_\sigma^2)+(\rho^2-r_\sigma^2)^2)^{1/2}}\\
&+ Cr \sup_{x\geq t}\left(|\lambda''(x)|\lambda(x)^{4\alpha-4}\right)\\
&\cdot \int_0^\infty \int_0^1 \frac{(\lambda(t)^{2-2\alpha}+\rho^2+r_\sigma^2)}{(1+2(\rho^2+r_\sigma^2)\lambda(t)^{2\alpha-2}+(\rho^2-r_\sigma^2)^2\lambda(t)^{4\alpha-4})^{3/2}} d\sigma d\rho
\end{aligned}$$

and where we obtained the second, third and fourth lines by switching the s and ρ integration order, and using

$$\int_{\rho+t}^\infty \frac{ds}{(s-t)\sqrt{(s-t)^2-\rho^2}} \leq \frac{C}{\rho}$$

So, it suffices to estimate

(3.132) $$B_3(y)$$
$$= \int_0^\infty \frac{d\rho}{(1+2(\rho^2+y^2)+(\rho^2-y^2)^2)^{1/2}} \leq C\int_0^{2y} \frac{d\rho}{y} + \int_{2y}^\infty \frac{d\rho}{(1+\rho^2)} \leq C$$

This gives

$$|\frac{1}{r}\int_t^\infty \int_0^{s-t} \frac{\rho \lambda''(s)}{(s-t)\sqrt{(s-t)^2-\rho^2}}$$
$$\left(\frac{-1-\rho^2+r^2}{\sqrt{(1+\rho^2-r^2)^2+4r^2}}+1+F_3(r,\rho,\lambda(s))-1\right)d\rho ds|$$
$$\leq C\frac{r}{t^2\log^{b+1}(t)}+C\frac{r}{t^2\log^{1+b\alpha}(t)}$$

We then consider the second line on the right-hand side of (3.131), and use the same procedure as above:

$$|\frac{-1}{r}\int_t^\infty \int_0^{s-t} \frac{\rho}{(s-t)}\frac{\lambda''(s)}{\sqrt{(s-t)^2-\rho^2}}\left(\frac{8\rho^2r^2}{(4r^2+(1+\rho^2-r^2)^2)^{3/2}}\right.$$
$$\left.-\frac{8\lambda(s)^{4\alpha-4}\rho^2r^2}{(4\lambda(s)^{2\alpha-2}r^2+(1+\lambda(s)^{2\alpha-2}(\rho^2-r^2))^2)^{3/2}}\right)d\rho ds|$$
$$\leq Cr\int_0^\infty \left(\frac{\rho^2\left(\sup_{x\geq t}|\lambda''(x)|\right)}{(4r^2+(1+\rho^2-r^2)^2)^{3/2}}+\frac{\rho^2\sup_{x\geq t}\left(|\lambda''(x)|\lambda(x)^{4\alpha-4}\right)}{(1+2\lambda(t)^{2\alpha-2}(r^2+\rho^2)+\lambda(t)^{4\alpha-4}(\rho^2-r^2)^2)^{3/2}}\right)d\rho$$
$$\leq Cr\int_0^\infty \frac{1}{t^2\log^{b+1}(t)(1+2(\rho^2+r^2)+(\rho^2-r^2)^2)^{1/2}}d\rho$$
$$+Cr\int_0^\infty \frac{\lambda(t)^{3-3\alpha}dq}{t^2\log^{b+1}(t)\log^{-4b+4\alpha b}(t)(1+2(q^2+r^2\lambda(t)^{2\alpha-2})+(q^2-r^2\lambda(t)^{2\alpha-2})^2)^{1/2}}$$
$$\leq Cr\frac{B_3(r)}{t^2\log^{b+1}(t)}+Cr\frac{B_3(r\lambda(t)^{\alpha-1})}{t^2\log^{b+1}(t)\log^{-b+\alpha b}(t)}$$

This gives

$$|\frac{-1}{r}\int_t^\infty \int_0^{s-t} \frac{\rho}{(s-t)}\frac{\lambda''(s)}{\sqrt{(s-t)^2-\rho^2}}\left(\frac{8\rho^2r^2}{(4r^2+(1+\rho^2-r^2)^2)^{3/2}}\right.$$
$$\left.-\frac{8\lambda(s)^{4\alpha-4}\rho^2r^2}{(4\lambda(s)^{2\alpha-2}r^2+(1+\lambda(s)^{2\alpha-2}(\rho^2-r^2))^2)^{3/2}}\right)d\rho ds|$$
$$\leq C\frac{r}{t^2\log^{b+1}(t)}+C\frac{r}{t^2\log^{1+b\alpha}(t)}$$

In total, we get (3.128).

Next, we have to estimate $\partial_t v_4$ in the region $r \leq \frac{t}{2}$. Again, we write the formula for v_4 as

$$v_4(t,r)$$
$$=\frac{-1}{2\pi}\int_t^\infty \int_0^{s-t} \frac{\rho}{\sqrt{(s-t)^2-\rho^2}}\int_0^{2\pi} \frac{v_{4,c}(s,\sqrt{r^2+2r\rho\cos(\theta)+\rho^2})}{\sqrt{r^2+\rho^2+2r\rho\cos(\theta)}}(r+\rho\cos(\theta))d\theta d\rho ds$$

Like done previously for v_4, we introduce $x = r\mathbf{e}_1 \in \mathbb{R}^2$ to ease notation, and get

(3.133)
$$\partial_t v_4(t,r) = -\int_t^\infty \frac{v_{4,s}(t,r)}{(s-t)}ds - \frac{1}{2\pi}\int_t^\infty \int_{B_{s-t}(0)} \frac{\text{integrand}_{v_{4,1}}}{\sqrt{(s-t)^2-|y|^2}}dA(y)ds$$

where we recall

$$v_{4,s}(t,r) = \frac{-1}{2\pi}\int_{B_{s-t}(0)} \frac{1}{\sqrt{(s-t)^2-|y|^2}} \frac{v_{4,c}(s,|x+y|)\left((x+y)\cdot\widehat{x}\right)}{|x+y|}dA(y)$$

$$\begin{aligned}\text{integrand}_{v_{4,1}} &= \frac{-\partial_2 v_{4,c}(s,|x+y|)}{|x+y|^2(s-t)}\left((x+y)\cdot y\right)\left(\hat{x}\cdot(x+y)\right)\\&+\frac{v_{4,c}(s,|x+y|)}{(s-t)|x+y|}\left(-y\cdot\hat{x}+\frac{\left(\hat{x}\cdot(x+y)\right)\left(y\cdot(y+x)\right)}{|x+y|^2}\right)\end{aligned}$$

So,

$$|\text{integrand}_{v_{4,1}}| \leq C|\partial_2 v_{4,c}(s,|x+y|)| + \frac{C|v_{4,c}(s,|x+y|)|}{|x+y|}$$

This is the same estimate obtained when estimating the integrand of v_4, except for a factor of $\frac{1}{r}$. So, the second term of (3.133) is estimated as follows:

$$|-\frac{1}{2\pi}\int_t^\infty\int_{B_{s-t}(0)}\frac{\text{integrand}_{v_{4,1}}}{\sqrt{(s-t)^2-|y|^2}}dA(y)ds| \leq \frac{C}{t^2\log^{3b+2N-1}(t)}, \quad r\leq\frac{t}{2}$$

For the first term in (3.133), we have

(3.134)

$$\begin{aligned}&|\int_t^\infty \frac{v_{4,s}(t,r)}{(s-t)}ds|\\&\leq C\int_t^\infty\frac{1}{(s-t)}\int_{B_{s-t}(0)}\frac{|v_{4,c}(s,|x+y|)|}{\sqrt{(s-t)^2-|y|^2}}\left(\mathbb{1}_{B_{\frac{s}{2}}(-x)}(y)+\mathbb{1}_{(B_{\frac{s}{2}}(-x))^c}(y)\right)dA(y)ds\\&\leq C\int_t^\infty\int_0^{s-t}\rho\int_0^{2\pi}\frac{1}{(s-t)\sqrt{(s-t)^2-\rho^2}}\frac{1}{s^2\log^{N+3b}(s)}\frac{d\theta d\rho ds}{(\log^{2N}(s)+r^2+\rho^2+2r\rho\cos(\theta))}\\&+C\int_t^\infty\int_0^{s-t}\rho\int_0^{2\pi}\frac{1}{(s-t)\sqrt{(s-t)^2-\rho^2}}\frac{\log(s)}{\log^{2b}(s)s^2t}\frac{1}{(s^2+\rho^2+r^2+2r\rho\cos(\theta))}d\theta d\rho ds\end{aligned}$$

We first estimate the third line of (3.134):

$$\begin{aligned}&|\int_t^\infty\int_0^{s-t}\rho\int_0^{2\pi}\frac{1}{(s-t)\sqrt{(s-t)^2-\rho^2}}\frac{1}{s^2\log^{N+3b}(s)}\frac{d\theta d\rho ds}{(\log^{2N}(s)+r^2+\rho^2+2r\rho\cos(\theta))}|\\&\leq C\int_0^\infty\rho\int_0^{2\pi}\frac{1}{(t+\rho)^2\log^{3b+N}(t)(\log^{2N}(t)+\rho^2+r^2+2r\rho\cos(\theta))}\int_{\rho+t}^\infty\frac{dsd\theta d\rho}{(s-t)\sqrt{(s-t)^2-\rho^2}}\\&\leq C\int_0^\infty\frac{1}{(t+\rho)^2\log^{3b+N}(t)}\int_0^{2\pi}\frac{d\theta d\rho}{(\log^{2N}(t)+\rho^2+r^2+2r\rho\cos(\theta))}\\&\leq C\int_0^\infty\frac{1}{(t+\rho)^2\log^{3b+N}(t)}\frac{d\rho}{\sqrt{(\log^{2N}(t)+(r+\rho)^2)(\log^{2N}(t)+(r-\rho)^2)}}\\&\leq\frac{C}{\log^{3b+N}(t)}\int_0^\infty\frac{1}{(t+\rho)^2}\frac{1}{\sqrt{\log^{2N}(t)+(r+\rho)^2}}\frac{d\rho}{\log^N(t)}\\&\leq\frac{C}{t^2\log^{3b+2N-1}(t)}\end{aligned}$$

Next, we have

$$\begin{aligned}
&|\int_t^\infty \int_0^{s-t} \rho \int_0^{2\pi} \frac{1}{(s-t)\sqrt{(s-t)^2-\rho^2}} \frac{\log(s) d\theta d\rho ds}{\log^{2b}(s)(s^2+\rho^2+r^2+2r\rho\cos(\theta))s^2 t}| \\
&\leq C \int_0^\infty \rho \int_0^{2\pi} \frac{1}{\log^{2b}(t)} \frac{\log(t)}{(\rho^2+r^2+2r\rho\cos(\theta)+t^2)} \frac{1}{(\rho+t)} \frac{1}{t^2} \int_{\rho+t}^\infty \frac{ds d\theta d\rho}{(s-t)\sqrt{(s-t)^2-\rho^2}} \\
&\leq \frac{C}{t^2 \log^{2b}(t)} \int_0^\infty \int_0^{2\pi} \frac{\log(t) d\theta d\rho}{(\rho^2+r^2+2r\rho\cos(\theta)+t^2)(\rho+t)} \\
&\leq \frac{C\log(t)}{t^3 \log^{2b}(t)} \int_0^\infty \frac{d\rho}{(\rho+t)\sqrt{t^2+\rho^2}} \\
&\leq \frac{C\log(t)}{t^4 \log^{2b}(t)}
\end{aligned}$$

Combining these, we get (3.129).

Now, we are ready to estimate $\partial_t v_5$. Again, we have

$$\begin{aligned}
&\partial_t v_5 \\
&= \frac{1}{2\pi} \int_t^\infty \int_0^{s-t} \frac{\rho}{(s-t)} \frac{1}{\sqrt{(s-t)^2-\rho^2}} \int_0^{2\pi} \frac{N_2(f_{v_5})(s, \sqrt{r^2+\rho^2+2r\rho\cos(\theta)})(r+\rho\cos(\theta))}{\sqrt{r^2+\rho^2+2r\rho\cos(\theta)}} d\theta d\rho ds \\
&- \frac{1}{2\pi} \int_t^\infty \int_{B_{s-t}(0)} \frac{\text{integrand}_{v_{5,1}}}{\sqrt{(s-t)^2-|y|^2}} dA(y) ds
\end{aligned}$$

with

$$|\text{integrand}_{v_{5,1}}| \leq C|\partial_2 N_2(f_{v_5})(s,|x+y|)| + C\frac{|N_2(f_{v_5})(s,|x+y|)}{|x+y|}$$

Again, this is the same estimate for the integrand which appeared in the v_5 pointwise estimates, aside from an extra factor of $\frac{1}{r}$. So, we get

$$|-\frac{1}{2\pi} \int_t^\infty \int_{B_{s-t}(0)} \frac{\text{integrand}_{v_{5,1}}}{\sqrt{(s-t)^2-|y|^2}} dA(y) ds| \leq \frac{C}{t^{7/2} \log^{3b-3+\frac{5N}{2}}(t)}, \quad r \leq \frac{t}{2}$$

Then, we need to estimate

$$\begin{aligned}
&\int_t^\infty \int_{B_{s-t}(0)} \frac{1}{\sqrt{(s-t)^2-|y|^2}} \frac{|N_2(f_{v_5})|(s,|x+y|)}{(s-t)} \left(\mathbb{1}_{B_{\frac{s}{2}}(-x)}(y) + \mathbb{1}_{(B_{\frac{s}{2}}(-x))^c}(y)\right) dA(y) ds \\
&\leq C \int_t^\infty \int_0^{s-t} \rho \int_0^{2\pi} \frac{1}{\sqrt{(s-t)^2-\rho^2}} \frac{1}{(s-t)} \left(\frac{r+\rho}{(\lambda(s)^2+r^2+\rho^2+2r\rho\cos(\theta))s^4 \log^{3b}(s)} \right. \\
&\qquad \left. + \frac{s}{s^6 \log^{3b}(s)} \right) d\theta d\rho ds \\
&+ C \int_t^\infty \int_0^{s-t} \rho \int_0^{2\pi} \frac{1}{\sqrt{(s-t)^2-\rho^2}} \frac{1}{(s-t)} \left(\frac{\log^3(s)}{s^2 t^3} + \frac{1}{s^6 \log^{3N+7b-2}(s)} \right) d\theta d\rho ds \\
&\leq C \int_0^\infty \rho \int_0^{2\pi} \frac{1}{\rho} \frac{(r+\rho)}{(1+r^2+\rho^2+2r\rho\cos(\theta))} \frac{1}{\log^b(t)(\rho+t)^4} d\theta d\rho + C \int_t^\infty \frac{ds}{s^5 \log^{3b}(s)} \\
&+ C \int_t^\infty \left(\frac{\log^3(s)}{s^2 t^3} + \frac{1}{s^6 \log^{3N+7b-2}(s)} \right) ds \\
&\leq C\frac{(r+1)\log(t)}{t^4 \log^b(t)} + C\frac{\log^3(t)}{t^4} \\
&\leq C\frac{r\log(t)}{t^4 \log^b(t)} + C\frac{\log^3(t)}{t^4}, \quad r \leq \frac{t}{2}
\end{aligned}$$

In total, we have (3.130).

Finally, we estimate $\partial_t v_1$. Using the same procedure used to estimate $\partial_t v_3$, we get

$$\partial_t v_1(t,r)$$
$$= \int_t^\infty \frac{-\lambda''(s)}{r} \int_0^{s-t} \frac{\rho}{(s-t)\sqrt{(s-t)^2-\rho^2}} \left(1+\frac{r^2-1-\rho^2}{\sqrt{(r^2-1-\rho^2)^2+4r^2}}\right) d\rho ds$$
$$+ \int_t^\infty \frac{\lambda''(s)}{r} \int_0^{s-t} \frac{\rho}{\sqrt{(s-t)^2-\rho^2}} \frac{8\rho^2 r^2}{(s-t)(4r^2+(1+\rho^2-r^2)^2)^{3/2}} d\rho ds$$

So, recalling (3.132), we have

$$|\partial_t v_1(t,r)| \leq \frac{Cr}{t^2 \log^{b+1}(t)} \int_0^1 B_3(r(1-\sigma))d\sigma + \frac{CrB_3(r)}{t^2\log^{b+1}(t)} \leq \frac{Cr}{t^2\log^{b+1}(t)} \square$$

Now, we return to (3.127), and get

$$|\partial_t \langle \left(\frac{\cos(2Q_{\frac{1}{\lambda(t)}})-1}{r^2}\right) v_3|_{r=R\lambda(t)}, \phi_0\rangle| \leq \frac{C\log(\log(t))}{t^3\log^2(t)} + \frac{C}{t^2\log^{-b+1+b\alpha}(t)}$$

$$|\partial_t \langle \left(\frac{\cos(2Q_{\frac{1}{\lambda(t)}})-1}{r^2}\right) v_4 \left(1-\chi_{\geq 1}(\frac{4r}{t})\right)|_{r=R\lambda(t)}, \phi_0\rangle|$$
$$\leq \frac{C}{t^3\log^{2b+2N}(t)} + \frac{C}{t^2\log^{b+2N-1}(t)}$$

$$|\partial_t \langle \left(\frac{\cos(2Q_{\frac{1}{\lambda(t)}})-1}{r^2}\right) v_5 \left(1-\chi_{\geq 1}(\frac{4r}{t})\right)|_{r=R\lambda(t)}, \phi_0\rangle|$$
$$\leq \frac{C}{t^{9/2}\log^{2b-2+\frac{5N}{2}}(t)} + \frac{C}{t^{7/2}\log^{b-3+\frac{5N}{2}}(t)}$$

$$|\partial_t \langle \left(\frac{\cos(2Q_{\frac{1}{\lambda(t)}})-1}{r^2}\right) \chi_{\geq 1}(\frac{2r}{\log^N(t)}) (v_1+v_2+v_3)|_{r=R\lambda(t)}, \phi_0\rangle|$$
$$\leq \frac{C}{t^3\log^{2b+2N+1}(t)} + \frac{C}{t^2\log^{1+b\alpha+b+2N}(t)}$$

Combining these estimates, we finally conclude

$$|RHS_2'(t)| \leq \frac{C}{t^2\log^{1-3b}(t)}$$

which implies

$$|\lambda'''(t)| \leq \frac{C}{t^2\log(\log(t))\log^{1-2b}(t)}$$

Since $\lambda(t) = \lambda_0(t) + e_0(t)$, we get

$$|e_0'''(t)| \leq \frac{C}{t^2\log(\log(t))\log^{1-2b}(t)} \tag{3.135}$$

Now that we have this preliminary estimate, we return to (3.123), and write

$$\lambda(t) = \lambda_{0,0}(t) + e(t)$$

Note that the function e defined in the above expression is different from the function e which appeared in the construction of λ as a solution to the modulation equation. It will suffice for our purposes to have an estimate of the form $|e'''(t)| \leq C|\lambda_{0,0}'''(t)|$, and this is why we will not need to use the slightly more complicated decomposition

$$\lambda(t) = \lambda_0(t) + e_0(t) = \lambda_{0,0}(t) + \lambda_{0,1}(t) + e_0(t)$$

Then, we differentiate (3.123), and write the equation in the following way, with any differentiations under the integral sign justified by the preliminary estimate on e'''. (Note that $e'''(t) = e_0'''(t) + \lambda_{0,1}'''(t)$).

(3.136)

$$\begin{aligned}
&-4\int_t^\infty \frac{e'''(s)ds}{\log(\lambda_{0,0}(s))(1+s-t)} - 4\int_t^\infty \frac{e'''(s)ds}{(1+s-t)^3\log(\lambda_{0,0}(s))(\lambda_{0,0}(t)^{1-\alpha}+s-t)}\\
&+4\alpha e'''(t)\\
&= -4\alpha\lambda_{0,0}'''(t) + \frac{-4(1-\alpha)\lambda(t)^{-\alpha}\lambda'(t)}{\log(\lambda_{0,0}(t))}\int_t^\infty \frac{e''(s)ds}{(\lambda(t)^{1-\alpha}+s-t)^2(1+s-t)^3}\\
&+\frac{1}{\log(\lambda_{0,0}(t))}\left(\frac{-4\alpha\lambda'(t)\lambda''(t)}{\lambda(t)} + \partial_t G(t,\lambda(t)) + 4\partial_t\left(\int_0^\infty \frac{\lambda_{0,0}''(t+w)dw}{(\lambda(t)^{1-\alpha}+w)(1+w)^3}\right)\right)\\
&+\frac{1}{\log(\lambda_{0,0}(t))}\left(-E_{\lambda_{0,0}}'(t) + 4\alpha(\log(\lambda_{0,0}(t)) - \log(\lambda(t)))\lambda'''(t)\right)\\
&-\frac{4}{\log(\lambda_{0,0}(t))}\int_t^\infty \frac{e'''(s)}{(1+s-t)^3}\left(\frac{1}{\lambda_{0,0}(t)^{1-\alpha}+s-t} - \frac{1}{\lambda(t)^{1-\alpha}+s-t}\right)ds\\
&-4\int_t^\infty \frac{e'''(s)}{1+s-t}\left(\frac{1}{\log(\lambda_{0,0}(s))} - \frac{1}{\log(\lambda_{0,0}(t))}\right)ds\\
&-4\int_t^\infty \frac{e'''(s)}{(1+s-t)^3(\lambda_{0,0}(t)^{1-\alpha}+s-t)}\left(\frac{1}{\log(\lambda_{0,0}(s))} - \frac{1}{\log(\lambda_{0,0}(t))}\right)ds\\
&:= RHS_3(t)
\end{aligned}$$

We now proceed to estimate RHS_3, starting with the terms which do not involve G.

$$|-4\alpha\lambda_{0,0}'''(t)| \leq \frac{C}{t^3\log^{b+1}(t)}$$

$$\begin{aligned}
&\left|\frac{-4(1-\alpha)\lambda(t)^{-\alpha}\lambda'(t)}{\log(\lambda_{0,0}(t))}\int_t^\infty \frac{e''(s)ds}{(\lambda(t)^{1-\alpha}+s-t)^2(1+s-t)^3}\right|\\
&\leq \frac{C}{t^3\log^{b+2}(t)(\log(\log(t)))^{3/2}}
\end{aligned}$$

$$\left|\frac{-4\alpha\lambda'(t)\lambda''(t)}{\log(\lambda_{0,0}(t))\lambda(t)}\right| \leq \frac{C}{t^3\log^{b+2}(t)\log(\log(t))}$$

$$\begin{aligned}
&\left|\frac{4}{\log(\lambda_{0,0}(t))}\partial_t\left(\int_0^\infty \frac{\lambda_{0,0}''(t+w)dw}{(\lambda(t)^{1-\alpha}+w)(1+w)^3}\right)\right|\\
&\leq \frac{C}{\log(\log(t))}\left(\int_0^\infty \frac{|\lambda_{0,0}'''(t+w)|dw}{(\lambda(t)^{1-\alpha}+w)(1+w)^3} + \int_0^\infty \frac{|\lambda_{0,0}''(t+w)||\lambda(t)^{-\alpha}||\lambda'(t)|dw}{(\lambda(t)^{1-\alpha}+w)^2(1+w)^3}\right)\\
&\leq \frac{C}{t^3\log^{b+1}(t)}
\end{aligned}$$

By the same procedure used to estimate $E_{\lambda_{0,0}}$, we have

$$\frac{|E'_{\lambda_{0,0}}(t)|}{|\log(\lambda_{0,0}(t))|} \leq \frac{C}{t^3 \log^{b+1}(t)\log(\log(t))}$$

In addition,

$$\begin{aligned}
&|\frac{4\alpha(\log(\lambda_{0,0}(t)) - \log(\lambda(t)))\lambda'''(t)}{\log(\lambda_{0,0}(t))}| \\
&\leq \frac{C}{\log(\log(t))}|\log(1+\frac{e(t)}{\lambda_{0,0}(t)})| \left(\frac{1}{t^3\log^{b+1}(t)} + |e'''(t)|\right) \\
&\leq \frac{C}{(\log(\log(t)))^{3/2}}\left(\frac{1}{t^3\log^{b+1}(t)} + |e'''(t)|\right)
\end{aligned}$$

$$\begin{aligned}
&|\frac{-4}{\log(\lambda_{0,0}(t))}\int_t^\infty \frac{e'''(s)}{(1+s-t)^3}\left(\frac{1}{\lambda_{0,0}(t)^{1-\alpha}+s-t} - \frac{1}{\lambda(t)^{1-\alpha}+s-t}\right)ds| \\
&\leq \frac{C}{\log(\log(t))}\int_t^\infty \frac{|e'''(s)|}{(1+s-t)^3}\frac{|\lambda(t)^{1-\alpha}-\lambda_{0,0}(t)^{1-\alpha}|}{(\lambda_{0,0}(t)^{1-\alpha}+s-t)^2}ds \\
&\leq \frac{C\sup_{x\geq t}|e'''(x)|}{\log(\log(t))}\int_t^\infty \frac{1}{\sqrt{\log(\log(t))}}\frac{ds}{(1+s-t)^3(\lambda_{0,0}(t)^{1-\alpha}+s-t)} \\
&\leq \frac{C\sup_{x\geq t}|e'''(x)|}{\sqrt{\log(\log(t))}}
\end{aligned}$$

Similarly,

$$\begin{aligned}
&|-4\int_t^\infty \frac{e'''(s)}{(1+s-t)}\left(\frac{1}{\log(\lambda_{0,0}(s))} - \frac{1}{\log(\lambda_{0,0}(t))}\right)ds| \\
&\leq \frac{C}{t(\log(\log(t)))^2\log(t)}\int_t^\infty |e'''(s)|ds \leq \frac{C\sup_{x\geq t}\left(|e'''(x)|x^{3/2}\right)}{t^{3/2}(\log(\log(t)))^2\log(t)}
\end{aligned}$$

Finally,

$$\begin{aligned}
&|-4\int_t^\infty \frac{e'''(s)}{(1+s-t)^3(\lambda_{0,0}(t)^{1-\alpha}+s-t)}\left(\frac{1}{\log(\lambda_{0,0}(s))} - \frac{1}{\log(\lambda_{0,0}(t))}\right)ds| \\
&\leq \frac{C\sup_{x\geq t}|e'''(x)|}{t\log(t)(\log(\log(t)))^2}
\end{aligned}$$

Now, we start to estimate the terms arising from $\partial_t G(t,\lambda(t))$:

$$|\partial_t\left(\lambda(t)E_{0,1}(\lambda(t),\lambda'(t),\lambda''(t))\right)| \leq \frac{C}{t^3\log^{b+1}(t)} + C|e'''(t)|$$

$$\begin{aligned}
&\partial_t\left(-16\int_t^\infty \lambda''(s)\left(K_3(s-t,\lambda(t)) - K_{3,0}(s-t,\lambda(t))\right)ds\right) \\
&= -16\int_t^\infty \lambda'''(s)\left(K_3(s-t,\lambda(t)) - K_{3,0}(s-t,\lambda(t))\right)ds \\
&\quad -16\int_t^\infty \lambda''(s)\left(\partial_2(K_3-K_{3,0})(s-t,\lambda(t))\right)\lambda'(t)ds
\end{aligned} \tag{3.137}$$

For the first line of the right-hand side of (3.137), we have

$$\begin{aligned}&|-16\int_t^\infty \lambda'''(s)\left(K_3(s-t,\lambda(t))-K_{3,0}(s-t,\lambda(t))\right)ds|\\&\leq C\sup_{x\geq t}|\lambda'''(x)|\int_t^\infty |K_3(s-t,\lambda(t))-K_{3,0}(s-t,\lambda(t))|ds\\&\leq \frac{C}{t^3\log^{b+1}(t)}+C\sup_{x\geq t}|e'''(x)|\end{aligned}$$

where we recall that $\int_t^\infty |K_3(s-t,\lambda(t))-K_{3,0}(s-t,\lambda(t))|ds$ was estimated in the v_3 inner product subsection. For the second line of the right-hand side of (3.137), we start with

$$\begin{aligned}&K_3(w,\lambda(t))-K_{3,0}(w,\lambda(t))\\&=\left(\frac{w}{1+w^2}-\frac{w}{\lambda(t)^{2-2\alpha}+w^2}\right)\frac{w^4}{4(w^2+36\lambda(t)^2)^2}+\frac{1}{4(\lambda(t)^{1-\alpha}+w)(1+w)^3}\end{aligned}$$

Then, we get

$$\begin{aligned}|\partial_2(K_3-K_{3,0})(w,\lambda(t))|\leq{}& \frac{C\lambda(t)^\alpha}{(1+w)^3(\lambda(t)+\lambda(t)^\alpha w)^2}\\&+\frac{C\lambda(t)^{1+2\alpha}w^5}{(36\lambda(t)^2+w^2)^2(\lambda(t)^2+\lambda(t)^{2\alpha}w^2)^2}\\&+\frac{C\lambda(t)w^5}{(36\lambda(t)^2+w^2)^3}\left|\frac{1}{1+w^2}-\frac{1}{(\lambda(t)^{2-2\alpha}+w^2)}\right|\end{aligned}$$

Then, we note that

$$\int_0^\infty \frac{\lambda(t)^{-\alpha}dw}{(1+w)^3(\lambda(t)^{1-\alpha}+w)^2}\leq C\log(\log(t))\log^b(t)$$

$$\int_0^\infty \frac{\lambda(t)^{1-2\alpha}w^5dw}{(36\lambda(t)^2+w^2)^2(\lambda(t)^{2-2\alpha}+w^2)^2}\leq\frac{C}{\lambda(t)}$$

$$\begin{aligned}&\int_0^\infty \frac{\lambda(t)w^5}{(36\lambda(t)^2+w^2)^3}\left|\frac{1}{1+w^2}-\frac{1}{(\lambda(t)^{2-2\alpha}+w^2)}\right|dw\\&\leq\frac{C\log(\log(t))}{\lambda(t)}\end{aligned}$$

This gives

$$\begin{aligned}&|\partial_t\left(-16\int_t^\infty \lambda''(s)\left(K_3(s-t,\lambda(t))-K_{3,0}(s-t,\lambda(t))\right)ds\right)|\\&\leq C\sup_{x\geq t}|e'''(x)|+\frac{C}{t^3\log^{b+1}(t)}\end{aligned}$$

As mentioned before, some terms arising in $\partial_t G(t,\lambda_0(t)+e(t))$ will be treated differently, now that we have the preliminary estimate on e'''. We start with the term

$$\begin{aligned}
&\partial_t\left(\frac{16}{\lambda(t)^2}\int_t^\infty \lambda''(s)\left(K_1(s-t,\lambda(t))-\frac{\lambda(t)^2}{4(1+s-t)}\right)ds\right)\\
&=\frac{-32\lambda'(t)}{\lambda(t)^3}\int_t^\infty \lambda''(s)\left(K_1(s-t,\lambda(t))-\frac{\lambda(t)^2}{4(1+s-t)}\right)ds\\
&+\frac{16}{\lambda(t)^2}\int_t^\infty \lambda'''(s)\left(K_1(s-t,\lambda(t))-\frac{\lambda(t)^2}{4(1+s-t)}\right)ds\\
&+\frac{16}{\lambda(t)^2}\int_t^\infty \lambda''(s)\left(\partial_2K_1(s-t,\lambda(t))\lambda'(t)-\frac{\lambda(t)\lambda'(t)}{2(1+s-t)}\right)ds
\end{aligned}$$

For $s-t\leq 1$, we estimate $\partial_2 K_1$ as follows:

$$\begin{aligned}
&|\partial_2K_1(s-t,\lambda(t))|\\
&\leq\int_0^\infty \frac{R}{(1+R^2)^3}\int_0^{s-t}\frac{\rho}{(s-t)}\frac{4\lambda(t)R^2\left(1+\rho^2+R^2\lambda(t)^2\right)}{(4\lambda(t)^2R^2+(1+\rho^2-R^2\lambda(t)^2)^2)^{3/2}}d\rho dR
\end{aligned}$$

We then note that

$$4R^2\lambda(t)^2+(1+\rho^2-R^2\lambda(t)^2)^2=(1+(\rho+R\lambda(t))^2)(1+(\rho-R\lambda(t))^2)$$

So,

$$|\frac{4\lambda(t)R^2\left(1+\rho^2+R^2\lambda(t)^2\right)}{(4\lambda(t)^2R^2+(1+\rho^2-R^2\lambda(t)^2)^2)^{3/2}}|\leq CR^2\lambda(t)$$

Then, we obtain

$$|\partial_2K_1(s-t,\lambda(t))|\leq C\lambda(t)(s-t),\quad s-t\leq 1$$

From here, we get

$$\int_t^{t+1}|\lambda''(x)||\partial_2K_1(x-t,\lambda(t))-\frac{\lambda(t)}{2(1+x-t)}||\lambda'(t)|dx\leq\frac{C}{t^3\log^{3b+2}(t)}$$

For $s-t\geq 1$, we use

$$\begin{aligned}
&\partial_2K_1(s-t,\lambda(t))\\
&=\int_0^\infty\frac{R}{(1+R^2)^3}\int_0^\infty\frac{\rho}{(s-t)}\frac{4\lambda(t)R^2(1+\rho^2+\lambda(t)^2R^2)}{(4\lambda(t)^2R^2+(1+\rho^2-R^2\lambda(t)^2)^2)^{3/2}}d\rho dR\\
&-\int_0^\infty\frac{R}{(1+R^2)^3}\int_{s-t}^\infty\frac{\rho}{(s-t)}\frac{4\lambda(t)R^2(1+\rho^2+R^2\lambda(t)^2)}{(4R^2\lambda(t)^2+(1+\rho^2-R^2\lambda(t)^2)^2)^{3/2}}d\rho dR\\
&=\frac{\lambda(t)}{2(s-t)}-\int_0^\infty\frac{R}{(1+R^2)^3}\int_{s-t}^\infty\frac{\rho}{(s-t)}\frac{4\lambda(t)R^2(1+\rho^2+R^2\lambda(t)^2)}{(4R^2\lambda(t)^2+(1+\rho^2-R^2\lambda(t)^2)^2)^{3/2}}d\rho dR
\end{aligned}$$

Then,

$$\begin{aligned}
&\int_{t+1}^{\infty} |\lambda''(s)||\partial_2 K_1(s-t,\lambda(t)) - \frac{\lambda(t)}{2(1+s-t)}|ds|\lambda'(t)|\\
&\leq \frac{C}{t^3\log^{2b+2}(t)} \int_{t+1}^{\infty} |\partial_2 K_1(s-t,\lambda(t)) - \frac{\lambda(t)}{2(1+s-t)}|ds\\
&\leq \frac{C}{t^3\log^{2b+2}(t)} \int_{1}^{\infty} \left(\frac{\lambda(t)}{2w} - \frac{\lambda(t)}{2(1+w)}\right) dw\\
&+ \frac{C}{t^3\log^{2b+2}(t)} \int_{0}^{\infty} \frac{R}{(1+R^2)^3} \int_{1}^{\infty} \rho\log(\rho) \frac{4\lambda(t)R^2(1+\rho^2+R^2\lambda(t)^2)}{(4R^2\lambda(t)^2+(1+\rho^2-R^2\lambda(t)^2)^2)^{3/2}} d\rho dR\\
&\leq \frac{C}{t^3\log^{3b+2}(t)} + \frac{C}{t^3\log^{3b+2}(t)} \int_{0}^{\infty} \frac{R^3}{(1+R^2)^3} \log(2+R\lambda(t))dR\\
&\leq \frac{C}{t^3\log^{3b+2}(t)}
\end{aligned}$$

We also have

$$\begin{aligned}
&|\frac{-32\lambda'(t)}{\lambda(t)^3} \int_{t}^{\infty} \lambda''(s) \left(K_1(s-t,\lambda(t)) - \frac{\lambda(t)^2}{4(1+s-t)}\right) ds|\\
&\leq \frac{C}{t^3\log^{2-b}(t)} \int_{t}^{\infty} |K_1(s-t,\lambda(t)) - \frac{\lambda(t)^2}{4(1+s-t)}|ds\\
&\leq \frac{C}{t^3\log^{b+2}(t)}
\end{aligned}$$

Finally, we have

$$\begin{aligned}
&|\frac{16}{\lambda(t)^2} \int_{t}^{\infty} \lambda'''(x) \left(K_1(x-t,\lambda(t)) - \frac{\lambda(t)^2}{4(1+x-t)}\right) dx|\\
&\leq \frac{C}{\lambda(t)^2} \sup_{x\geq t}(|\lambda'''(x)|) \int_{t}^{\infty} |K_1(x-t,\lambda(t)) - \frac{\lambda(t)^2}{4(1+x-t)}|dx\\
&\leq C\sup_{x\geq t} |\lambda'''(x)|
\end{aligned}$$

Combining these, we get

$$\begin{aligned}
&|\partial_t \left(\frac{16}{\lambda(t)^2} \int_{t}^{\infty} \lambda''(s) \left(K_1(s-t,\lambda(t)) - \frac{\lambda(t)^2}{4(1+s-t)}\right) ds\right)|\\
&\leq \frac{C}{t^3\log^{b+1}(t)} + C\sup_{x\geq t} |e'''(x)|
\end{aligned}$$

Next, we consider

$$\begin{aligned}
&\partial_t \left(\frac{16}{\lambda(t)^2} \int_{t}^{\infty} \lambda''(x)K(x-t,\lambda(t))dx\right)\\
&= \frac{-32\lambda'(t)}{\lambda(t)^3} \int_{t}^{\infty} \lambda''(x)K(x-t,\lambda(t))dx + \frac{16}{\lambda(t)^2} \int_{t}^{\infty} \lambda'''(x)K(x-t,\lambda(t))dx\\
&+ \frac{16}{\lambda(t)^2} \int_{t}^{\infty} \lambda''(x)\partial_2 K(x-t,\lambda(t))\lambda'(t)dx
\end{aligned}$$

Note that

$$|\frac{-32\lambda'(t)}{\lambda(t)^3}\int_t^\infty \lambda''(x)K(x-t,\lambda(t))dx| \leq \frac{C\log^{3b}(t)}{t^3\log^{2b+2}(t)}\int_t^\infty |K(x-t,\lambda(t))|dx \leq \frac{C}{t^3\log^{b+2}(t)}$$

Next,

$$|\frac{16}{\lambda(t)^2}\int_t^\infty \lambda'''(x)K(x-t,\lambda(t))dx| \leq C\sup_{x\geq t}|\lambda'''(x)| \leq \frac{C}{t^3\log^{b+1}(t)} + C\sup_{x\geq t}|e'''(x)|$$

Finally, the integral

$$\frac{16}{\lambda(t)^2}\int_t^\infty \lambda''(s)\partial_2 K(s-t,\lambda(t))\lambda'(t)ds$$

was treated during the preliminary estimates (it is equal to $\lambda(t)\cdot I$ in the notation of that section). So, in total, we get

$$|\partial_t\left(\frac{16}{\lambda(t)^2}\int_t^\infty \lambda''(x)K(x-t,\lambda(t))dx\right)| \leq \frac{C}{t^3\log^{b+1}(t)} + C\sup_{x\geq t}|e'''(x)|$$

Now, we return to (3.77) to prove an estimate on $\partial_t v_3$ which will be sufficient to estimate $\partial_t v_4$. We start with

(3.138)

$$\begin{aligned}&\partial_t v_3(t,r)\\ &= \frac{-1}{r}\int_0^\infty\int_0^w \frac{\rho}{\sqrt{w^2-\rho^2}}\lambda'''(w+t)\left(\frac{-1-\rho^2+r^2}{\sqrt{(-1-\rho^2+r^2)^2+4r^2}} + F_3(r,\rho,\lambda(t+w))\right)d\rho dw\\ &-\frac{1}{r}\int_0^\infty\int_0^w \frac{\rho}{\sqrt{w^2-\rho^2}}\lambda''(t+w)\partial_3 F_3(r,\rho,\lambda(t+w))\lambda'(t+w)d\rho dw\end{aligned}$$

The first line of (3.138) is treated in the same manner as v_3 was, while obtaining (3.77). In particular, the analog of $v_{3,2}$ which appeared in an intermediate step in obtaining (3.77) is estimated by

$$\begin{aligned}&Cr\sup_{x\geq t}|\lambda'''(x)| + Cr\sup_{x\geq t}\left(|\lambda'''(x)|\lambda(x)^{\alpha-1}\left(\lambda(x)^{\alpha-1}-\lambda(t)^{\alpha-1}\right)\right)\lambda(t)^{2-2\alpha}\\ &\leq \frac{Cr}{t^3\log^{b+1}(t)} + \frac{Cr}{t\log^{b-b\alpha}(t)}\sup_{x\geq t}\left(\frac{|e'''(x)|x}{\lambda(x)^{1-\alpha}}\right)\end{aligned}$$

while the analog of $v_{3,1}$ is estimated by

$$Cr\sup_{x\geq t}\left(\frac{|\lambda'''(x)|}{\lambda(x)^{2-2\alpha}}\right)\log(\log(t))\lambda(t)^{2-2\alpha}$$

To treat the second line of (3.138), we start with

$$|\partial_3 F_3(r,\rho,\lambda(s))| \leq \frac{Cr^2\lambda(s)^{2\alpha-3}}{(1+\lambda(s)^{4\alpha-4}(\rho^2-r^2)^2+2\lambda(s)^{2\alpha-2}(\rho^2+r^2))}$$

Then, we get

$$\begin{aligned}
&|\frac{-1}{r}\int_t^\infty \int_0^{s-t} \frac{\rho}{(s-t)}\lambda''(s)\partial_3 F_3(r,\rho,\lambda(s))\lambda'(s)d\rho ds| \\
&\leq \frac{C}{r}\int_t^{t+1} \frac{1}{(s-t)}\frac{1}{s^3\log^{2b+2}(s)}\int_0^{s-t}\rho\frac{r^2\lambda(s)^{2\alpha-3}}{(1+2\lambda(s)^{2\alpha-2}\rho^2)}d\rho ds \\
&+\frac{C}{r}\int_{t+1}^\infty \frac{1}{s^3\log^{2b+2}(s)(s-t)}\int_0^{s-t}\rho|\partial_3 F_3(r,\rho,\lambda(s))|d\rho ds \\
&\leq \frac{Cr}{t^3\log^{b+1}(t)}
\end{aligned}$$

and

$$\begin{aligned}
&|\frac{-1}{r}\int_t^\infty \int_0^{s-t}\rho\left(\frac{1}{\sqrt{(s-t)^2-\rho^2}}-\frac{1}{(s-t)}\right)\lambda''(s)\partial_3 F_3(r,\rho,\lambda(s))\lambda'(s)d\rho ds| \\
&\leq Cr\int_0^\infty \frac{\rho d\rho}{(1+\lambda(t)^{4\alpha-4}(\rho^2-r^2)^2+2\lambda(t)^{2\alpha-2}(\rho^2+r^2))}\frac{\log^{(3-2\alpha)b}(t)}{t^3\log^{2b+2}(t)} \\
&\leq \frac{Cr}{t^3\log^{b+2}(t)}
\end{aligned}$$

Combining these, we get

$$|\partial_t v_3(t,r)| \leq \frac{Cr\log(\log(t))}{t^3\log^{b+1}(t)}+\frac{Cr}{t}\sup_{x\geq t}\left(\frac{x|e'''(x)|}{\lambda(x)^{2-2\alpha}}\right)\log(\log(t))\lambda(t)^{2-2\alpha} \tag{3.139}$$

Next, we estimate $\partial_t E_5$. First, we recall the definition of E_5:

$$\begin{aligned}
&E_5(t,r) \\
&=-\frac{1}{r}\int_t^{t+6r}\int_0^{s-t}\frac{\rho}{(s-t)}\lambda''(s)\left(\frac{-1-\rho^2+r^2}{\sqrt{(-1-\rho^2+r^2)^2+4r^2}}+F_3(r,\rho,\lambda(s))\right)d\rho ds \\
&-\frac{1}{r}\int_{6r}^\infty\left(\int_0^w \frac{\rho}{w}\lambda''(t+w)\left(\frac{-1-\rho^2+r^2}{\sqrt{(-1-\rho^2+r^2)^2+4r^2}}+F_3(r,\rho,\lambda(t+w))\right)d\rho\right. \\
&\left.-\frac{\lambda''(t+w)}{w}r^2w^2\left(\frac{1}{1+w^2}-\frac{1}{(\lambda(t+w)^{2-2\alpha}+w^2)}\right)\right)dw \\
&-r\int_{t+6r}^\infty \lambda''(s)(s-t)\left(\frac{1}{(\lambda(t)^{2-2\alpha}+(s-t)^2)}-\frac{1}{\lambda(s)^{2-2\alpha}+(s-t)^2}\right)ds \\
&+v_{3,2}(t,r)
\end{aligned} \tag{3.140}$$

We then estimate the time derivative of each of the lines of (3.140). For the first line, we have

$$\begin{aligned}&|\partial_t\left(-\frac{1}{r}\int_0^{6r}\int_0^w \frac{\rho}{w}\lambda''(t+w)\left(\frac{-1-\rho^2+r^2}{\sqrt{(-1-\rho^2+r^2)^2+4r^2}}+F_3(r,\rho,\lambda(t+w))\right)d\rho dw\right)|\\&\leq \frac{C}{r}\int_0^{6r}\int_0^w \frac{\rho}{w}|\lambda'''(t+w)|\cdot|\frac{-1-\rho^2+r^2}{\sqrt{(-1-\rho^2+r^2)^2+4r^2}}+F_3(r,\rho,\lambda(t+w))|d\rho dw\\&+\frac{C}{r}\int_0^{6r}\int_0^w \frac{\rho}{w}|\lambda''(t+w)\partial_3F_3(r,\rho,\lambda(t+w))\lambda'(t+w)|d\rho dw\\&\leq \frac{C}{r}\int_0^{6r}\int_0^w \frac{\rho}{w}|\lambda'''(t+w)|d\rho dw+\frac{Cr}{t^3\log^{b+1}(t)}\end{aligned}$$

where we note that the third line of the inequality above has been estimated already. Next, we have

(3.141)

$$\begin{aligned}&\partial_t\left(-\frac{1}{r}\int_{6r}^{\infty}\left(\left(\int_0^w \frac{\rho}{w}\lambda''(t+w)\left(\frac{-1-\rho^2+r^2}{\sqrt{(-1-\rho^2+r^2)^2+4r^2}}+F_3(r,\rho,\lambda(t+w))\right)d\rho\right.\right.\right.\\&\left.\left.\left.-\frac{\lambda''(t+w)}{w}r^2w^2\left(\frac{1}{1+w^2}-\frac{1}{(\lambda(t+w)^{2-2\alpha}+w^2)}\right)\right)dw\right)\\&=\frac{1}{r}\int_{6r}^{\infty}\lambda''(t+w)r^2w\left(\frac{(2-2\alpha)}{(\lambda(t+w)^{2-2\alpha}+w^2)^2}\lambda(t+w)^{1-2\alpha}\lambda'(t+w)\right)dw\\&-\frac{1}{r}\int_{6r}^{\infty}\int_0^w \frac{\rho}{w}\lambda'''(t+w)\left(\frac{-1-\rho^2+r^2}{\sqrt{(-1-\rho^2+r^2)^2+4r^2}}+F_3(r,\rho,\lambda(t+w))\right)d\rho dw\\&+\frac{1}{r}\int_{6r}^{\infty}\lambda'''(t+w)r^2w\left(\frac{1}{1+w^2}-\frac{1}{(\lambda(t+w)^{2-2\alpha}+w^2)}\right)dw\\&-\frac{1}{r}\int_{6r}^{\infty}\int_0^w \frac{\rho}{w}\lambda''(t+w)\partial_3F_3(r,\rho,\lambda(t+w))\lambda'(t+w)d\rho dw\end{aligned}$$

We get

$$\begin{aligned}&|\frac{1}{r}\int_{6r}^{\infty}\lambda''(t+w)r^2w\left(\frac{(2-2\alpha)}{(\lambda(t+w)^{2-2\alpha}+w^2)^2}\lambda(t+w)^{1-2\alpha}\lambda'(t+w)\right)dw|\\&\leq Cr\int_{6r}^{\infty}\frac{|\lambda''(t+w)|w}{(\lambda(t+w)^{2-2\alpha}+w^2)^2}\lambda(t+w)^{1-2\alpha}|\lambda'(t+w)|dw\\&\leq \frac{Cr}{t^3\log^{b+2}(t)}\end{aligned}$$

By the identical procedure used to estimate E_4 in the v_3 subsection, we have

$$\begin{aligned}&|-\frac{1}{r}\int_{6r}^{\infty}\int_0^w \frac{\rho}{w}\lambda'''(t+w)\left(\frac{-1-\rho^2+r^2}{\sqrt{(-1-\rho^2+r^2)^2+4r^2}}+F_3(r,\rho,\lambda(t+w))\right)d\rho dw\\&+\frac{1}{r}\int_{6r}^{\infty}\lambda'''(t+w)r^2w\left(\frac{1}{1+w^2}-\frac{1}{(\lambda(t+w)^{2-2\alpha}+w^2)}\right)dw|\\&\leq \frac{Cr}{t^3\log^{b+1}(t)}+Cr\sup_{x\geq t}|e'''(x)|\end{aligned}$$

We then note that the last line of (3.141) was already estimated, and is bounded above in absolute value by

$$\frac{Cr}{t^3\log^{b+1}(t)}$$

Then, we estimate

$$\begin{aligned}&|\partial_t\left(r\int_{t+6r}^{\infty}\lambda''(s)(s-t)\left(\frac{1}{(\lambda(t)^{2-2\alpha}+(s-t)^2)}-\frac{1}{\lambda(s)^{2-2\alpha}+(s-t)^2)}\right)ds\right)|\\&\leq Cr\int_{6r}^{\infty}|\lambda'''(t+w)|\frac{|\lambda(t+w)^{2-2\alpha}-\lambda(t)^{2-2\alpha}|}{(\lambda(t+w)^{2-2\alpha}+w^2)^2}wdw\\&+Cr|\int_{6r}^{\infty}\lambda''(t+w)w\partial_t\left(\frac{1}{(\lambda(t)^{2-2\alpha}+w^2)}-\frac{1}{\lambda(t+w)^{2-2\alpha}+w^2}\right)dw|\end{aligned}$$

The second line of the above expression is estimated by

$$\begin{aligned}&r\int_{6r}^{\infty}|\lambda'''(t+w)|\frac{|\lambda(t+w)^{2-2\alpha}-\lambda(t)^{2-2\alpha}|}{(\lambda(t+w)^{2-2\alpha}+w^2)^2}wdw\\&\leq Cr\int_{6r}^{\infty}|\lambda'''(t+w)|\frac{\lambda(t)^{1-2\alpha}|\lambda'(t)|w^2}{(\lambda(t+w)^{2-2\alpha}+w^2)^2}dw\\&\leq\frac{Cr}{t\log^{3b+1-3b\alpha}(t)}\left(\frac{1}{t^3\log^{1-b+2\alpha b}(t)}+\sup_{x\geq t}\left(\frac{|e'''(x)|}{\lambda(x)^{2-2\alpha}}\right)\right)\end{aligned}$$

On the other hand, we have

$$\begin{aligned}&|r\int_{6r}^{\infty}\lambda''(t+w)w\partial_t\left(\frac{1}{(\lambda(t)^{2-2\alpha}+w^2)}-\frac{1}{\lambda(t+w)^{2-2\alpha}+w^2}\right)dw|\\&\leq\frac{Cr}{t^3\log^{b+2}(t)}\end{aligned}$$

Finally, we recall

$$\begin{aligned}v_{3,2}(t,r)=\frac{-1}{r}\int_t^{\infty}\int_0^{s-t}\rho&\left(\frac{1}{\sqrt{(s-t)^2-\rho^2}}-\frac{1}{(s-t)}\right)\lambda''(s)\\&\left(\frac{-1-\rho^2+r^2}{\sqrt{(-1-\rho^2+r^2)^2+4r^2}}+F_3(r,\rho,\lambda(s))\right)d\rho ds\end{aligned}$$

After taking the time derivative, we estimate the following term with the same argument used for $v_{3,2}$.

$$\begin{aligned}&|\frac{-1}{r}\int_0^{\infty}\int_0^{w}\rho\left(\frac{1}{\sqrt{w^2-\rho^2}}-\frac{1}{w}\right)\lambda'''(t+w)\left(\frac{-1-\rho^2+r^2}{\sqrt{(-1-\rho^2+r^2)^2+4r^2}}+F_3(r,\rho,\lambda(t+w))\right)d\rho dw|\\&\leq\frac{Cr}{t^3\log^{b+1}(t)}+\frac{Cr\sup_{x\geq t}\left(\frac{|e'''(x)|x}{\lambda(x)^{3-2\alpha}}\right)}{t\log^{3b-2b\alpha}(t)}\end{aligned}$$

Finally,

$$\begin{aligned}&|\frac{-1}{r}\int_0^{\infty}\int_0^{w}\rho\left(\frac{1}{\sqrt{w^2-\rho^2}}-\frac{1}{w}\right)\lambda''(t+w)\partial_3F_3(r,\rho,\lambda(t+w))\lambda'(t+w)d\rho dw|\\&\leq\frac{Cr}{t^3\log^{b+2}(t)}\end{aligned}$$

where we note that the integral has already been estimated while studying $\partial_t v_3$.

In total, we get

$$|\partial_t E_5(t,r)| \leq \frac{Cr}{t^3 \log^{b+1}(t)} + \frac{Cr \sup_{x\geq t}\left(\frac{|e'''(x)|x}{\lambda(x)^{3-2\alpha}}\right)}{t \log^{(3-2\alpha)b}(t)}$$

Now, we will obtain an estimate on $\partial_t v_4$ which is better than the preliminary one, (3.129). In particular, now that we have the preliminary estimate on λ''', we can use the same procedure used to estimate $\partial_t v_3$, to see that $\partial_t v_1$ solves, with 0 Cauchy data at infinity, the same equation as v_1 does, except with λ'' replaced by λ'''. This observation, combined with the estimates on v_1 gives

$$|\partial_t v_1(t,r)| \leq \frac{Cr\left(\log(t)+\log(3+2r)\right)}{t} \sup_{x\geq t}\left(|\lambda'''(x)|x\right) \tag{3.142}$$

(Note that we can not directly use the $\partial_t v_1$ analog of (3.12), because the preliminary estimate on $|e'''|$ is not good enough to justify the steps which would produce such an estimate).

Now, we use the estimate (3.139), combined with the above estimate on $\partial_t v_1$, and the previous estimate on $\partial_t v_2$, to get

$$\begin{aligned}|\partial_t v_{4,c}(t,r)| \leq& \frac{C|\chi_{\geq 1}'(\frac{2r}{\log^N(t)})|}{t\log(t)}\left(\frac{\lambda(t)^2}{(\lambda(t)^2+r^2)^2}|v_1+v_2+v_3| + |F_{0,2}(t,r)|\right)\\ &+\frac{C\chi_{\geq 1}(\frac{2r}{\log^N(t)})\lambda(t)|\lambda'(t)|}{(r^2+\lambda(t)^2)^2}|v_1+v_2+v_3|\\ &+\frac{C\chi_{\geq 1}(\frac{2r}{\log^N(t)})\lambda(t)^2|\partial_t(v_1+v_2+v_3)|}{(\lambda(t)^2+r^2)^2}\\ &+C\chi_{\geq 1}(\frac{2r}{\log^N(t)})|\partial_t F_{0,2}(t,r)|\end{aligned}$$

Using $|\chi_{\geq 1}'(\frac{2r}{\log^N(t)})| \cdot \frac{r}{t\log^{N+1}(t)} \leq C\frac{\mathbb{1}_{\{r\geq \frac{\log^N(t)}{2}\}}}{t\log(t)}$, we get

$$\begin{aligned}|\partial_t v_{4,c}(t,r)| \leq& \frac{C\mathbb{1}_{\{r\geq\frac{\log^N(t)}{2}\}}}{\log^{2b}(t)r^4}\begin{cases}\frac{r}{t^3\log^b(t)}, & r\leq \frac{t}{2}\\ \frac{\log(r)}{|t-r|}\left(\frac{1}{|t-r|}+\frac{1}{t\log(t)}\right), & t>r>\frac{t}{2}\end{cases}\\ &+\frac{C\mathbb{1}_{\{r\geq\frac{\log^N(t)}{2}\}}}{r^3t^3}\left(\frac{\log(t)+\log(r)}{\log^{3b+1}(t)}\right)\\ &+C\mathbb{1}_{\{r\geq\frac{\log^N(t)}{2}\}}\frac{\lambda(t)^{2-2\alpha}(\log(t)+\log(r))}{r^3t\log^{2b}(t)}\sup_{x\geq t}\left(\frac{|e'''(x)|x}{\lambda(x)^{2-2\alpha}}\right)\end{aligned}$$

Then, we note that

$$\begin{aligned}&\partial_t v_4(t,r)\\ &= \frac{-1}{2\pi}\int_0^\infty\int_0^w \frac{\rho}{\sqrt{w^2-\rho^2}}\int_0^{2\pi}\frac{\partial_1 v_{4,c}(t+w,\sqrt{r^2+\rho^2+2r\rho\cos(\theta)})}{\sqrt{r^2+\rho^2+2r\rho\cos(\theta)}}(r+\rho\cos(\theta))d\theta d\rho dw\end{aligned}$$

and, carry out the identical procedure done to obtain the estimate on $v_4^{\lambda_1}-v_4^{\lambda_2}$. This results in

$$|\partial_t v_4(t,r)| \leq \frac{C}{t^3\log^{3b+N-2}(t)} + \frac{C\lambda(t)^{2-2\alpha}}{t\log^{2b+N-3}(t)}\sup_{x\geq t}\left(\frac{x|e'''(x)|}{\lambda(x)^{2-2\alpha}}\right), \quad r\leq\frac{t}{2}$$

Like previously, in order to estimate $\partial_t v_4$ in the region $r \geq \frac{t}{2}$, we first record a slightly different estimate on $\partial_t v_{4,c}$. To obtain this, we use (3.50) in the region $t-\sqrt{t} \leq r \leq t+\sqrt{t}$, and (3.51) for the other parts of the region $r \geq \frac{t}{2}$ to estimate v_2. For $\partial_t v_2$, we use (3.50) in the region $t-t^{1/4} \leq r \leq t+t^{1/4}$, and (3.51) in the other parts of the region $r \geq \frac{t}{2}$.

This leads to

$$\begin{aligned}|\partial_t v_{4,c}(t,r)| \leq & \frac{C\mathbb{1}_{\{r\geq \frac{\log^N(t)}{2}\}}}{\log^{2b}(t) r^4}\\ & \cdot \begin{cases} \frac{r}{t^3 \log^b(t)}, & r \leq \frac{t}{2} \\ \frac{\log(r)}{|t-r|}\left(\frac{1}{|t-r|}+\frac{1}{t\log(t)}\right), & \frac{t}{2} \leq r \leq t-\sqrt{t}, \text{ or } r \geq t+\sqrt{t} \\ \frac{\log(r)}{(t-r)^2}+\frac{1}{\sqrt{rt}\log(t)}, & t-\sqrt{t} \leq r \leq t-t^{1/4}, \text{ or } t+t^{1/4} \leq r \leq t+\sqrt{t} \\ \frac{1}{\sqrt{r}}, & t-t^{1/4} \leq r \leq t+t^{1/4} \end{cases}\\ & + \frac{C\mathbb{1}_{\{r\geq \frac{\log^N(t)}{2}\}}}{r^3 t^3}\left(\frac{\log(t)+\log(r)}{\log^{3b+1}(t)}\right)\\ & + C\mathbb{1}_{\{r\geq \frac{\log^N(t)}{2}\}} \frac{\lambda(t)^{2-2\alpha}(\log(t)+\log(r))}{r^3 t\log^{2b}(t)} \sup_{x\geq t}\left(\frac{|e'''(x)|x}{\lambda(x)^{2-2\alpha}}\right)\end{aligned}$$

First, we obtain

$$||\partial_t v_{4,c}||_{L^2(rdr)} \leq \frac{C}{t^3 \log^{3b+2N}(t)} + \frac{C\sup_{x\geq t}\left(\frac{x|e'''(x)|}{\lambda(x)^{2-2\alpha}}\right)}{t\log^{4b-2b\alpha+2N-1}(t)} \tag{3.143}$$

Then, we use the same procedure used to estimate $v_4^{\lambda_1}-v_4^{\lambda_2}$ in the region $r \geq \frac{t}{2}$, and get

$$|\partial_t v_4(t,r)| \leq \frac{C}{t^2\log^{3b+2N}(t)} + \frac{C\sup_{x\geq t}\left(\frac{x|e'''(x)|}{\lambda(x)^{2-2\alpha}}\right)}{\log^{4b-2\alpha b+2N-2}(t)}, \quad r \geq \frac{t}{2} \tag{3.144}$$

Now, we recall (3.127), and prove new estimates on the terms involving $v_4, E_5, v_1+v_2+v_3$, and $F_{0,2}$

$$\begin{aligned}&|\partial_t\langle\left(\frac{\cos(2Q_{\frac{1}{\lambda(t)}})-1}{r^2}\right)v_4\left(1-\chi_{\geq 1}(\frac{4r}{t})\right)|_{r=R\lambda(t)}, \phi_0\rangle|\\ &\leq \frac{C}{t^3\log^{b+N-2}(t)} + \frac{C\sup_{x\geq t}\left(\frac{x|e'''(x)|}{\lambda(x)^{2-2\alpha}}\right)}{t\log^{N-3+2b-2b\alpha}(t)}\end{aligned}$$

$$|\partial_t\langle\left(\frac{\cos(2Q_{\frac{1}{\lambda(t)}})-1}{r^2}\right)E_5|_{r=R\lambda(t)}, \phi_0\rangle| \leq \frac{C}{t^3\log(t)} + \frac{C\sup_{x\geq t}\left(\frac{x|e'''(x)|}{\lambda(x)^{3-2\alpha}}\right)}{\log^{(2-2\alpha)b}(t)t}$$

$$\begin{aligned}&|\partial_t\langle\left(\frac{\cos(2Q_{\frac{1}{\lambda(t)}})-1}{r^2}\right)\chi_{\geq 1}(\frac{2r}{\log^N(t)})\left(v_1+v_2+v_3\right)|_{r=R\lambda(t)}, \phi_0\rangle|\\ &\leq \frac{C}{t^3\log^{2b+2N}(t)} + \frac{C\lambda(t)^{2-2\alpha}\sup_{x\geq t}\left(\frac{|e'''(x)|x}{\lambda(x)^{2-2\alpha}}\right)}{t\log^{2N+b-1}(t)}\end{aligned}$$

$$|\partial_t\langle\chi_{\geq 1}(\frac{2r}{\log^N(t)})F_{0,2}|_{r=R\lambda(t)}, \phi_0\rangle| \leq \frac{C}{t^3\log^{2b+1-2b\alpha+2N}(t)} + \frac{C|e'''(t)|}{\log^{b-2b\alpha+2N}(t)}$$

From our previous estimates, we also have

$$\begin{aligned}
&|\lambda'(t)||\langle\left(\frac{\cos(2Q_{\frac{1}{\lambda(t)}})-1}{r^2}\right)\left((v_4+v_5)\left(1-\chi_{\geq 1}(\frac{4r}{t})\right)+E_5\right)|_{r=R\lambda(t)},\phi_0\rangle| \\
&+|\lambda'(t)||\langle\left(\frac{\cos(2Q_{\frac{1}{\lambda(t)}})-1}{r^2}\right)\chi_{\geq 1}(\frac{2r}{\log^N(t)})\,(v_1+v_2+v_3)\,|_{r=R\lambda(t)},\phi_0\rangle| \\
&+|\lambda'(t)||\langle\chi_{\geq 1}(\frac{2r}{\log^N(t)})F_{0,2}|_{r=R\lambda(t)},\phi_0\rangle| \\
&\leq \frac{C}{t^3\log^{b+2}(t)}
\end{aligned}$$

Combining these, we get

$$\begin{aligned}
&|\partial_t\left(\lambda(t)\langle\left(\frac{\cos(2Q_{\frac{1}{\lambda(t)}})-1}{r^2}\right)\left((v_4+v_5)\left(1-\chi_{\geq 1}(\frac{4r}{t})\right)+E_5\right)|_{r=R\lambda(t)},\phi_0\rangle\right)| \\
&+|\partial_t\left(\lambda(t)\langle\left(\frac{\cos(2Q_{\frac{1}{\lambda(t)}})-1}{r^2}\right)\chi_{\geq 1}(\frac{2r}{\log^N(t)})\,(v_1+v_2+v_3)\,|_{r=R\lambda(t)},\phi_0\rangle\right)| \\
&+|\partial_t\left(\lambda(t)\langle\chi_{\geq 1}(\frac{2r}{\log^N(t)})F_{0,2}|_{r=R\lambda(t)},\phi_0\rangle\right)| \\
&\leq \frac{C}{t^3\log^{b+1}(t)}+\frac{C\sup_{x\geq t}\left(\frac{x|e'''(x)|}{\lambda(x)^{3-2\alpha}}\right)}{t\log^{(3-2\alpha)b}(t)}
\end{aligned}$$

where we use (3.130) to estimate $\partial_t v_5$. Finally, for all remaining terms in RHS_3, we use the same estimate that was used in obtaining the preliminary estimate on e''', and conclude

$$|RHS_3(t)|\leq \frac{C}{t^3\log^{b+1}(t)}+\frac{C\sup_{x\geq t}\left(x^{3/2}|e'''(x)|\right)}{t^{3/2}\sqrt{\log(\log(t))}}$$

Note that the left-hand side of (3.136) is the same as that of (3.66), except with e'' replaced with e''', and λ_0 replaced with $\lambda_{0,0}$. This will not cause any major differences in the study of (3.136) relative to (3.66), because the key estimates (3.69) and (3.71) are invariant under multiplication of $K(t,s)$ by any non-negative function of s, and (3.69) is still true with the replacement of $\frac{1}{(\lambda_0(t)^{1-\alpha}+s-t)}$ by $\frac{1}{(\lambda_{0,0}(t)^{1-\alpha}+s-t)}$. In particular, if we define

$$K_2(t,s):=\frac{\mathbb{1}_{\leq 0}(s-t)}{\alpha|\log(\lambda_{0,0}(-s))|}\left(\frac{1}{1-s+t}+\frac{1}{(1-s+t)^3(\lambda_{0,0}(-t)^{1-\alpha}-s+t)}\right)$$

then, by the above discussion, the resolvent kernel associated to K_2 in the same way that r was associated to K, exists, and satisfies the same estimate as r: (3.72). Let us denote this resolvent kernel by r_2.

So far, we have that e''' is a solution to (3.136), which can be re-cast into the form

$$K_2 * x + x = H_2$$

similarly to (3.66). Next, we carry out the following computation, noting that the step which requires Fubini's theorem is justified by the preliminary estimate on e''',

(3.135).

$$\begin{aligned} r_2 * (K_2 * x) + r_2 * x &= r_2 * H_2 \\ (r_2 * K_2) * x + r_2 * x &= r_2 * H_2 \\ H_2 - x = K_2 * x &= r_2 * H_2 \end{aligned}$$

where we used the equation

$$r_2 + r_2 * K_2 = K_2$$

(Recall that r_2 solves the r_2 and K_2 analogs of (3.70)). Translating x and H_2 back to e''' and RHS_3, we have

$$e'''(t) = \frac{RHS_3(t)}{4\alpha} - \int_t^\infty \frac{RHS_3(z)}{4\alpha} r_2(-t,-z)dz, \quad \text{a.e. } t \geq T_0 \tag{3.145}$$

e''', RHS_3 are continuous functions, and the above equation can be rearranged to yield

$$\int_t^\infty \frac{RHS_3(z)}{4\alpha} r_2(-t,-z)dz = \frac{RHS_3(t)}{4\alpha} - e'''(t), \quad \text{a.e. } t \geq T_0$$

So,

$$t \mapsto \int_t^\infty \frac{RHS_3(z)}{4\alpha} r_2(-t,-z)dz$$

agrees with a continuous function almost everywhere, and can thus be extended to a continuous function. Moreover, using (3.72), for r_2 instead of r, we have

$$|\int_t^\infty \frac{RHS_3(z)}{4\alpha} r_2(-t,-z)dz| \leq ||\frac{RHS_3}{2\alpha}||_{L^\infty([t,\infty))}, \quad \text{a.e. } t \geq T_0$$

So, the same estimate holds for all $t \geq T_0$, where we now identify

$$t \mapsto \int_t^\infty \frac{RHS_3(z)}{4\alpha} r_2(-t,-z)dz$$

with its continuous extension described above. Such an identification will be performed without further mention. Returning to (3.145), we get

$$\begin{aligned} |e'''(t)| &\leq C \sup_{x \geq t} |RHS_3(x)| \\ &\leq \frac{C}{t^3 \log^{b+1}(t)} + \frac{C \sup_{x\geq t} \left(x^{3/2}|e'''(x)|\right)}{t^{3/2}\sqrt{\log(\log(t))}} \end{aligned}$$

Thus,

$$t^{3/2}|e'''(t)| \leq \frac{C}{t^{3/2}\log^{b+1}(t)} + \frac{C \sup_{x\geq t} \left(x^{3/2}|e'''(x)|\right)}{\sqrt{\log(\log(t))}}$$

But, as mentioned earlier, e''' is a continuous function on $[T_0, \infty)$, so $t \mapsto t^{3/2}|e'''(t)|$ is also continuous on $[T_0, \infty)$, and

$$|e'''(t)|t^{3/2} \to 0, \quad t \to \infty$$

by the preliminary estimate on e'''. So, for all $t \geq T_0$, there exists $y(t) \geq t$ such that

$$\sup_{x \geq t} \left(x^{3/2}|e'''(x)|\right) = y(t)^{3/2}|e'''(y(t))|$$

But, then,

$$\begin{aligned}\sup_{x\geq t}\left(x^{3/2}|e'''(x)|\right) &= y(t)^{3/2}|e'''(y(t))|\\ &\leq \frac{C}{y(t)^{3/2}\log^{b+1}(y(t))}+\frac{C\sup_{x\geq y(t)}\left(x^{3/2}|e'''(x)|\right)}{\sqrt{\log(\log(y(t))}}\\ &\leq \frac{C}{t^{3/2}\log^{b+1}(t)}+\frac{C\sup_{x\geq t}\left(x^{3/2}|e'''(x)|\right)}{\sqrt{\log(\log(t))}}\end{aligned}$$

So, there exists some absolute constants $C_{p_2}, C_{p_1} > 0$ such that, for all

$$t \geq C_{p_1} + T_0$$

we have

$$\sup_{x\geq t}\left(x^{3/2}|e'''(x)|\right) \leq \frac{C_{p2}}{t^{3/2}\log^{b+1}(t)}$$

But, $e''' \in C([T_0,\infty))$, so, there exists some absolute constant $C_{p_3} > 0$ such that

$$|e'''(t)| \leq \frac{C_{p3}}{t^3\log^{b+1}(t)}, \quad t \geq T_0$$

Recalling $\lambda(t) = \lambda_{0,0}(t) + e(t)$, we have

$$|\lambda'''(t)| \leq \frac{C}{t^3\log^{b+1}(t)}, \quad t \geq T_0 \tag{3.146}$$

To finish, we only need to establish the estimates on $\partial_t\partial_r^j v_k$ in the proposition statement. We recall that $\partial_t v_1$ solves the same equation as v_1, also with 0 Cauchy data at infinity, except with λ'' on the right-hand side replaced with λ'''. Now that we have established (3.146), we can justify the steps leading to the $\partial_t v_1$ analog of (3.12), which gives

$$|\partial_t v_1(t,r)| \leq \frac{C}{rt\log^{b+1}(t)}, \quad r > \frac{t}{2}$$

Combined with (3.142), this gives (3.118).

A similar large r estimate can be also proven for $\partial_t v_3$: Starting with (3.138), we get

$$\begin{aligned}&|\partial_t v_3(t,r)|\\ &\leq \frac{C}{r}\int_t^\infty |\lambda'''(s)|(s-t)ds\\ &+\frac{C}{r}\int_t^\infty\int_0^{s-t}\frac{\rho}{\sqrt{(s-t)^2-\rho^2}}\frac{|\lambda''(s)||\lambda'(s)|r^2\lambda(s)^{2\alpha-3}}{(1+\lambda(s)^{4\alpha-4}(\rho^2-r^2)^2+2\lambda(s)^{2\alpha-2}(\rho^2+r^2))}d\rho ds\\ &\leq \frac{C}{rt\log^{b+1}(t)}, \quad r > \frac{t}{2}\end{aligned}$$

Combining this with (3.139) gives (3.119).

Next, we estimate $\partial_{tr} v_k, \quad k = 1, 3$. We start with $\partial_{tr} v_3$:

(3.147)

$$\partial_t v_3(t, r)$$
$$= \frac{-1}{r} \int_t^\infty \int_0^{s-t} \frac{\rho}{\sqrt{(s-t)^2 - \rho^2}} \lambda'''(s) \left(\frac{-1 - \rho^2 + r^2}{\sqrt{(1 + \rho^2 - r^2)^2 + 4r^2}} + F_3(r, \rho, \lambda(s)) \right) d\rho ds$$
$$- \frac{1}{r} \int_t^\infty \int_0^{s-t} \frac{\rho}{\sqrt{(s-t)^2 - \rho^2}} \lambda''(s) \partial_3 F_3(r, \rho, \lambda(s)) \lambda'(s) d\rho ds$$

Since the first line of (3.147) is the same expression as v_3, except with λ''' replacing λ'', we use the same procedure for this line, as was used to estimate $\partial_r v_3$. For the second line of (3.147), we start by noting that

$$|\partial_r \left(\frac{\partial_3 F_3(r, \rho, \lambda(s))}{r} \right)|$$
$$\leq \frac{C\lambda(s)^{2\alpha+1}(\lambda(s)^6 + (\rho^2 + r^2)\lambda(s)^{4+2\alpha} + (\rho^2 + r^2)^2 \lambda(s)^{2+4\alpha})}{\lambda(s)^{10}(1 + (\rho^2 - r^2)^2 \lambda(s)^{4\alpha-4} + 2(\rho^2 + r^2)\lambda(s)^{2\alpha-2})^{5/2}}$$
$$+ \frac{C\lambda(s)^{8\alpha+1}(\rho^2 - r^2)^2(\rho^2 + r^2)}{\lambda(s)^{10}(1 + (\rho^2 - r^2)^2 \lambda(s)^{4\alpha-4} + 2(\rho^2 + r^2)\lambda(s)^{2\alpha-2})^{5/2}}$$

We then estimate the partial r derivative of the second line of (3.147) using the same procedure used before to estimate $\partial_r v_3$, and get

$$|\partial_r \left(-\frac{1}{r} \int_t^\infty \int_0^{s-t} \frac{\rho}{\sqrt{(s-t)^2 - \rho^2}} \lambda''(s) \partial_3 F_3(r, \rho, \lambda(s)) \lambda'(s) d\rho ds \right)| \leq \frac{C}{t^3 \log^{1+b}(t)}$$

(Note that we have the factor $|\lambda'(s)| \leq \frac{C}{s \log^{b+1}(s)}$ in the integrand of the second line of (3.147), which explains the gain relative to the analogous term arising in the $\partial_r v_3$ estimates). In total, we get (3.120).

As observed before, $\partial_t v_1$ has the same exact representation formulae as v_1, except with an extra derivative on λ''. Therefore, we can use the identical procedure used to estimate $\partial_r v_1$, to get (3.121).

This gives

(3.148)

$$|\partial_{tr} v_{4,c}(t, r)| \leq \frac{C|\chi''_{\geq 1}(\frac{2r}{\log^N(t)})|}{r^3 t^3 \log^{1+N+3b}(t)} + \frac{C|\chi'_{\geq 1}(\frac{2r}{\log^N(t)})|}{r^3 t^3 \log^{3b+N}(t)}$$
$$+ C\chi_{\geq 1}(\frac{2r}{\log^N(t)}) \begin{cases} \frac{1}{r^4 t^3 \log^{3b}(t)}, & r \leq \frac{t}{2} \\ \frac{\log(r)}{\log^{2b}(t) r^4} \left(\frac{1}{\log(t) r|t-r|t} + \frac{1}{(t-r)^2 t} + \frac{1}{(t-r)^3} \right) \\ + \frac{1}{t^3 r^4 \log^{3b}(t)}, & t > r > \frac{t}{2} \end{cases}$$

Using the final estimates on λ''', we also get

$$\frac{|\partial_t v_{4,c}(t, r)|}{r} \leq \frac{C \mathbb{1}_{\{r \geq \frac{\log^N(t)}{2}\}}}{r^4 t^3 \log^{3b+1-2\alpha b}(t)}$$
$$+ \frac{C \mathbb{1}_{\{r \geq \frac{\log^N(t)}{2}\}}}{\log^{2b}(t) r^5} \begin{cases} \frac{r}{t^3 \log^b(t)}, & r \leq \frac{t}{2} \\ \frac{\log(r)}{|t-r|} \left(\frac{1}{t \log(t)} + \frac{1}{|t-r|} \right) + \frac{1}{rt \log^{b+1}(t)}, & t > r \geq \frac{t}{2} \end{cases}$$

Now, we can prove (3.122) in the region $r \leq \frac{t}{2}$, by writing

$$\partial_t v_4(t,r)$$
$$= \frac{-r}{2\pi} \int_t^\infty \int_0^{s-t} \frac{\rho}{\sqrt{(s-t)^2-\rho^2}} \int_0^1 \int_0^{2\pi} \partial_r$$
$$\left(\frac{\partial_1 v_{4,c}(s, \sqrt{r^2+\rho^2+2r\rho\cos(\theta)})}{\sqrt{r^2+\rho^2+2r\rho\cos(\theta)}}(r+\rho\cos(\theta))\right)(s,\rho,\beta r) d\theta d\beta d\rho ds$$

and using the same procedure used for v_4. In total, we get

$$|\partial_t v_4(t,r)| \leq \frac{Cr}{t^3 \log^{3b+2N-2}(t)}, \quad r \leq \frac{t}{2} \tag{3.149}$$

Combining this with (3.144) (and the final estimate on λ''') gives (3.122). □

3.0.8.11. *Estimating λ''''.*

PROPOSITION 3.4. *$\lambda \in C^4([T_0,\infty))$ and we have the following estimates:*

$$|\lambda''''(t)| \leq \frac{C}{t^4 \log^{b+1}(t)}, \quad t \geq T_0$$

$$|\partial_t^2 v_1(t,r)| \leq \begin{cases} \frac{Cr}{t^4 \log^b(t)}, & r \leq \frac{t}{2} \\ \frac{C}{rt^2 \log^{b+1}(t)}, & r > \frac{t}{2} \end{cases} \tag{3.150}$$

$$|\partial_t^2 v_3(t,r)| \leq \frac{Cr\log(\log(t))}{t^4 \log^{b+1}(t)}$$

$$|\partial_t^2 v_4(t,r)| \leq \begin{cases} \frac{C}{t^4 \log^{3b-2+N}(t)}, & r \leq \frac{t}{2} \\ \frac{C}{t^{35/12} \log^{2b-1}(t)}, & r > \frac{t}{2} \end{cases} \tag{3.151}$$

$$|\partial_t^2 v_5(t,r)| \leq \frac{C}{t^4 \log^{3N+b-2}(t)}, \quad r \leq \frac{t}{2} \tag{3.152}$$

PROOF. Recalling that $\lambda(t) = \lambda_{0,0}(t) + e(t)$, we will show that $e \in C^4([T_0,\infty))$, and estimate e''''. Returning to (3.123), we have

$$\lambda'''(t) = \frac{RHS_2'(t)}{g_2(t)} - \frac{g_2'(t)}{g_2(t)^2} RHS_2(t)$$

Because $\lambda''' \in C^0([T_0,\infty))$, an inspection of the definition of RHS_2 shows that $RHS_2 \in C^2([T_0,\infty))$; so, $\lambda''' \in C^1([T_0,\infty))$, with

$$\lambda''''(t) = \partial_t \left(\frac{RHS_2'(t)}{g_2(t)} - \frac{g_2'(t)}{g_2(t)^2} RHS_2(t)\right)$$

The previously obtained estimates on RHS_2, RHS_2' then show that

$$|\lambda''''(t)| \leq \frac{C|RHS_2''(t)|}{\log(\log(t))\log^b(t)} + \frac{C}{t^3 \log^{2-2b}(t)\log(\log(t))} \tag{3.153}$$

Again, we will first obtain a preliminary estimate on λ'''', which will be improved afterwards. We start by recalling the definition of RHS_2:

$$\begin{aligned}RHS_2(t) = &-\frac{16}{\lambda(t)^3}\int_t^\infty \lambda''(s)\left(K_1(s-t,\lambda(t))+K(s-t,\lambda(t))\right)ds + 2\frac{(\lambda'(t))^2}{\lambda(t)^2}\\
&+\frac{4b}{\lambda(t)t^2\log^b(t)} + E_{v_2,ip}(t,\lambda(t))\\
&+\langle\left(\frac{\cos(2Q_{\frac{1}{\lambda(t)}})-1}{r^2}\right)\left(v_3+(v_4+v_5)\left(1-\chi_{\geq 1}(\frac{4r}{t})\right)\right)|_{r=R\lambda(t)},\phi_0\rangle\\
&-\langle\left(\frac{\cos(2Q_{\frac{1}{\lambda(t)}})-1}{r^2}\right)\chi_{\geq 1}\left(\frac{2r}{\log^N(t)}\right)(v_1+v_2+v_3)\,|_{r=R\lambda(t)},\phi_0\rangle\\
&-4\int_0^\infty \chi_{\geq 1}(\frac{2R\lambda(t)}{\log^N(t)})\frac{(\lambda'(t))^2R^2\phi_0(R)dR}{\lambda(t)^2(R^2+1)^2}\end{aligned}$$

We then note

$$|\partial_t^2\left(\frac{2\lambda'(t)^2}{\lambda(t)^2}\right)| \leq \frac{C}{t^4\log^2(t)}$$

$$|\partial_t^2\left(\frac{4b}{\lambda(t)t^2\log^b(t)}\right)| \leq \frac{C}{t^4}$$

$$|\partial_t^2\left(\frac{-4\lambda'(t)^2}{\lambda(t)^2}\int_0^\infty \chi_{\geq 1}(\frac{2R\lambda(t)}{\log^N(t)})\frac{R^2\phi_0(R)dR}{(R^2+1)^2}\right)| \leq \frac{C}{t^4\log^{2+2b+2N}(t)}$$

To continue estimating RHS_2'', we start with

$$\begin{aligned}&\partial_t^2\left(\frac{16}{\lambda(t)^2}\int_t^\infty \lambda''(x)K_1(x-t,\lambda(t))dx\right)\\
&=\left(\frac{96}{\lambda(t)^4}\lambda'(t)^2-\frac{32\lambda''(t)}{\lambda(t)^3}\right)\int_t^\infty \lambda''(x)K_1(x-t,\lambda(t))dx\\
&-\frac{64\lambda'(t)}{\lambda(t)^3}\int_t^\infty \lambda'''(x)K_1(x-t,\lambda(t))dx\\
&-\frac{64\lambda'(t)}{\lambda(t)^3}\int_t^\infty \lambda''(x)\partial_2K_1(x-t,\lambda(t))\lambda'(t)dx\\
&-\frac{16}{\lambda(t)^2}\int_t^\infty \lambda'''(x)\partial_1K_1(x-t,\lambda(t))dx\\
&+\frac{32}{\lambda(t)^2}\int_t^\infty \lambda'''(x)\partial_2K_1(x-t,\lambda(t))\lambda'(t)dx\\
&+\frac{16}{\lambda(t)^2}\int_t^\infty \lambda''(x)\left(\partial_2^2K_1(x-t,\lambda(t))\lambda'(t)^2+\partial_2K_1(x-t,\lambda(t))\lambda''(t)\right)dx\end{aligned} \tag{3.154}$$

The only term in (3.154) which we can not immediately estimate from previous estimates is the one involving $\partial_2^2K_1$. The analogous term involving K is also present in $RHS_3''(t)$. We start with

$$\begin{aligned}&|\partial_{22}K(x,\lambda(t))(\lambda'(t))^2+\partial_2K(x,\lambda(t))\lambda''(t)|\\
&\leq \int_0^\infty \frac{Cr\lambda(t)^2\left(\lambda'(t)^2+\lambda(t)\lambda''(t)\right)}{(r^2+\lambda(t)^2)^3}\int_0^x\rho\left(\frac{1}{\sqrt{x^2-\rho^2}}-\frac{1}{x}\right)\left(1+\frac{r^2-1-\rho^2}{\sqrt{(r^2-1-\rho^2)^2+4r^2}}\right)d\rho dr\end{aligned}$$

which gives

$$
\begin{aligned}
&|\frac{16}{\lambda(t)^2}\int_t^\infty \lambda''(x)\left(\partial_2^2 K(x-t,\lambda(t))\lambda'(t)^2+\partial_2 K(x-t,\lambda(t))\lambda''(t)\right)dx|\\
&\leq \frac{C}{t^4\log^{3b+2}(t)}\int_t^\infty\int_0^\infty \frac{r}{(r^2+\lambda(t)^2)^3}\int_0^{x-t}\rho\\
&\qquad\left(\frac{1}{\sqrt{(x-t)^2-\rho^2}}-\frac{1}{x-t}\right)\left(1+\frac{r^2-1-\rho^2}{\sqrt{(r^2-1-\rho^2)^2+4r^2}}\right)d\rho dr dx\\
&\leq \frac{C}{t^4\log^{3b+2}(t)}\int_0^\infty \frac{r}{(r^2+\lambda(t)^2)^3}\int_0^\infty \rho\\
&\qquad\left(1+\frac{r^2-1-\rho^2}{\sqrt{(r^2-1-\rho^2)^2+4r^2}}\right)\int_{\rho+t}^\infty\left(\frac{1}{\sqrt{(x-t)^2-\rho^2}}-\frac{1}{x-t}\right)dx d\rho dr\\
&\leq \frac{C}{t^4\log^{3b+2}(t)}\int_0^\infty \frac{r^3}{(r^2+\lambda(t)^2)^3}dr\\
&\leq \frac{C}{t^4\log^{2+b}(t)}
\end{aligned}
$$

Then, we treat the K_1 term:

$$K_1(x,\lambda(t))=\int_0^\infty \frac{r}{\lambda(t)^2(1+\frac{r^2}{\lambda(t)^2})^3}\int_0^x \frac{\rho}{x}\left(1+\frac{r^2-1-\rho^2}{\sqrt{(r^2-1-\rho^2)^2+4r^2}}\right)d\rho dr$$

Proceeding as for K, we get

$$
\begin{aligned}
&|\frac{16}{\lambda(t)^2}\int_t^\infty \lambda''(x)\left(\partial_2^2 K_1(x-t,\lambda(t))\lambda'(t)^2+\partial_2 K_1(x-t,\lambda(t))\lambda''(t)\right)dx|\\
&\leq C\int_t^\infty |\lambda''(x)|\frac{1}{t^2\log^{2b+1}(t)}\int_0^\infty \frac{r}{(r^2+\lambda(t)^2)^3}\frac{1}{(x-t)}\int_0^{x-t}\rho\left(1+\frac{r^2-1-\rho^2}{\sqrt{(r^2-1-\rho^2)^2+4r^2}}\right)d\rho dr dx\\
&\leq C\int_t^\infty |\lambda''(x)|\frac{1}{t^2\log^{2b+1}(t)}\int_0^\infty \frac{r}{(r^2+\lambda(t)^2)^3}\frac{1}{(x-t)}\begin{cases}\int_0^{x-t}2\rho d\rho, & x-t\leq 1\\ \int_0^\infty \rho\left(1+\frac{r^2-1-\rho^2}{\sqrt{(r^2-1-\rho^2)^2+4r^2}}\right)d\rho, & x-t\geq 1\end{cases} dr dx\\
&\leq C\int_t^{t+1}\frac{(x-t)}{x^2\log^{b+1}(x)t^2\log^{2b+1}(t)}\frac{dx}{\lambda(t)^4}\\
&+C\int_{t+1}^\infty \frac{dx}{x^2\log^{b+1}(x)(x-t)t^2\log(t)}\\
&\leq \frac{C}{t^4\log^{1-b}(t)}
\end{aligned}
$$

Combining this with estimates for the other terms in (3.154) (which are deduced from the procedure used in obtaining preliminary estimates on e'''), we get

$$
\begin{aligned}
&|\partial_t^2\left(\frac{16}{\lambda(t)^2}\int_t^\infty \lambda''(x)K_1(x-t,\lambda(t))dx+\frac{16}{\lambda(t)^2}\int_t^\infty \lambda''(x)K(x-t,\lambda(t))dx\right)|\\
&\leq \frac{C}{t^3\log^{1-2b}(t)}
\end{aligned}
$$

Next, we consider terms arising in $\partial_t^2\left(\lambda(t)E_{v_2,ip}(t,\lambda(t))\right)$.

$$\begin{aligned}
&\partial_t^2\left(2c_b\lambda(t)\int_0^\infty \frac{\sin(t\xi)}{t^2}\psi_{v_2}(\xi,\lambda(t))d\xi\right)\\
&= 2c_b\lambda''(t)\int_0^\infty \frac{\sin(t\xi)}{t^2}\psi_{v_2}(\xi,\lambda(t))d\xi + 4c_b\lambda'(t)\int_0^\infty \frac{-2\sin(t\xi)}{t^3}\psi_{v_2}(\xi,\lambda(t))d\xi\\
&+ 4c_b\lambda'(t)\int_0^\infty \frac{\xi\cos(t\xi)}{t^2}\psi_{v_2}(\xi,\lambda(t))d\xi + 4c_b\lambda'(t)\int_0^\infty \frac{\sin(t\xi)}{t^2}\partial_2\psi_{v_2}\lambda'(t)d\xi\\
&+ 4c_b\lambda(t)\int_0^\infty \frac{-2\xi\cos(t\xi)}{t^3}\psi_{v_2}(\xi,\lambda(t))d\xi + 2c_b\lambda(t)\int_0^\infty \frac{6\sin(t\xi)}{t^4}\psi_{v_2}(\xi,\lambda(t))d\xi\\
&+ 4c_b\lambda(t)\int_0^\infty \frac{-2\sin(t\xi)}{t^3}\partial_2\psi_{v_2}\lambda'(t)d\xi\\
&- 2c_b\lambda(t)\int_0^\infty \frac{\xi^2\sin(t\xi)}{t^2}\psi_{v_2}(\xi,\lambda(t))d\xi + 4c_b\lambda(t)\int_0^\infty \frac{\xi\cos(t\xi)}{t^2}\partial_2\psi_{v_2}\lambda'(t)d\xi\\
&+ 2c_b\lambda(t)\int_0^\infty \frac{\sin(t\xi)}{t^2}\left(\partial_{22}\psi_{v_2}\lambda'(t)^2 + \partial_2\psi_{v_2}\lambda''(t)\right)d\xi
\end{aligned}$$

Only three integrals can not be immediately estimated based on our previous estimates. These three integrals are estimated using the same procedure used before, by estimating $K_1(x)$ and various of its derivatives, using the fact that K_1 appears in $\psi_{v_2}(\xi\lambda(t))$ with its argument $x=\xi\lambda(t)$ satisfying $0<x<\frac{1}{4}$. We have

$$-2c_b\lambda(t)\int_0^\infty \frac{\xi^2\sin(t\xi)}{t^2}\psi_{v_2}(\xi,\lambda(t))d\xi = \frac{2c_b\lambda(t)}{t^5}\int_0^\infty \cos(t\xi)\partial_\xi^3\left(\xi^2\psi_{v_2}(\xi,\lambda(t))\right)d\xi$$

$$4c_b\lambda(t)\int_0^\infty \frac{\xi\cos(t\xi)}{t^2}\partial_2\psi_{v_2}\lambda'(t)d\xi = -4c_b\lambda(t)\int_0^\infty \frac{\cos(t\xi)}{t^4}\partial_\xi^2\left(\xi\partial_2\psi_{v_2}\right)\lambda'(t)d\xi$$

and

$$2c_b\lambda(t)\int_0^\infty \frac{\sin(t\xi)}{t^2}\left(\partial_{22}\psi_{v_2}\lambda'(t)^2\right)d\xi = 2c_b\lambda(t)\int_0^\infty \frac{\cos(t\xi)}{t^3}\partial_{122}\psi_{v_2}(\xi,\lambda(t))\lambda'(t)^2d\xi$$

Using these, and the procedure described above, we get

$$\begin{aligned}
&|-2c_b\lambda(t)\int_0^\infty \frac{\xi^2\sin(t\xi)}{t^2}\psi_{v_2}(\xi,\lambda(t))d\xi| + |4c_b\lambda(t)\int_0^\infty \frac{\xi\cos(t\xi)}{t^2}\partial_2\psi_{v_2}\lambda'(t)d\xi|\\
&+ |2c_b\lambda(t)\int_0^\infty \frac{\sin(t\xi)}{t^2}\left(\partial_{22}\psi_{v_2}\lambda'(t)^2\right)d\xi|\\
&\leq \frac{C}{t^5}
\end{aligned}$$

Every other term arising in

$$\partial_t^2\left(2c_b\lambda(t)\int_0^\infty \frac{\sin(t\xi)}{t^2}\psi_{v_2}(\xi,\lambda(t))d\xi\right)$$

can be estimated by using estimates for $\lambda,\lambda',\lambda''$, and the terms estimated when considering

$$\partial_t\left(2c_b\lambda(t)\int_0^\infty \frac{\sin(t\xi)}{t^2}\psi_{v_2}(\xi,\lambda(t))d\xi\right)$$

In total, we then get

$$|\partial_t^2\left(2c_b\lambda(t)\int_0^\infty \frac{\sin(t\xi)}{t^2}\psi_{v_2}(\xi,\lambda(t))d\xi\right)| \leq \frac{C}{t^5}$$

For the term

$$\partial_t^2\left(2c_b\lambda(t)\int_0^\infty \frac{\sin(t\xi)}{t^2}\chi_{\leq\frac{1}{4}}(\xi)F_{v_2}(\xi,\lambda(t))d\xi\right)$$

we use the identical procedure as above, noting that the only difference is that $\psi_{v_2}(\xi,\lambda(t))$ is replaced here by $\chi_{\leq\frac{1}{4}}(\xi)F_{v_2}(\xi,\lambda(t))$. Then, we get

$$\begin{aligned}
&|-2c_b\lambda(t)\int_0^\infty \frac{\xi^2\sin(t\xi)}{t^2}\chi_{\leq\frac{1}{4}}(\xi)F_{v_2}(\xi,\lambda(t))d\xi|\\
&+|4c_b\lambda(t)\int_0^\infty \frac{\xi\cos(t\xi)}{t^2}\chi_{\leq\frac{1}{4}}(\xi)\partial_2\left(F_{v_2}(\xi,\lambda(t))\right)\lambda'(t)d\xi|\\
&+|2c_b\lambda(t)\int_0^\infty \frac{\sin(t\xi)}{t^2}\left(\chi_{\leq\frac{1}{4}}(\xi)\partial_{22}\left(F_{v_2}(\xi,\lambda(t))\right)\lambda'(t)^2\right)d\xi|\\
&\leq \frac{C\log(\log(t))}{t^5\log^{2b}(t)}
\end{aligned}$$

In total, we have

$$|\partial_t^2\left(2c_b\lambda(t)\int_0^\infty \frac{\sin(t\xi)}{t^2}\chi_{\leq\frac{1}{4}}(\xi)F_{v_2}(\xi,\lambda(t))d\xi\right)|\leq \frac{C\log(\log(t))}{t^5\log^{2b}(t)}$$

The next term to consider is

$$\begin{aligned}
&\partial_t^2\left(\int_0^{\frac{1}{2}}\left(\chi_{\leq\frac{1}{4}}(\xi)-1\right)\frac{\sin(t\xi)}{t^2}\left(\frac{b-1}{\xi\log^b(\frac{1}{\xi})}+\frac{b(b-1)}{\xi\log^{b+1}(\frac{1}{\xi})}\right)d\xi\right)\\
&=\frac{4}{t^4}\left(\frac{b-1}{\log^b(2)}+\frac{b(b-1)}{\log^{b+1}(2)}\right)\sin(\frac{t}{2})+\mathrm{Err}\\
&-\left(\frac{2(b-1)}{\log^b(2)}+\frac{2b(b-1)}{\log^{b+1}(2)}\right)\frac{\cos(\frac{t}{2})}{4t^3}+\frac{\sin(\frac{t}{2})}{t^4}\left(\partial_\xi\left(\frac{\xi(b-1)}{\log^b(\frac{1}{\xi})}+\frac{\xi b(b-1)}{\log^{b+1}(\frac{1}{\xi})}\right)|_{\xi=\frac{1}{2}}\right)\\
&+E_2
\end{aligned}$$

where we integrate by parts after differentiating the integral, and

$$|\mathrm{Err}|\leq\frac{C}{t^5},\quad |E_2|\leq\frac{C}{t^5}$$

On the other hand, we have

$$\begin{aligned}
&\partial_t^2\left(\int_0^{\frac{t}{2}}\frac{\sin(u)b(b-1)du}{t^2u\log^{b+1}(\frac{t}{u})}\right)\\
&=\frac{\cos(\frac{t}{2})b(b-1)}{2t^3\log^{b+1}(2)}-\frac{3\sin(\frac{t}{2})b(b-1)}{t^4\log^{b+1}(2)}\\
&+\frac{\sin(\frac{t}{2})b(b-1)}{t}\left(\frac{-2}{t^3\log^{b+1}(2)}-\frac{(b+1)}{t^3\log^{b+2}(2)}\right)\\
&+\int_0^{\frac{t}{2}}\frac{\sin(u)b(b-1)}{u}\left(\frac{6}{t^4\log^{b+1}(\frac{t}{u})}+\frac{5(b+1)}{t^4\log^{b+2}(\frac{t}{u})}+\frac{(b+2)(b+1)}{t^4\log^{b+3}(\frac{t}{u})}\right)du
\end{aligned}$$

So,

$$\begin{aligned}
&\partial_t^2\left(\int_0^{\frac{1}{2}}\left(\chi_{\leq\frac{1}{4}}(\xi)-1\right)\frac{\sin(t\xi)}{t^2}\left(\frac{b-1}{\xi\log^b(\frac{1}{\xi})}+\frac{b(b-1)}{\xi\log^{b+1}(\frac{1}{\xi})}\right)d\xi\right)\\
&+\partial_t^2\left(\int_0^{\frac{t}{2}}\frac{\sin(u)(b-1)}{t^2u}\left(\frac{b}{\log^{b+1}(\frac{t}{u})}+\frac{1}{\log^b(\frac{t}{u})}\right)du\right)\\
&=O\left(\frac{1}{t^5}\right)+\int_0^{\frac{t}{2}}\frac{\sin(u)(b-1)}{u}\left(\frac{6}{t^4\log^b(\frac{t}{u})}+\frac{5b}{t^4\log^{b+1}(\frac{t}{u})}+\frac{b(b+1)}{t^4\log^{b+2}(\frac{t}{u})}\right)du\\
&+\int_0^{\frac{t}{2}}\frac{\sin(u)b(b-1)}{u}\left(\frac{6}{t^4\log^{b+1}(\frac{t}{u})}+\frac{5(b+1)}{t^4\log^{b+2}(\frac{t}{u})}+\frac{(b+1)(b+2)}{t^4\log^{b+3}(\frac{t}{u})}\right)du\\
&=O\left(\frac{1}{t^5}\right)+\frac{3(b-1)\pi}{t^4\log^b(t)}+O\left(\frac{1}{t^4\log^{b+1}(t)}\right)
\end{aligned}$$

where we used the computations from the section which constructed v_2, to obtain the $\frac{1}{t^4\log^b(t)}$ term above. We conclude the following estimate on one of the terms arising in $\partial_t^2\left(\lambda(t)E_{v_2,ip}(t,\lambda(t))\right)$:

$$\begin{aligned}
&|\partial_t^2\left(\int_0^{\frac{1}{2}}\left(\chi_{\leq\frac{1}{4}}(\xi)-1\right)\frac{\sin(t\xi)}{t^2}\left(\frac{b-1}{\xi\log^b(\frac{1}{\xi})}+\frac{b(b-1)}{\xi\log^{b+1}(\frac{1}{\xi})}\right)d\xi\right)\\
&+\partial_t^2\left(\int_0^{\frac{t}{2}}\frac{\sin(u)(b-1)}{t^2u}\left(\frac{b}{\log^{b+1}(\frac{t}{u})}+\frac{1}{\log^b(\frac{t}{u})}\right)du-\left(\frac{(b-1)\pi}{2t^2\log^b(t)}\right)\right)|\\
&\leq\frac{C}{t^4\log^{b+1}(t)}
\end{aligned}$$

In total, we finally get

$$|\partial_t^2\left(\lambda(t)E_{v_2,ip}(t,\lambda(t))\right)|\leq\frac{C}{t^4\log^{b+1}(t)}$$

By the same procedure, this estimate is also true for the case $b=1$.

Now, we record preliminary estimates on $\partial_t^2 v_k,\quad k=1,3,4,5$

LEMMA 3.15 (Preliminary estimates on $\partial_t^2 v_k$). *We have the following* preliminary *estimates on* $\partial_t^2 v_k$*:*

$$|\partial_t^2 v_1(t,r)|\leq\frac{Cr}{(1+r)}\frac{1}{t^3\log^{b+1}(t)}\tag{3.155}$$

$$|\partial_t^2 v_3(t,r)|\leq\frac{Cr}{t^4\log^{b+1}(t)}+\frac{C}{t^3\log^{1+2\alpha b-b}(t)}\tag{3.156}$$

$$|\partial_t^2 v_4(t,r)|\leq\begin{cases}\frac{C}{t^3\log^{b+2N-2}(t)}, & r\leq\frac{t}{2}\\ \frac{C}{t^2\log^{3N+b}(t)}, & r\geq\frac{t}{2}\end{cases}\tag{3.157}$$

$$|\partial_t^2 v_5(t,r)|\leq\frac{C}{t^4\log^{3N+b-2}(t)},\quad r\leq\frac{t}{2}\tag{3.158}$$

PROOF. First, we estimate $\partial_t^2 v_3$. We start with the formula for $\partial_t v_3$ used in the process of proving the final estimate on λ''':

(3.159)

$$\begin{aligned}&\partial_t v_3(t,r)\\&=\frac{-1}{r}\int_t^\infty\int_0^{s-t}\frac{\rho}{\sqrt{(s-t)^2-\rho^2}}\lambda'''(s)\left(\frac{-1-\rho^2+r^2}{\sqrt{(1+\rho^2-r^2)^2+4r^2}}+F_3(r,\rho,\lambda(s))\right)d\rho ds\\&-\frac{1}{r}\int_0^\infty\int_0^w\frac{\rho}{\sqrt{w^2-\rho^2}}\lambda''(t+w)\partial_3F_3(r,\rho,\lambda(t+w))\lambda'(t+w)d\rho dw\end{aligned}$$

Denote the first line on the right-hand side of (3.159) by $v_{3,1,t}$. Then, we have

(3.160)

$$\begin{aligned}&\partial_t v_{3,1,t}(t,r)\\&=\frac{1}{r}\int_t^\infty\int_0^{s-t}\frac{\rho}{(s-t)\sqrt{(s-t)^2-\rho^2}}\lambda'''(s)\left(\frac{-1-\rho^2+r^2}{\sqrt{(1+\rho^2-r^2)^2+4r^2}}+F_3(r,\rho,\lambda(s))\right)d\rho ds\\&-\frac{1}{r}\int_t^\infty\int_0^{s-t}\frac{\rho}{(s-t)\sqrt{(s-t)^2-\rho^2}}\lambda'''(s)\left(\frac{8\rho^2r^2}{(4r^2+(1+\rho^2-r^2)^2)^{3/2}}\right.\\&+\left.\frac{-8\lambda(s)^{4\alpha-4}\rho^2r^2}{(4\lambda(s)^{2\alpha-2}r^2+(1+\lambda(s)^{2\alpha-2}(\rho^2-r^2))^2)^{3/2}}\right)d\rho ds\end{aligned}$$

We will need to use a more complicated procedure for (3.160) than what was used to treat the analogous term which arose in the course of obtaining (3.128). For the first line on the right-hand side of (3.160), we start with

$$\begin{aligned}&|\frac{-1-\rho^2+r^2}{\sqrt{(1+\rho^2-r^2)^2+4r^2}}+F_3(r,\rho,\lambda(s))|\\&\leq\begin{cases}2, & \rho\leq 4r\\ |\frac{-1-\rho^2+r^2}{\sqrt{(1+\rho^2-r^2)^2+4r^2}}+1|+|-1+F_3(r,\rho,\lambda(s))|, & \rho>4r\end{cases}\\&\leq\begin{cases}2, & \rho\leq 4r\\ \frac{r^2}{\rho^2}, & \rho>4r\end{cases}\end{aligned}$$

Then, we have

$$\begin{aligned}&|\frac{1}{r}\int_t^\infty\int_0^{s-t}\frac{\rho}{(s-t)\sqrt{(s-t)^2-\rho^2}}\lambda'''(s)\left(\frac{-1-\rho^2+r^2}{\sqrt{(1+\rho^2-r^2)^2+4r^2}}+F_3(r,\rho,\lambda(s))\right)d\rho ds|\\&\leq\frac{C}{r}\int_0^\infty\frac{\rho}{t^3\log^{b+1}(t)}\left(\begin{cases}2, & \rho\leq 4r\\ \frac{r^2}{\rho^2}, & \rho>4r\end{cases}\right)\int_{\rho+t}^\infty\frac{dsd\rho}{(s-t)\sqrt{(s-t)^2-\rho^2}}\\&\leq\frac{C}{r}\int_0^{4r}\frac{d\rho}{t^3\log^{b+1}(t)}+\frac{C}{r}\int_{4r}^\infty\frac{r^2d\rho}{\rho^2t^3\log^{b+1}(t)}\\&\leq\frac{C}{t^3\log^{b+1}(t)}\end{aligned}$$

We then consider the second line on the right-hand side of (3.160). First, note that, because $\lambda'(x)\leq 0,\quad x\geq T_0$, we have $\lambda(s)^{2\alpha-2}\geq\lambda(t)^{2\alpha-2},\quad s\geq t$. Then, we start with the case $r\leq\frac{\lambda(t)^{1-\alpha}}{4}\leq\frac{1}{4}$. We have $1+\rho^2\geq 16r^2$, and

$1+\lambda(t)^{2\alpha-2}\rho^2 \geq 16\lambda(t)^{2\alpha-2}r^2$. Whence, we have

$$\frac{1}{(4r^2+(1-r^2+\rho^2)^2)^{3/2}} \leq \frac{C}{(1+\rho^2)^3}$$

$$\frac{1}{(4\lambda(t)^{2\alpha-2}r^2+(1+\lambda(t)^{2\alpha-2}\rho^2-\lambda(t)^{2\alpha-2}r^2)^2)^{3/2}} \leq \frac{C}{(1+\rho^2\lambda(t)^{2\alpha-2})^3}$$

So,

$$\begin{aligned}
&|-\frac{1}{r}\int_t^\infty \int_0^{s-t} \frac{\rho}{(s-t)\sqrt{(s-t)^2-\rho^2}}\lambda'''(s)\left(\frac{8\rho^2r^2}{(4r^2+(1+\rho^2-r^2)^2)^{3/2}}\right.\\
&\left.+\frac{-8\lambda(s)^{4\alpha-4}\rho^2r^2}{(4\lambda(s)^{2\alpha-2}r^2+(1+\lambda(s)^{2\alpha-2}(\rho^2-r^2))^2)^{3/2}}\right)d\rho ds|\\
&\leq \frac{C}{r}\int_0^\infty \frac{1}{t^3\log^{b+1}(t)}\left(\frac{\rho^2r^2}{(1+\rho^2)^3}+\frac{\rho^2r^2}{(1+\lambda(t)^{2\alpha-2}\rho^2)^3\log^{4\alpha b-4b}(t)}\right)d\rho\\
&\leq \frac{Cr}{t^3\log^{1+b\alpha}(t)}, \quad r \leq \frac{\lambda(t)^{1-\alpha}}{4}
\end{aligned}$$

If $r \geq \frac{\lambda(t)^{1-\alpha}}{4}$, then,

$$\begin{aligned}
&|-\frac{1}{r}\int_t^\infty \int_0^{s-t} \frac{\rho}{(s-t)\sqrt{(s-t)^2-\rho^2}}\lambda'''(s)\left(\frac{8\rho^2r^2}{(4r^2+(1+\rho^2-r^2)^2)^{3/2}}\right.\\
&\left.+\frac{-8\lambda(s)^{4\alpha-4}\rho^2r^2}{(4\lambda(s)^{2\alpha-2}r^2+(1+\lambda(s)^{2\alpha-2}(\rho^2-r^2))^2)^{3/2}}\right)d\rho ds|\\
&\leq \frac{C}{r}\left(\frac{r^2}{t^3\log^{b+1}(t)}\int_0^{4r}\frac{\rho^2 d\rho}{(4r^2+(1-r^2+\rho^2)^2)^{3/2}}\right.\\
&\left.+\frac{r^2}{t^3\log^{b+1}(t)\log^{4\alpha b-4b}(t)}\int_0^{4r}\frac{\rho^2 d\rho}{(4\lambda(t)^{2\alpha-2}r^2+(1+\lambda(t)^{2\alpha-2}(\rho^2-r^2))^2)^{3/2}}\right)\\
&+\frac{C}{rt^3\log^{b+1}(t)}\int_{4r}^\infty\left(\frac{\rho^2r^2}{\rho^6}+\frac{\rho^2r^2}{\log^{4\alpha b-4b}(t)\lambda(t)^{6\alpha-6}\rho^6}\right)d\rho\\
&\leq \frac{C}{r}\left(\frac{r^3}{t^3\log^{b+1}(t)}\int_0^\infty\frac{\rho d\rho}{(4r^2+(1-r^2+\rho^2)^2)^{3/2}}\right.\\
&\left.+\frac{r^3}{t^3\log^{b+1}(t)\log^{4\alpha b-4b}(t)}\int_0^\infty\frac{\rho d\rho}{(4\lambda(t)^{2\alpha-2}r^2+(1+\lambda(t)^{2\alpha-2}(\rho^2-r^2))^2)^{3/2}}\right)\\
&+\frac{C}{rt^3\log^{b+1}(t)}\left(\frac{1}{r}+\frac{1}{r\log^{2b-2\alpha b}(t)}\right)\\
&\leq \frac{C}{t^3\log^{-b+2\alpha b+1}(t)}
\end{aligned}$$

In total, we get

$$|\partial_t v_{3,1,t}(t,r)| \leq \frac{C}{t^3\log^{1+2\alpha b-b}(t)}$$

For the t derivative of the second line of (3.159), we start with

$$|\partial_3^2 F_3(r,\rho,\lambda(s))| \leq \frac{C|\partial_3 F_3(r,\rho,\lambda(s))|}{\lambda(s)} + \frac{C\lambda(s)^{4\alpha-6}r^2}{\lambda(s)^{2\alpha-2}\left(1+(\rho^2-r^2)^2\lambda(s)^{4\alpha-4}+2(\rho^2+r^2)\lambda(s)^{2\alpha-2}\right)}$$

Then, we use our estimate for $\partial_3 F_3$ proven while estimating $\partial_t v_3$ previously, and note that the estimate above for $\partial_3^2 F_3$ gives rise to an estimate of the t derivative of the integrand of the second line of (3.159) which is of the same form as the estimate on the integrand used to obtain the final estimate on $\partial_t v_3$. We can therefore read off an estimate from our previous computations, and get that the partial t derivative of the second line of (3.159) is bounded above in absolute value by

$$\frac{Cr}{t^4\log^{b+1}(t)}$$

This gives (3.156).

For $\partial_t^2 v_1$, we have

(3.161)

$$\begin{aligned}
&\partial_{tt}v_1(t,r)\\
&= \int_t^\infty \frac{-\lambda'''(s)}{r}\int_0^{s-t}\frac{\rho}{\sqrt{(s-t)^2-\rho^2}(s-t)}\left(1+\frac{r^2-1-\rho^2}{\sqrt{(r^2-1-\rho^2)^2+4r^2}}\right)d\rho ds\\
&+\int_t^\infty \frac{\lambda'''(s)}{r}\int_0^{s-t}\frac{\rho}{\sqrt{(s-t)^2-\rho^2}}\frac{8\rho^2r^2}{(s-t)(4r^2+(1+\rho^2-r^2)^2)^{3/2}}d\rho ds
\end{aligned}$$

For the first line of (3.161),we get

$$\begin{aligned}
&\int_t^\infty \frac{-\lambda'''(s)}{r}\int_0^{s-t}\frac{\rho}{(s-t)\sqrt{(s-t)^2-\rho^2}}\left(1+\frac{r^2-1-\rho^2}{\sqrt{(r^2-1-\rho^2)^2+4r^2}}\right)d\rho ds\\
&=-\int_0^\infty \rho\left(1+\frac{r^2-1-\rho^2}{\sqrt{(r^2-1-\rho^2)^2+4r^2}}\right)\int_{\rho+t}^\infty \frac{\lambda'''(s)}{r(s-t)\sqrt{(s-t)^2-\rho^2}}ds d\rho
\end{aligned}$$

Next, if

$$f_{v_1,t}(x,y)=1+\frac{x-y}{\sqrt{(x-y)^2+4x}},\quad x\geq 0, y\geq 1$$

then,

$$|f_{v_1,t}(x,y)|\leq \frac{Cx}{y^2},\quad x\leq \frac{y}{16}$$

Also, we have

$$|f_{v_1,t}(x,y)|\leq 2$$

Note that if $r \leq \frac{1}{4}$, then, $r^2 \leq \frac{1+\rho^2}{16}$. So, we get

$$\begin{aligned}
&|\int_t^\infty \frac{-\lambda'''(s)}{r} \int_0^{s-t} \frac{\rho}{(s-t)\sqrt{(s-t)^2-\rho^2}} \left(1+\frac{r^2-1-\rho^2}{\sqrt{(r^2-1-\rho^2)^2+4r^2}}\right) d\rho ds| \\
&\leq C \frac{\sup_{x\geq t}|\lambda'''(x)|}{r} \begin{cases} \int_0^\infty |f_{v_1,t}(r^2, 1+\rho^2)| d\rho, & r < \frac{1}{4} \\ \int_0^{4r} 2 d\rho + C\int_{4r}^\infty \frac{r^2 d\rho}{\rho^4}, & r > \frac{1}{4} \end{cases} \\
&\leq C \sup_{x\geq t}|\lambda'''(x)| \begin{cases} r, & r < \frac{1}{4} \\ 1, & r > \frac{1}{4} \end{cases} \\
&\leq C\frac{r}{1+r} \sup_{x\geq t}|\lambda'''(x)|
\end{aligned}$$

For the second line of (3.161), we use a similar procedure as above, again treating first the case $r \leq \frac{1}{4}$, and get

$$\begin{aligned}
&|\int_t^\infty \frac{\lambda'''(s)}{r} \int_0^{s-t} \frac{\rho}{\sqrt{(s-t)^2-\rho^2}} \frac{8\rho^2 r^2}{(s-t)(4r^2+(1-r^2+\rho^2)^2)^{3/2}} d\rho ds| \\
&\leq \frac{C\sup_{x\geq t}|\lambda'''(x)|}{r} \int_0^\infty \frac{\rho^3 r^2}{(4r^2+(1-r^2+\rho^2)^2)^{3/2}} \int_{\rho+t}^\infty \frac{1}{\sqrt{(s-t)^2-\rho^2}} \frac{ds d\rho}{(s-t)} \\
&\leq C\frac{\sup_{x\geq t}|\lambda'''(x)|}{r} r^2 \int_0^\infty \frac{\rho^2 d\rho}{(1+\rho^2)^3} \\
&\leq Cr \sup_{x\geq t}|\lambda'''(x)|, \quad r \leq \frac{1}{4}
\end{aligned}$$

where we used the fact that

$$r \leq \frac{1}{4} \implies 1+\rho^2-r^2 \geq C(1+\rho^2)$$

Next, we consider the second line of (3.161) for $r > \frac{1}{4}$. Here, we get

$$\begin{aligned}
&|\int_t^\infty \frac{\lambda'''(s)}{r} \int_0^{s-t} \frac{\rho}{\sqrt{(s-t)^2-\rho^2}} \frac{8\rho^2 r^2}{(s-t)(4r^2+(1-r^2+\rho^2)^2)^{3/2}} d\rho ds| \\
&\leq C\frac{\sup_{x\geq t}|\lambda'''(x)|}{r} \left(\int_0^{2r} \frac{\rho^2 r^2 d\rho}{(4r^2+(1-r^2+\rho^2)^2)^{3/2}} + \int_{2r}^\infty \frac{\rho^2 r^2 d\rho}{(4r^2+(1-r^2+\rho^2)^2)^{3/2}}\right) \\
&\leq \frac{C\sup_{x\geq t}|\lambda'''(x)|}{r} \left(2r\int_0^{2r} \frac{r^2\rho d\rho}{(4r^2+(1-r^2+\rho^2)^2)^{3/2}} + r^2 \int_{2r}^\infty \frac{\rho^2 d\rho}{\rho^6}\right) \\
&\leq C\sup_{x\geq t}|\lambda'''(x)|, \quad r > \frac{1}{4}
\end{aligned}$$

In total, we get (3.155).

Now, we will obtain an estimate on $\partial_t^2 v_4$. We write

$$v_4 = v_4^0(t,r) + v_4^1(t,r)$$

where v_4^0 solves the same equation as v_4, with 0 Cauchy data at infinity, except with right-hand side equal to $v_{4,c}^0$, which is given by

$$v_{4,c}^0(t,r) = \chi_{\geq 1}(\frac{2r}{\log^N(t)}) \left(\left(\frac{\cos(2Q_1(\frac{r}{\lambda(t)}))-1}{r^2}\right)(v_1+v_2+v_3)\right)$$

and similarly for v_4^1, where

$$v_{4,c}^1(t,r) = \chi_{\geq 1}(\frac{2r}{\log^N(t)})F_{0,2}(t,r)$$

We start with

$$\begin{aligned}&\partial_t^2 v_4^0(t,r)\\&=\frac{-1}{2\pi}\int_t^\infty \int_0^{s-t} \frac{\rho}{\sqrt{(s-t)^2-\rho^2}}\\&\qquad\int_0^{2\pi} \frac{\partial_1^2 v_{4,c}^0(s,\sqrt{r^2+\rho^2+2r\rho\cos(\theta)})\,(r+\rho\cos(\theta))}{\sqrt{r^2+\rho^2+2r\rho\cos(\theta)}}d\theta d\rho ds\end{aligned}$$

Then, we note that

$$\begin{aligned}|\partial_t^2 v_{4,c}^0(t,r)| &\leq C\frac{\mathbb{1}_{\{r\geq \frac{\log^N(t)}{2}\}}|v_1+v_2+v_3|}{t^2\log^{2b+1}(t)r^4}+\frac{C\mathbb{1}_{\{r\geq \frac{\log^N(t)}{2}\}}|\partial_t(v_1+v_2+v_3)|}{t\log^{2b+1}(t)r^4}\\&+\frac{C\mathbb{1}_{\{r\geq \frac{\log^N(t)}{2}\}}|\partial_t^2(v_1+v_2+v_3)|}{r^4\log^{2b}(t)}\end{aligned}$$

For the purposes of obtaining a preliminary estimate on $\partial_t^2 v_4^0$ in the region $r \leq \frac{t}{2}$, we combine all of our previous estimates in the following way:

$$|\partial_t^2 v_{4,c}^0(t,r)| \leq \mathbb{1}_{\{r\geq \frac{\log^N(t)}{2}\}}\begin{cases}\frac{C}{r^4\log^{2b}(t)t^3}\left(\frac{1}{\log^{1+2b\alpha-b}(t)}+\frac{1}{\log^b(t)}\right), & r\leq \frac{t}{2}\\ \frac{C\log(r)}{r^4\log^{2b}(t)}\left(\frac{1}{|t-r|^3}+\frac{1}{t\log(t)(t-r)^2}+\frac{1}{t^2\log(t)|t-r|}\right)\\ +\frac{C}{r^4t^3\log^{1+b+2b\alpha}(t)}, & t>r>\frac{t}{2}\end{cases}$$

Directly inserting this estimate into our previous formula for $\partial_t^2 v_4^0(t,r)$, and estimating as in previous sections, we get

$$|\partial_t^2 v_4^0(t,r)| \leq \frac{C}{t^3\log^{2N+b-2}(t)}, \quad r\leq \frac{t}{2}$$

We will use the same procedure used for $\partial_t v_4$ to estimate $\partial_t^2 v_4^0$ in the region $r \geq \frac{t}{2}$. As in previous cases, we will use a different combination of $\partial_t^j v_2$ estimates to get a different estimate for $\partial_t^2 v_{4,c}^0$:

$$|\partial_t^2 v_{4,c}^0(t,r)| \leq \begin{cases}\frac{C\mathbb{1}_{\{r\geq \frac{\log^N(t)}{2}\}}}{r^4\log^b(t)t^3}, & r\leq\frac{t}{2}\\ \frac{C}{\log^{2b}(t)r^{9/2}}, & t-t^{1/6}\leq r\leq t+t^{1/6}\\ \frac{C\log(r)}{r^4\log^{2b}(t)}\left(\frac{1}{|t-r|^3}+\frac{1}{t\log(t)(t-r)^2}+\frac{1}{t^2\log(t)|t-r|}\right)\\ +\frac{C}{r^4t^3\log^{1+b+2b\alpha}(t)}, & t-t^{1/6}>r>\frac{t}{2}, \text{ or } r>t+t^{1/6}\end{cases}$$

We then get

$$|\partial_t^2 v_4^0(t,r)| \leq \frac{C}{t^2\log^{3N+b}(t)}, \quad r\geq\frac{t}{2}$$

which gives

$$|\partial_t^2 v_4^0(t,r)| \leq \begin{cases}\frac{C}{t^3\log^{2N+b-2}(t)}, & r\leq\frac{t}{2}\\ \frac{C}{t^2\log^{3N+b}(t)}, & r\geq\frac{t}{2}\end{cases}$$

Now, we estimate $\partial_t^2 v_4^1$. This time, we follow the procedure used to obtain the preliminary estimate on $\partial_t v_4$:

(3.162)

$$\begin{aligned}\partial_t^2 v_4^1(t,r) &= \frac{1}{2\pi}\int_t^\infty \frac{1}{(s-t)}\int_0^{s-t}\frac{\rho}{\sqrt{(s-t)^2-\rho^2}}\int_0^{2\pi}\frac{\partial_1 v_{4,c}^1(s,\sqrt{r^2+\rho^2+2r\rho\cos(\theta)})(r+\rho\cos(\theta))}{\sqrt{r^2+\rho^2+2r\rho\cos(\theta)}}d\theta d\rho ds\\ &\quad -\frac{1}{2\pi}\int_t^\infty\int_{B_{s-t}(0)}\frac{\text{integrand}_{v_{4,2}^1}}{\sqrt{(s-t)^2-|y|^2}}dA(y)ds\end{aligned}$$

where

$$\begin{aligned}\text{integrand}_{v_{4,2}^1} &= \frac{-\partial_{12}v_{4,c}^1(s,|x+y|)}{|x+y|^2(s-t)}((x+y)\cdot y)(\widehat{x}\cdot(x+y))\\ &\quad + \frac{\partial_1 v_{4,c}^1(s,|x+y|)}{(s-t)|x+y|}\left(-y\cdot\widehat{x}+\frac{(\widehat{x}\cdot(x+y))(y\cdot(y+x))}{|x+y|^2}\right)\end{aligned}$$

which gives

$$|\text{integrand}_{v_{4,2}^1}| \leq C|\partial_{12}v_{4,c}^1| + C\frac{|\partial_1 v_{4,c}^1|}{|x+y|}$$

We start with the term on the first line of (3.162):

$$\begin{aligned}&|\frac{1}{2\pi}\int_t^\infty \frac{1}{(s-t)}\int_0^{s-t}\frac{\rho}{\sqrt{(s-t)^2-\rho^2}}\int_0^{2\pi}\frac{\partial_1 v_{4,c}^1(s,\sqrt{r^2+\rho^2+2r\rho\cos(\theta)})(r+\rho\cos(\theta))}{\sqrt{r^2+\rho^2+2r\rho\cos(\theta)}}d\theta d\rho ds|\\ &\leq C\int_t^\infty\frac{1}{(s-t)}\int_0^{s-t}\frac{\rho}{\sqrt{(s-t)^2-\rho^2}}\int_0^{2\pi}\frac{d\theta d\rho ds}{s^3\log^{3b+1+N-2b\alpha}(s)(\log^{2N}(s)+\rho^2+r^2+2r\rho\cos(\theta))}\\ &\leq C\int_0^\infty\rho\int_0^{2\pi}\frac{1}{(\rho+t)^3\log^{3b+1+N-2b\alpha}(t)(\log^{2N}(t)+r^2+\rho^2+2r\rho\cos(\theta))}\int_{\rho+t}^\infty\frac{ds d\theta d\rho}{(s-t)\sqrt{(s-t)^2-\rho^2}}\\ &\leq \frac{C}{t^3\log^{3b+2N-2b\alpha}(t)}\end{aligned}$$

For the second line of (3.162), we use the same procedure used to estimate v_4 previously, and we get

$$|-\frac{1}{2\pi}\int_t^\infty\int_{B_{s-t}(0)}\frac{\text{integrand}_{v_{4,2}^1}}{\sqrt{(s-t)^2-|y|^2}}dA(y)ds| \leq \frac{C}{t^3\log^{3b-1+2N-2\alpha b}(t)}$$

In total, we get

$$|\partial_t^2 v_4^1(t,r)| \leq \frac{C}{t^3\log^{3b+2N-1-2\alpha b}(t)}, \quad r\geq 0$$

Combining with the estimate on $\partial_t^2 v_4^0$ above, we get (3.157).

Next, we estimate $\partial_t^2 v_5$. We start with

(3.163)

$$\begin{aligned}|\partial_t^2 N_2(f)(t,r)| &\leq \left(\frac{C\lambda(t)\lambda'(t)^2}{r(r^2+\lambda(t)^2)^2}+\frac{C|\lambda''(t)|}{r(r^2+\lambda(t)^2)}\right)|f(t,r)|^2\\ &\quad + \frac{C|\lambda'(t)|}{r(r^2+\lambda(t)^2)}|f(t,r)\partial_t f(t,r)|\\ &\quad + \frac{C\lambda(t)}{r(r^2+\lambda(t)^2)}\left((\partial_t f(t,r))^2+|f(t,r)\partial_{tt}f(t,r)|\right)\\ &\quad + \frac{C\left((\lambda'(t))^2+\lambda(t)|\lambda''(t)|\right)}{(r^2+\lambda(t)^2)^2}|f(t,r)^3|+\frac{C\lambda(t)|\lambda'(t)|}{(r^2+\lambda(t)^2)^2}|f(t,r)^2\partial_t f(t,r)|\\ &\quad + \frac{C}{r^2}\left(|f(t,r)|(\partial_t f(t,r))^2+(f(t,r))^2|\partial_{tt}f(t,r)|\right)\end{aligned}$$

where $f = v_1 + v_2 + v_3 + v_4$.

For the purposes of estimating $\partial_t^2 v_5$ in the region $r \leq \frac{t}{2}$, we return to (3.163), and use (3.51) for all quantities involving v_2 in the region $r \geq \frac{t}{2}$, with the following exceptions. For the term involving $|f| \cdot |\partial_t f|^2$ on the last line of (3.163), we take the following average of (3.51) and (3.50) to estimate $(\partial_t v_2)^2$ in the region $r \geq \frac{t}{2}$:

$$(\partial_t v_2(t,r))^2 \leq C \left(\frac{\log(r)}{|t-r|^2}\right)^{3/2} \cdot \frac{1}{r^{1/4}}$$

Similarly, for the term involving $f^2|\partial_{tt} f|$, we estimate $v_2^2|\partial_{tt} v_2|$ by

$$(v_2(t,r))^2|\partial_{tt} v_2| \leq \frac{C\log(r)^2}{|t-r|^4\sqrt{r}}$$

(This is so that we will not have to have an extra argument for a term analogous to the last line of (3.95)). This procedure gives

$$|\partial_t^2 N_2(f)(t,r)| \leq C \begin{cases} \frac{1}{t^5 \log^b(t)(r^2+\lambda(t)^2)}, & r \leq \frac{t}{2} \\ \frac{\log(r)}{r^3 t^2 |t-r| \log^{3N+2b}(t)} + \frac{\log^2(r)}{r^2(t-r)^2 t^2 \log^{3N+b}(t)} + \frac{\log^3(r)}{t^2 r^2 \log(t)|t-r|^3} \\ + \frac{\log^{5/2}(r)}{r^{9/4}(t-r)^4}, & \frac{t}{2} \leq r < t \end{cases}$$

Using these estimates in the formula

$$\partial_t^2 v_5(t,r) = \frac{-1}{2\pi}\int_t^\infty \int_0^{s-t} \frac{\rho}{\sqrt{(s-t)^2-\rho^2}} \int_0^{2\pi} \frac{\partial_1^2 N_2(f)(s, \sqrt{r^2+\rho^2+2r\rho\cos(\theta)})}{\sqrt{r^2+\rho^2+2r\rho\cos(\theta)}} (r+\rho\cos(\theta)) d\theta d\rho ds$$

gives (3.158). □

Then, we use all of our estimates above to get

$$|\partial_t^2 \left(\langle \left(\frac{\cos(2Q_{\frac{1}{\lambda(t)}}(r))-1}{r^2}\right) v_3|_{r=R\lambda(t)}, \phi_0\rangle\right)| \leq \frac{C}{t^3 \log^{1+2\alpha b-3b}(t)}$$

$$|\partial_t^2 \left(\langle \left(\frac{\cos(2Q_{\frac{1}{\lambda(t)}}(r))-1}{r^2}\right) v_4 \left(1-\chi_{\geq 1}(\frac{4r}{t})\right)|_{r=R\lambda(t)}, \phi_0\rangle\right)| \leq \frac{C}{t^3 \log^{2N-2-b}(t)}$$

$$|\partial_t^2 \left(\langle \left(\frac{\cos(2Q_{\frac{1}{\lambda(t)}}(r))-1}{r^2}\right) v_5 \left(1-\chi_{\geq 1}(\frac{4r}{t})\right)|_{r=R\lambda(t)}, \phi_0\rangle\right)| \leq \frac{C}{t^4 \log^{3N-b-2}(t)}$$

$$|\partial_t^2 \left(\langle \left(\frac{\cos(2Q_{\frac{1}{\lambda(t)}}(r))-1}{r^2}\right) \left(\chi_{\geq 1}(\frac{2r}{\log^N(t)})(v_1+v_2+v_3)\right)|_{r=R\lambda(t)}, \phi_0\rangle\right)| \leq \frac{C}{t^3 \log^{1+3N+2\alpha b}(t)}$$

Finally, assembling together all of our previous estimates, we conclude

$$|RHS_2''(t)| \leq \frac{C}{t^3 \log^{1-3b}(t)}$$

Returning to (3.153), we get

$$|\lambda''''(t)| \leq \frac{C}{t^3 \log^{1-2b}(t)\log(\log(t))}$$

Now that we have this preliminary estimate, we return to (3.63), substitute

$$\lambda(t) = \lambda_{0,0}(t) + e(t)$$

and differentiate twice in t, to get

(3.164)

$$\begin{aligned}
&-4\int_t^\infty \frac{e''''(s)ds}{\log(\lambda_{0,0}(s))(1+s-t)} + 4\alpha e''''(t)\\
&-4\int_t^\infty \frac{e''''(s)ds}{\log(\lambda_{0,0}(s))(\lambda_{0,0}(t)^{1-\alpha}+s-t)(1+s-t)^3}\\
&= \frac{-E''_{\lambda_{0,0}}(t)}{\log(\lambda_{0,0}(t))} - \frac{4\alpha}{\log(\lambda_{0,0}(t))}\left(\frac{\lambda''(t)}{\lambda(t)} - \frac{\lambda'(t)^2}{\lambda(t)^2}\right)\lambda''(t) - \frac{8\alpha\lambda'(t)\lambda'''(t)}{\lambda(t)\log(\lambda_{0,0}(t))}\\
&-4\alpha\left(\frac{\log(\lambda(t))}{\log(\lambda_{0,0}(t))} - 1\right)e''''(t) - \frac{4\alpha\log(\lambda(t))}{\log(\lambda_{0,0}(t))}\lambda_{0,0}''''(t)\\
&+\frac{4}{\log(\lambda_{0,0}(t))}\partial_t^2\left(\int_0^\infty \frac{\lambda_{0,0}''(t+w)}{(\lambda(t)^{1-\alpha}+w)(1+w^3)}dw\right)\\
&+4\int_0^\infty \frac{e''''(t+w)}{\log(\lambda_{0,0}(t))}\left(\frac{1}{(\lambda(t)^{1-\alpha}+w)} - \frac{1}{(\lambda_{0,0}(t)^{1-\alpha}+w)}\right)\frac{dw}{(1+w)^3}\\
&-\frac{8(1-\alpha)\lambda(t)^{-\alpha}\lambda'(t)}{\log(\lambda_{0,0}(t))}\int_0^\infty \frac{e'''(t+w)}{(\lambda(t)^{1-\alpha}+w)^2(1+w)^3}dw\\
&-\frac{4(1-\alpha)(-\alpha\lambda(t)^{-\alpha-1}(\lambda'(t))^2 + \lambda(t)^{-\alpha}\lambda''(t))}{\log(\lambda_{0,0}(t))}\int_0^\infty \frac{e''(t+w)dw}{(1+w)^3(\lambda(t)^{1-\alpha}+w)^2}\\
&+\frac{8}{\log(\lambda_{0,0}(t))}\int_0^\infty \frac{e''(t+w)}{(1+w)^3}\frac{(1-\alpha)^2\lambda(t)^{-2\alpha}\lambda'(t)^2dw}{(\lambda(t)^{1-\alpha}+w)^3} + \frac{\partial_t^2 G(t,\lambda_{0,0}(t)+e(t))}{\log(\lambda_{0,0}(t))}\\
&+4\int_t^\infty e''''(s)\left(\frac{1}{\log(\lambda_{0,0}(t))} - \frac{1}{\log(\lambda_{0,0}(s))}\right)\frac{ds}{(1+s-t)}\\
&+4\int_t^\infty \frac{e''''(s)}{(1+s-t)^3(\lambda_{0,0}(t)^{1-\alpha}+s-t)}\left(\frac{1}{\log(\lambda_{0,0}(t))} - \frac{1}{\log(\lambda_{0,0}(s))}\right)ds\\
&:= RHS_4(t)
\end{aligned}$$

Like before, we start by estimating all the terms on the right-hand side of (3.164) which do not involve G:

$$\left|\frac{-E''_{\lambda_{0,0}}(t)}{\log(\lambda_{0,0}(t))}\right| \leq \frac{C}{t^4\log^{b+1}(t)\log(\log(t))}$$

$$\left|\frac{-4\alpha}{\log(\lambda_{0,0}(t))}\left(\frac{\lambda''(t)}{\lambda(t)} - \frac{\lambda'(t)^2}{\lambda(t)^2}\right)\lambda''(t)\right| \leq \frac{C}{t^4\log^{b+2}(t)\log(\log(t))}$$

$$\left|\frac{-8\alpha\lambda'(t)\lambda'''(t)}{\lambda(t)\log(\lambda_{0,0}(t))}\right| \leq \frac{C}{t^4\log^{b+2}(t)\log(\log(t))}$$

$$\left|-4\alpha\left(\frac{\log(\lambda_{0,0}(t)+e(t))}{\log(\lambda_{0,0}(t))} - 1\right)e''''(t)\right| \leq \frac{C|e''''(t)|}{(\log(\log(t)))^{3/2}}$$

$$\left|\frac{-4\alpha\log(\lambda(t))}{\log(\lambda_{0,0}(t))}\lambda_{0,0}''''(t)\right| \leq \frac{C}{t^4\log^{b+1}(t)}$$

Then, we note that the following estimate:

$$|\partial_t^2 \left(\frac{\lambda_{0,0}''(t+w)}{w+\lambda(t)^{1-\alpha}}\right)| \leq \frac{C}{t^2(t+w)^2 \log^{b+1}(t+w)(w+\lambda(t)^{1-\alpha})}$$

implies

$$|\frac{4}{\log(\lambda_{0,0}(t))} \int_0^\infty \partial_t^2 \left(\frac{\lambda_{0,0}''(t+w)}{w+\lambda(t)^{1-\alpha}}\right) \frac{dw}{(1+w)^3}| \leq \frac{C}{t^4 \log^{b+1}(t)}$$

Next, we have

$$\begin{aligned}&|4 \int_0^\infty \frac{e''''(t+w)}{\log(\lambda_{0,0}(t))} \left(\frac{\lambda_{0,0}(t)^{1-\alpha} - \lambda(t)^{1-\alpha}}{(\lambda(t)^{1-\alpha}+w)(\lambda_{0,0}(t)^{1-\alpha}+w)}\right) \frac{dw}{(1+w)^3}| \\ &\leq \frac{C \sup_{x\geq t} |e''''(x)|}{\sqrt{\log(\log(t))}}\end{aligned}$$

$$|\frac{-8(1-\alpha)\lambda(t)^{-\alpha}\lambda'(t)}{\log(\lambda_{0,0}(t))} \int_0^\infty \frac{e'''(t+w)dw}{(\lambda(t)^{1-\alpha}+w)^2(1+w)^3}| \leq \frac{C}{t^4 \log^{b+2}(t)}$$

$$\begin{aligned}&|\frac{-4(1-\alpha)(-\alpha\lambda(t)^{-\alpha-1}(\lambda'(t))^2 + \lambda(t)^{-\alpha}\lambda''(t))}{\log(\lambda_{0,0}(t))} \int_0^\infty \frac{e''(t+w)dw}{(1+w)^3(\lambda(t)^{1-\alpha}+w)^2}| \\ &\leq \frac{C}{t^4 \log^{b+2}(t)}\end{aligned}$$

$$|\frac{8}{\log(\lambda_{0,0}(t))} \int_0^\infty \frac{e''(t+w)}{(1+w)^3} \frac{(1-\alpha)^2\lambda(t)^{-2\alpha}\lambda'(t)^2 dw}{(\lambda(t)^{1-\alpha}+w)^3}| \leq \frac{C}{t^4 \log^{b+3}(t)}$$

$$|4 \int_t^\infty e''''(s) \left(\frac{1}{\log(\lambda_{0,0}(t))} - \frac{1}{\log(\lambda_{0,0}(s))}\right) \frac{ds}{1+s-t}| \leq \frac{C \sup_{x\geq t} \left(x^2 |e''''(x)|\right)}{t^2 \log(t)(\log(\log(t)))^2}$$

and

$$\begin{aligned}&|4 \int_t^\infty \frac{e''''(s)}{(1+s-t)^3(\lambda_{0,0}(t)^{1-\alpha}+s-t)} \left(\frac{1}{\log(\lambda_{0,0}(t))} - \frac{1}{\log(\lambda_{0,0}(s))}\right) ds| \\ &\leq \frac{C \sup_{x\geq t} \left(|e''''(x)| x^2\right)}{t^3 \log(t) \log(\log(t))}\end{aligned}$$

Combining these, we get that all of the terms on the right-hand side of (3.164) which do not involve G are bounded above in absolute value by

$$\frac{C}{t^4 \log^{b+1}(t)} + \frac{C \sup_{x\geq t} \left(x^2 |e''''(x)|\right)}{\sqrt{\log(\log(t))} t^2}$$

We now estimate terms in $\partial_t^2 G(t, \lambda(t))$, starting with the term involving K_1. We again use the preliminary estimate on e'''' to justify any differentiations under the integral sign, to get

$$
\begin{aligned}
&\partial_t^2\left(\frac{16}{\lambda(t)^2}\int_t^\infty \lambda''(x)\left(K_1(x-t,\lambda(t))-\frac{\lambda(t)^2}{4(1+x-t)}\right)dx\right)\\
&=\left(\frac{96\lambda'(t)^2}{\lambda(t)^4}-\frac{32\lambda''(t)}{\lambda(t)^3}\right)\int_0^\infty \lambda''(t+w)\left(K_1(w,\lambda(t))-\frac{\lambda(t)^2}{4(1+w)}\right)dw\\
&-\frac{64\lambda'(t)}{\lambda(t)^3}\int_0^\infty \lambda'''(w+t)\left(K_1(w,\lambda(t))-\frac{\lambda(t)^2}{4(1+w)}\right)dw\\
&-\frac{64\lambda'(t)}{\lambda(t)^3}\int_0^\infty \lambda''(w+t)\left(\partial_2K_1(w,\lambda(t))-\frac{\lambda(t)}{2(1+w)}\right)\lambda'(t)dw\\
&+\frac{16}{\lambda(t)^2}\int_0^\infty \lambda''''(w+t)\left(K_1(w,\lambda(t))-\frac{\lambda(t)^2}{4(1+w)}\right)dw\\
&+\frac{32}{\lambda(t)^2}\int_0^\infty \lambda'''(w+t)\left(\partial_2K_1(w,\lambda(t))-\frac{\lambda(t)}{2(1+w)}\right)\lambda'(t)dw\\
&+\frac{16}{\lambda(t)^2}\int_0^\infty \lambda''(w+t)\left(\left(\partial_2^2K_1(w,\lambda(t))\lambda'(t)-\frac{\lambda'(t)}{2(1+w)}\right)\lambda'(t)\right.\\
&\qquad\left.+\left(\partial_2K_1(w,\lambda(t))-\frac{\lambda(t)}{2(1+w)}\right)\lambda''(t)\right)dw
\end{aligned}
\tag{3.165}
$$

The only term not involving λ'''' which has not already been estimated is the first term on the last line of the above expression. For $w\geq 1$, we start with

$$K_1(w,\lambda(t))=\int_0^\infty \frac{r}{\lambda(t)^2(1+\frac{r^2}{\lambda(t)^2})^3}\int_0^w \frac{\rho}{w}\left(1+\frac{r^2-1-\rho^2}{\sqrt{(r^2-1-\rho^2)^2+4r^2}}\right)d\rho dr$$

whence

$$
\begin{aligned}
&\partial_2^2K_1(w,\lambda(t))\\
&=\int_0^\infty \frac{6\lambda(t)^2r\left(\lambda(t)^4-5\lambda(t)^2r^2+2r^4\right)}{(\lambda(t)^2+r^2)^5}\left(\frac{r^2}{w}-\int_w^\infty \frac{\rho}{w}\left(1+\frac{r^2-1-\rho^2}{\sqrt{(r^2-1-\rho^2)^2+4r^2}}\right)d\rho\right)dr\\
&=\frac{1}{2w}-\int_0^\infty \frac{6\lambda(t)^2r(\lambda(t)^4-5\lambda(t)^2r^2+2r^4)}{(\lambda(t)^2+r^2)^5}\int_w^\infty \frac{\rho}{w}\left(1+\frac{r^2-1-\rho^2}{\sqrt{(r^2-1-\rho^2)^2+4r^2}}\right)d\rho dr
\end{aligned}
$$

Then,

$$
\begin{aligned}
&|\frac{16}{\lambda(t)^2}\int_1^\infty \lambda''(w+t)\left(\partial_{22}K_1(w,\lambda(t))\lambda'(t)-\frac{\lambda'(t)}{2(1+w)}\right)\lambda'(t)dw|\\
&\leq \frac{C|\lambda'(t)|}{t^3\log^2(t)}\int_1^\infty\left(\frac{1}{2w}-\frac{1}{2(1+w)}\right)dw\\
&+\frac{C|\lambda'(t)|}{t^3\log^2(t)}\int_0^\infty \frac{r\lambda(t)^2}{(\lambda(t)^2+r^2)^3}\int_1^\infty\int_1^\rho \frac{\rho}{w}\left(1+\frac{r^2-1-\rho^2}{\sqrt{(r^2-1-\rho^2)^2+4r^2}}\right)dwd\rho dr\\
&\leq \frac{C}{t^4\log^{b+3}(t)}+\frac{C\lambda(t)^2}{t^4\log^{b+3}(t)}\int_0^\infty \frac{r^3\log(2+r^2)dr}{(\lambda(t)^2+r^2)^3}\\
&\leq \frac{C}{t^4\log^{b+3}(t)}
\end{aligned}
$$

For $w \leq 1$, we first change variables, and start with

$$K_1(w,\lambda(t))$$
$$= \int_0^\infty \frac{R}{(1+R^2)^3} \int_0^w \frac{\rho}{w} \left(1 + \frac{R^2\lambda(t)^2 - 1 - \rho^2}{\sqrt{(R^2\lambda(t)^2 - 1 - \rho^2)^2 + 4R^2\lambda(t)^2}}\right) d\rho dR$$

Then,

$$|\partial_2^2 K_1(w,\lambda(t))| \leq C \int_0^\infty \frac{R}{(1+R^2)^3} \int_0^w \frac{\rho}{w} R^2 d\rho dR$$

which gives

$$|\frac{16}{\lambda(t)^2} \int_0^1 \lambda''(w+t) \left(\partial_{22} K_1(w,\lambda(t))\lambda'(t) - \frac{\lambda'(t)}{2(1+w)}\right) \lambda'(t) dw|$$
$$\leq \frac{C}{t^3 \log^2(t)} \int_0^1 \left(w + \frac{1}{1+w}\right) \frac{dw}{t \log^{b+1}(t)} \leq \frac{C}{t^4 \log^{b+3}(t)}$$

We now consider the expression (3.165), except with all instances of

$$K_1(x-t,\lambda(t)) - \frac{\lambda(t)^2}{4(1+x-t)}$$

replaced with

$$K(x-t,\lambda(t))$$

(This expression also appears in $\partial_t^2(G(t,\lambda(t)))$). All of the integrals involving K which do not involve λ'''' have already been estimated, and we get

$$|\partial_t^2 \left(\frac{16}{\lambda(t)^2} \int_t^\infty \lambda''(x) \left(K_1(x-t,\lambda(t)) - \frac{\lambda(t)^2}{4(1+x-t)}\right) dx\right)|$$
$$+ |\partial_t^2 \left(\frac{16}{\lambda(t)^2} \int_t^\infty \lambda''(x) K(x-t,\lambda(t)) dx\right)|$$
$$\leq \frac{C}{t^4 \log^{b+1}(t)} + C \sup_{x \geq t} |e''''(x)|$$

From the explicit expressions, we also get

$$|\partial_t^2 \left(-\lambda(t) E_{0,1}(\lambda(t),\lambda'(t),\lambda''(t))\right)| \leq \frac{C}{t^4 \log^{b+1}(t)} + C|e''''(t)|$$

Next, we start with

$$|\partial_t^2 \left(K_3(w,\lambda(t))\right)| \leq \frac{C}{t^2 \log(t)} |\frac{w}{1+w^2} - \frac{w}{\lambda(t)^{2-2\alpha} + w^2}|$$
$$+ \frac{Cw}{t^2 \log^{(2-2\alpha)b+1}(t) (\lambda(t)^{2-2\alpha} + w^2)^2}$$

and

$$|\partial_t^2 K_{3,0}(w,\lambda(t))| \leq \frac{C}{t^2 \log^{b+1-b\alpha}(t)} \frac{1}{(\lambda(t)^{1-\alpha} + w)^2 (1+w)^3}$$

and use our previous estimates on $K_3 - K_{3,0}$, and $\partial_t \left((K_3 - K_{3,0})(w,\lambda(t))\right)$ to get

$$|\partial_t^2 \left(-16 \int_t^\infty \lambda''(s) \left(K_3(s-t,\lambda(t)) - K_{3,0}(s-t,\lambda(t))\right) ds\right)|$$
$$\leq C \sup_{x \geq t} |e''''(x)| + \frac{C}{t^4 \log^{b+1}(t)}$$

Next, we will need an estimate on $\partial_t^2 v_3$ which is different from that recorded while obtaining the preliminary estimate on λ''''. We start with

(3.166)

$$\begin{aligned}
&\partial_t^2 v_3(t,r)\\
&= \frac{-1}{r}\int_0^\infty \int_0^w \frac{\rho}{\sqrt{w^2-\rho^2}}\lambda''''(t+w)\left(\frac{-1-\rho^2+r^2}{\sqrt{(1+\rho^2-r^2)^2+4r^2}}+F_3(r,\rho,\lambda(t+w))\right)d\rho dw\\
&-\frac{2}{r}\int_0^\infty \int_0^w \frac{\rho}{\sqrt{w^2-\rho^2}}\lambda'''(t+w)\partial_t(F_3(r,\rho,\lambda(t+w)))d\rho dw\\
&-\frac{1}{r}\int_0^\infty \int_0^w \frac{\rho}{\sqrt{w^2-\rho^2}}\lambda''(t+w)\partial_t^2(F_3(r,\rho,\lambda(t+w)))d\rho dw
\end{aligned}$$

The second and third lines of (3.166) were already estimated in the course of obtaining the preliminary estimate on λ'''', and do not need to be estimated any differently here. The first line of (3.166), with the replacement of λ'''' with λ''' has already been estimated while obtaining the final estimate on λ'''. Therefore, we can read off that

$$(3.167)\quad |\partial_t^2 v_3(t,r)| \leq Cr\sup_{x\geq t}\left(\frac{x|e''''(x)|}{\lambda(x)^{2-2\alpha}}\right)\frac{\log(\log(t))\lambda(t)^{2-2\alpha}}{t}+\frac{Cr\log(\log(t))}{t^4\log^{b+1}(t)}$$

Next, we estimate $\partial_t^2 E_5(t,r)$ by proceeding line-by-line on the expression (3.140). Denoting, by $E_{5,t,t,i}$, the second partial t derivative of the ith line of (3.140), we get

(3.168)

$$\begin{aligned}
&E_{5,t,t,1}(t,r)\\
&= \frac{-1}{r}\int_0^{6r} \int_0^w \frac{\rho}{w}\lambda''''(t+w)\left(\frac{-1-\rho^2+r^2}{\sqrt{(1+\rho^2-r^2)^2+4r^2}}+F_3(r,\rho,\lambda(t+w))\right)d\rho dw\\
&-\frac{2}{r}\int_0^{6r} \int_0^w \frac{\rho}{w}\lambda'''(t+w)\partial_t(F_3(r,\rho,\lambda(t+w)))d\rho dw\\
&-\frac{1}{r}\int_0^{6r} \int_0^w \frac{\rho}{w}\lambda''(t+w)\partial_t^2(F_3(r,\rho,\lambda(t+w)))d\rho dw
\end{aligned}$$

Note that, with the replacement of λ'''' with λ''', the first line of (3.168) was already estimated in the course of obtaining the final estimate on λ'''. The same is true for the second line of (3.168), except with the replacement of λ''' with λ''. On the other hand, for the third line of (3.168), we use our previous estimate of

$$\partial_t^2(F_3(r,\rho,\lambda(t+w)))$$

to get

$$|E_{5,t,t,1}(t,r)| \leq \frac{Cr}{t^4\log^{b+1}(t)}+Cr\sup_{x\geq t}\left(|e''''(x)|\right)$$

Next, we have

(3.169)

$$\begin{aligned}
&E_{5,t,t,2}(t,r)\\
&= \frac{-1}{r}\int_{6r}^{\infty}\int_0^w \frac{\rho}{w}\lambda''''(t+w)\left(\frac{-1-\rho^2+r^2}{\sqrt{(1+\rho^2-r^2)^2+4r^2}}+F_3(r,\rho,\lambda(t+w))\right)d\rho dw\\
&-\frac{2}{r}\int_{6r}^{\infty}\int_0^w \frac{\rho}{w}\lambda'''(t+w)\partial_t(F_3(r,\rho,\lambda(t+w)))d\rho dw\\
&-\frac{1}{r}\int_{6r}^{\infty}\int_0^w \frac{\rho}{w}\lambda''(t+w)\partial_t^2(F_3(r,\rho,\lambda(t+w)))d\rho dw\\
&+\frac{1}{r}\int_{6r}^{\infty}\lambda''''(t+w)r^2w\left(\frac{1}{1+w^2}-\frac{1}{\lambda(t+w)^{2-2\alpha}+w^2}\right)dw\\
&+\frac{2}{r}\int_{6r}^{\infty}\frac{\lambda'''(t+w)r^2w}{(\lambda(t+w)^{2-2\alpha}+w^2)^2}(2-2\alpha)\lambda(t+w)^{1-2\alpha}\lambda'(t+w)dw\\
&+\frac{1}{r}\int_{6r}^{\infty}\frac{\lambda''(t+w)r^2w(2-2\alpha)\left((1-2\alpha)\lambda(t+w)^{-2\alpha}(\lambda'(t+w))^2+\lambda(t+w)^{1-2\alpha}\lambda''(t+w)\right)}{(\lambda(t+w)^{2-2\alpha}+w^2)^2}dw\\
&-\frac{2}{r}\int_{6r}^{\infty}\frac{\lambda''(t+w)r^2w\left((2-2\alpha)\lambda(t+w)^{1-2\alpha}\lambda'(t+w)\right)^2}{(\lambda(t+w)^{2-2\alpha}+w^2)^3}dw
\end{aligned}$$

Only two lines in (3.169) can not be estimated simply by comparing to an analogous term arising in the $\partial_t E_5$ estimates. They are estimated as follows:

$$\begin{aligned}
&|\frac{-1}{r}\int_{6r}^{\infty}\int_0^w \frac{\rho}{w}\lambda''(t+w)\partial_t^2(F_3(r,\rho,\lambda(t+w)))d\rho dw|\\
&\leq \frac{C}{r}\int_{6r}^{\infty}\int_0^w \frac{\rho}{w(t+w)^4\log^{2b+2}(t+w)}\left(\frac{Cr^2\lambda(t+w)^{2\alpha-3}}{(1+\lambda(t+w)^{4\alpha-4}(\rho^2-r^2)^2+2\lambda(t+w)^{2\alpha-2}(\rho^2+r^2))}\right)d\rho dw\\
&\leq \frac{C}{r}\int_0^{\lambda(t)^{1-\alpha}}\frac{1}{wt^4\log^{2b+2}(t)}\int_0^w r^2\lambda(t+w)^{2\alpha-3}\rho d\rho dw\\
&+\frac{C}{r}\int_{\lambda(t)^{1-\alpha}}^{\infty}\frac{1}{w}\int_0^w \frac{\rho}{(t+w)^4\log^{2b+2}(t+w)}\frac{r^2\lambda(t+w)^{2\alpha-3}d\rho dw}{(1+\lambda(t+w)^{4\alpha-4}(\rho^2-r^2)^2+2\lambda(t+w)^{2\alpha-2}(\rho^2+r^2))}\\
&\leq \frac{Cr}{t^4\log^{b+1}(t)}
\end{aligned}$$

$$\begin{aligned}
&|\frac{-1}{r}\int_{6r}^{\infty}\frac{\lambda''(t+w)r^2w(2-2\alpha)\left((1-2\alpha)\lambda(t+w)^{-2\alpha}(\lambda'(t+w))^2+\lambda(t+w)^{1-2\alpha}\lambda''(t+w)\right)dw}{(\lambda(t+w)^{2-2\alpha}+w^2)^2}|\\
&\leq \frac{Cr}{t^4\log^{b+2}(t)}
\end{aligned}$$

where we use the identical procedure used to estimate an analogous term arising while obtaining the final estimate on λ'''. Then, we get

$$|E_{5,t,t,2}(t,r)|\leq \frac{Cr}{t^4\log^{b+1}(t)}+Cr\sup_{x\geq t}|e''''(x)|$$

Next, we have

(3.170)

$$\begin{aligned}
&E_{5,t,t,3}(t,r)\\
&= -r\int_{6r}^{\infty}\lambda''''(t+w)w\left(\frac{1}{\lambda(t)^{2-2\alpha}+w^2}-\frac{1}{\lambda(t+w)^{2-2\alpha}+w^2)}\right)dw\\
&-2r\int_{6r}^{\infty}\lambda'''(t+w)w\left(\frac{-(2-2\alpha)\lambda(t)^{1-2\alpha}\lambda'(t)}{(\lambda(t)^{2-2\alpha}+w^2)^2}+\frac{(2-2\alpha)\lambda(t+w)^{1-2\alpha}\lambda'(t+w)}{(\lambda(t+w)^{2-2\alpha}+w^2)^2}\right)dw\\
&-r\int_{6r}^{\infty}\lambda''(t+w)w\partial_t\left(\frac{-(2-2\alpha)\lambda(t)^{1-2\alpha}\lambda'(t)}{(\lambda(t)^{2-2\alpha}+w^2)^2}+\frac{(2-2\alpha)\lambda(t+w)^{1-2\alpha}\lambda'(t+w)}{(\lambda(t+w)^{2-2\alpha}+w^2)^2}\right)dw
\end{aligned}$$

The first and second lines of (3.170) can be estimated based on estimates of analogous terms arising in the estimation of $\partial_t E_5$ done while obtaining the final estimate on λ'''. On the other hand, by appropriately using

$$|f(t+w)-f(t)| \leq \sup_{\sigma\in[0,1]} |f'(t+\sigma w)| \cdot w$$

we estimate the third line on the right-hand side of (3.170), and in total, we get

$$|E_{5,t,t,3}(t,r)| \leq \frac{Cr}{t^4\log^{b+2}(t)} + \frac{Cr}{t\log^{3b+1-3b\alpha}(t)} \sup_{x\geq t}\left(\frac{|e''''(x)|}{\lambda(x)^{2-2\alpha}}\right)$$

Finally, we have

(3.171)

$$\begin{aligned}
&E_{5,t,t,4}(t,r)\\
&= \frac{-1}{r}\int_0^\infty\int_0^w \rho\left(\frac{1}{\sqrt{w^2-\rho^2}}-\frac{1}{w}\right)\lambda''''(t+w)\left(\frac{-1-\rho^2+r^2}{\sqrt{(1+\rho^2-r^2)^2+4r^2}}+F_3(r,\rho,\lambda(t+w))\right)d\rho dw\\
&-\frac{2}{r}\int_0^\infty\int_0^w \rho\left(\frac{1}{\sqrt{w^2-\rho^2}}-\frac{1}{w}\right)\lambda'''(t+w)\partial_t\left(F_3(r,\rho,\lambda(t+w))\right)d\rho dw\\
&-\frac{1}{r}\int_0^\infty\int_0^w \rho\left(\frac{1}{\sqrt{w^2-\rho^2}}-\frac{1}{w}\right)\lambda''(t+w)\partial_t^2\left(F_3(r,\rho,\lambda(t+w))\right)d\rho dw
\end{aligned}$$

The only term in (3.171) which can not be estimated simply by reading off estimates of analogous terms arising in $\partial_t E_5$ estimates from before is the one involving $\partial_t^2\left(F_3(r,\rho,\lambda(t+w))\right)$. For this term, we have

$$\begin{aligned}
&|-\frac{1}{r}\int_0^\infty\int_0^w \rho\left(\frac{1}{\sqrt{w^2-\rho^2}}-\frac{1}{w}\right)\lambda''(t+w)\partial_t^2\left(F_3(r,\rho,\lambda(t+w))\right)d\rho dw|\\
&\leq Cr\int_0^\infty \rho\int_\rho^\infty\left(\frac{1}{\sqrt{w^2-\rho^2}}-\frac{1}{w}\right)\frac{\log^{3b-2\alpha b}(t)}{t^4\log^{2b+2}(t)(1+\lambda(t)^{4\alpha-4}(\rho^2-r^2)^2+2\lambda(t)^{2\alpha-2}(\rho^2+r^2))}dwd\rho\\
&\leq \frac{Cr}{t^4\log^{b+2}(t)}
\end{aligned}$$

Combining these, we get

$$|\partial_t^2 E_5(t,r)| \leq \frac{Cr}{t^4\log^{b+1}(t)} + \frac{Cr}{t\log^{(3-2\alpha)b}(t)}\sup_{x\geq t}\left(\frac{|e''''(x)|x}{\lambda(x)^{3-2\alpha}}\right)$$

Next, we need to record new estimates on $\partial_t^2 v_4$. First, we note that, by the same procedure used to obtain estimates on $\partial_t v_1$, we have

$$|\partial_t^2 v_1(t,r)| \leq Cr\left(\frac{\log(t)+\log(2+r^2)}{t}\right)\sup_{x\geq t}\left(|\lambda''''(x)|x\right) \tag{3.172}$$

Note that this time, the preliminary estimate on λ'''' is indeed strong enough to justify the steps leading up to a $\partial_t^2 v_1$ analog of (3.12), but this will be unnecessary for our purposes. We then use (3.167), and the same estimates for all other $\partial_t^j v_k$ used to obtain the preliminary estimate on $\partial_t^2 v_{4,c}$, and get

(3.173)

$$\begin{aligned}
&|\partial_t^2 v_{4,c}(t,r)|\\
&\leq \frac{C\mathbb{1}_{\geq 1}\left(\frac{2r}{\log^N(t)}\right)}{r^4\log^{2b}(t)}\\
&\cdot\begin{cases}\frac{r(\log(t)+\log(2+r^2))}{t}\left(\frac{1}{t^3\log^{b+1}(t)}+\sup_{x\geq t}\left(\frac{|e''''(x)|x}{\lambda(x)^{2-2\alpha}}\right)\lambda(t)^{2-2\alpha}\right), & r\leq\frac{t}{2}\\
\frac{r(\log(t)+\log(2+r^2))}{t}\left(\frac{1}{t^3\log^{b+1}(t)}+\sup_{x\geq t}\left(\frac{|e''''(x)|x}{\lambda(x)^{2-2\alpha}}\right)\lambda(t)^{2-2\alpha}\right) & \\
+\frac{\log(r)}{|t-r|^3}+\frac{\log(r)}{t\log(t)(t-r)^2}+\frac{\log(r)}{t^2|t-r|\log(t)}, & t>r>\frac{t}{2}\end{cases}
\end{aligned}$$

which leads, via the same procedure used to obtain the preliminary estimate on $\partial_t^2 v_4$, to

$$|\partial_t^2 v_4(t,r)| \leq \frac{C}{t^4 \log^{3b+N-2}(t)} + \frac{C\lambda(t)^{2-2\alpha}}{t \log^{2b-3+N}(t)} \sup_{x\geq t}\left(\frac{|e''''(x)|x}{\lambda(x)^{2-2\alpha}}\right), \quad r \leq \frac{t}{2}$$

As usual, for the following estimate, we modify (3.173) by using $\frac{C}{r^{9/2}\log^{2b}(t)}$ in the region $t - t^{1/6} \leq r \leq t + t^{1/6}$. (Note that the preliminary estimate on λ'''' implies the following).

$$\sup_{x\geq t}\left(\frac{|e''''(x)|x}{\lambda(x)^{2-2\alpha}}\right)\lambda(t)^{2-2\alpha}\frac{1}{r^4\log^{2b-1}(t)} \leq \frac{C}{r^{9/2}\log^{2b}(t)}, \quad t - t^{1/6} \leq r \leq t + t^{1/6}$$

This gives

$$||\partial_t^2 v_{4,c}||_{L^2(rdr)} \leq \frac{C}{t^{47/12}\log^{2b-1}(t)} + \frac{C\lambda(t)^{2-2\alpha}}{\log^{2N+2b-1}(t) t} \sup_{x\geq t}\left(\frac{|e''''(x)|x}{\lambda(x)^{2-2\alpha}}\right)$$

In total, this gives

$$|\partial_t^2 v_4(t,r)| \leq \begin{cases} \frac{C}{t^4\log^{3b+N-2}(t)} + \frac{C\lambda(t)^{2-2\alpha}}{t\log^{2b-3+N}(t)} \sup_{x\geq t}\left(\frac{|e''''(x)|x}{\lambda(x)^{2-2\alpha}}\right), & r \leq \frac{t}{2} \\ \frac{C}{t^{35/12}\log^{2b-1}(t)} + \frac{C\lambda(t)^{2-2\alpha}}{\log^{2N+2b-2}(t)} \sup_{x\geq t}\left(\frac{|e''''(x)|x}{\lambda(x)^{2-2\alpha}}\right), & r \geq \frac{t}{2} \end{cases}$$

Using the same estimates that were used for $\partial_t^j v_5$ in the course of proving the preliminary estimate on λ'''', we get

$$|\partial_t^2\left(-\lambda(t)\langle\left(\frac{\cos(2Q_{\frac{1}{\lambda(t)}}(r))-1}{r^2}\right)E_5|_{r=R\lambda(t)},\phi_0\rangle\right)|$$
$$\leq \frac{C}{t^4\log^{b+1}(t)} + \frac{C\sup_{x\geq t}\left(\frac{|e''''(x)|x}{\lambda(x)^{3-2\alpha}}\right)}{t\log^{(3-2\alpha)b}(t)}$$

$$|\partial_t^2\left(-\lambda(t)\langle\left(\frac{\cos(2Q_{\frac{1}{\lambda(t)}}(r))-1}{r^2}\right)v_4\left(1-\chi_{\geq 1}(\frac{4r}{t})\right)|_{r=R\lambda(t)},\phi_0\rangle\right)|$$
$$\leq \frac{C}{t^4\log^{2b+N-2}(t)} + \frac{C\lambda(t)^{2-2\alpha}\sup_{x\geq t}\left(\frac{|e''''(x)|x}{\lambda(x)^{2-2\alpha}}\right)}{t\log^{b-3+N}(t)}$$

$$|\partial_t^2\left(-\lambda(t)\langle\left(\frac{\cos(2Q_{\frac{1}{\lambda(t)}}(r))-1}{r^2}\right)v_5\left(1-\chi_{\geq 1}(\frac{4r}{t})\right)|_{r=R\lambda(t)},\phi_0\rangle\right)|$$
$$\leq \frac{C}{t^4\log^{3N-2}(t)}$$

$$|\partial_t^2\left(-\lambda(t)\langle\left(\frac{\cos(2Q_{\frac{1}{\lambda(t)}}(r))-1}{r^2}\right)\chi_{\geq 1}(\frac{2r}{\log^N(t)})\left(v_1+v_2+v_3\right)|_{r=R\lambda(t)},\phi_0\rangle\right)|$$
$$\leq \frac{C}{t^4\log^{3b+2N}(t)} + \frac{C\lambda(t)^{2-2\alpha}\sup_{x\geq t}\left(\frac{|e''''(x)|x}{\lambda(x)^{2-2\alpha}}\right)}{t\log^{2N+2b-1}(t)}$$

$$|\partial_t^2\left(\lambda(t)\langle\chi_{\geq 1}(\frac{2r}{\log^N(t)})F_{0,2}|_{r=R\lambda(t)},\phi_0\rangle\right)| \leq \frac{C}{t^4\log^{3b+1+2N-2b\alpha}(t)} + \frac{C|e''''(t)|}{\log^{2b+2N-2b\alpha}(t)}$$

Combining these, we get

$$|\partial_t^2 G(t,\lambda_{0,0}(t)+e(t))| \leq \frac{C}{t^4\log^{b+1}(t)} + C\sup_{x\geq t}\left(\frac{|e''''(x)|x}{\lambda(x)^{3-2\alpha}}\right)\frac{1}{t\log^{(3-2\alpha)b}(t)}$$

and

$$|RHS_4(t)| \leq \frac{C}{t^4\log^{b+1}(t)} + \frac{C\sup_{x\geq t}\left(\frac{x^2|e''''(x)|}{\lambda(x)^{3-2\alpha}}\right)}{\sqrt{\log(\log(t))}t^2\log^{(3-2\alpha)b}(t)}$$

We now return to (3.164), and note that it is of the form

$$\begin{aligned}&-4\int_t^\infty \frac{e''''(s)ds}{\log(\lambda_{0,0}(s))(1+s-t)} + 4\alpha e''''(t)\\ &-4\int_t^\infty \frac{e''''(s)ds}{\log(\lambda_{0,0}(s))(\lambda_{0,0}(t)^{1-\alpha}+s-t)(1+s-t)^3}\\ &= RHS_4(t)\end{aligned}$$

with

$$|RHS_4(t)| \leq \frac{C}{t^4\log^{b+1}(t)} + C\frac{\sup_{x\geq t}\left(\frac{x^2|e''''(x)|}{\lambda(x)^{3-2\alpha}}\right)}{\sqrt{\log(\log(t))}t^2\log^{(3-2\alpha)b}(t)}$$

We are now in the same situation as for e''', and repeating the procedure used there, we get

$$|e''''(t)| \leq \frac{C}{t^4\log^{b+1}(t)}$$

Using the explicit formula for $\lambda_{0,0}$, we finally conclude

$$|\lambda''''(t)| \leq \frac{C}{t^4\log^{b+1}(t)}, \quad t\geq T_0$$

To finish the proof of the proposition, we recall (3.12), and the fact that $\partial_t^2 v_1$ has the same formula as v_1, except for λ'''' replacing λ''. Using our estimate on λ'''', we get

$$|\partial_t^2 v_1(t,r)| \leq \frac{C}{rt^2\log^{b+1}(t)}$$

Combining this with (3.172), we get (3.150). □

3.0.9. Estimates on $\partial_r^k\partial_t^j F_4$. Later on, when we start to construct the exact solution to (1.1), we will utilize the orthogonality

$$\langle F_4, \phi_0(\frac{\cdot}{\lambda(t)})\rangle = 0, \quad t\geq T_0$$

by integrating by parts in various oscillatory integrals involving F_4. Therefore, we will need estimates on certain derivatives of F_4 in order to control the integrands which will arise in this process:

PROPOSITION 3.5. *For* $0 \leq k \leq 2, \quad 0 \leq j \leq 1, \quad j+k \leq 2$, *we have*
(3.174)
$$t^j r^k |\partial_r^k \partial_t^j F_4(t,r)| \leq \frac{C \mathbb{1}_{\{r \leq \log^N(t)\}} r}{t^2 \log^{3b+1-2\alpha b}(t)(r^2+\lambda(t)^2)^2} + \frac{C \mathbb{1}_{\{r \leq \frac{t}{2}\}} r}{t^2 \log^{5b+2N-2}(t)(r^2+\lambda(t)^2)^2}$$

In addition, we have
(3.175)
$$|\partial_t^2 F_4(t,r)| \leq \frac{C \mathbb{1}_{\{r \leq \log^N(t)\}} r}{t^4 \log^{3b+1-2\alpha b}(t)(r^2+\lambda(t)^2)^2} + \frac{C \mathbb{1}_{\{r \leq \frac{t}{2}\}} r}{t^4 \log^{5b+2N-2}(t)(r^2+\lambda(t)^2)^2} + \frac{C \mathbb{1}_{\{r \leq \frac{t}{2}\}}}{t^4 \log^{5b+N-2}(t)(r^2+\lambda(t)^2)^2}$$

PROOF. We start with a lemma which captures a delicate leading order cancellation near the origin between v_1 and v_2:

LEMMA 3.16 (Near origin cancellation between v_1 and v_2). *For* $0 \leq k, j \leq 2, \quad k+j \leq 2$, *we have*
$$t^j r^k |\partial_r^k \partial_t^j (v_1+v_2)| \leq \frac{C r \log(\log(t))}{t^2 \log^{b+1}(t)}, \quad r \leq \log^N(t)$$

PROOF. We use (3.11) and (3.49), to get
$$v_1(t,r)+v_2(t,r) = r\left(\int_t^\infty \frac{\lambda''(x)}{1+x-t}dx - \frac{b}{t^2 \log^b(t)}\right) + \text{Err}(t,r) + E_{v_2}(t,r), \quad 0 \leq r \leq \frac{t}{2}$$
Then, using the modulation equation, and the previous estimates on its terms, we get
(3.176)
$$|v_1(t,r)+v_2(t,r)| \leq \frac{C r \log(\log(t))}{t^2 \log^{b+1}(t)} + \frac{C r \log(3+2r)}{t^2 \log^{b+1}(t)} + \frac{Cr}{t^2 \log^{b+1}(t)} + \frac{Cr^2}{t^3 \log^b(t)}, \quad r \leq \frac{t}{2}$$
Note that, for (e.g) $r \leq \log^N(t)$, this estimate is slightly better than what would result from estimating v_1 and v_2 separately.

We use (3.13) and (3.49) to obtain the analogous estimate after taking an r derivative:
$$\partial_r v_1(t,r) + \partial_r v_2(t,r) = \int_t^\infty \frac{\lambda''(s)ds}{1+s-t} - \frac{b}{t^2 \log^b(t)} + E_{\partial_r v_1}(t,r) + E_{\partial_r v_2}(t,r)$$
Again, the modulation equation, combined with the estimates for $E_{\partial_r v_1}$ and $E_{\partial_r v_2}$ from the previous subsections, give
(3.177)
$$|\partial_r v_1(t,r) + \partial_r v_2(t,r)| \leq \frac{C \log(\log(t))}{t^2 \log^{b+1}(t)} + \frac{C \log(3+2r)}{t^2 \log^{b+1}(t)} + \frac{C}{t^2 \log^{b+1}(t)} + \frac{Cr}{t^3 \log^b(t)}, \quad r \leq \frac{t}{2}$$

We first note that, for $j=1,2$, $\partial_t^j v_1$ solves the same equation with 0 Cauchy data at infinity as v_1 does, except with $\lambda^{(2+j)}(t)$ on the right-hand side. Then, we use the fact that λ solves

$$\begin{aligned}&-4\int_t^\infty \frac{\lambda''(x)dx}{1+x-t}+\frac{4b}{t^2\log^b(t)}+4\alpha\log(\lambda(t))\lambda''(t)\\&-4\int_t^\infty \frac{\lambda''(s)ds}{(\lambda(t)^{1-\alpha}+s-t)(1+s-t)^3}\\&=G(t,\lambda(t))\end{aligned}$$

and differentiate j times for $j=1,2$, to get

$$|-4\int_t^\infty \frac{\lambda'''(s)}{1+s-t}ds-\frac{8b}{t^3\log^b(t)}|\leq \frac{C\log(\log(t))}{t^3\log^{b+1}(t)}$$

and

$$|\int_t^\infty \frac{\lambda''''(s)ds}{1+s-t}-\frac{6b}{t^4\log^b(t)}|\leq \frac{C\log(\log(t))}{t^4\log^{b+1}(t)}$$

Then, using (3.11),(3.49), we get

$$|\partial_t(v_1+v_2)(t,r)|\leq C\frac{r\log(\log(t))}{t^3\log^{b+1}(t)},\quad r\leq \log^N(t) \tag{3.178}$$

and

$$|\partial_t^2(v_1+v_2)(t,r)|\leq C\frac{r\log(\log(t))}{t^4\log^{b+1}(t)},\quad r\leq \log^N(t) \tag{3.179}$$

Next, we use the equations solved by v_1, v_2, to get

$$r^2\partial_{rr}(v_1+v_2)=r^2\left(\frac{-2r\lambda''(t)}{1+r^2}\right)+r^2\partial_{tt}(v_1+v_2)-r\partial_r(v_1+v_2)+(v_1+v_2)$$

which gives

$$r^2|\partial_{rr}(v_1+v_2)|\leq C\frac{r\log(\log(t))}{t^2\log^{b+1}(t)},\quad r\leq \log^N(t) \tag{3.180}$$

Finally, we study $\partial_{tr}(v_1+v_2)$. Because $\partial_t v_1$ has the same representation formula as v_1, except with λ'' replaced by λ''', we can use the same procedure used for $\partial_r v_1$, to get

$$\partial_{tr}v_1(t,r)=\int_t^\infty \frac{\lambda'''(s)ds}{1+s-t}+E_{\partial_{tr}v_1}(t,r)$$

with

$$|E_{\partial_{tr}v_1}(t,r)|\leq \frac{C\log(3+2r)}{t^3\log^{b+1}(t)}$$

Then, we use (3.49) to conclude

$$|\partial_{tr}(v_1+v_2)(t,r)|\leq \frac{C\log(\log(t))}{t^3\log^{b+1}(t)},\quad r\leq \log^N(t) \tag{3.181}$$

□

Next, we note that our previous estimate for $\partial_r v_3$, (3.79), sufficed for all of our purposes up to now. For estimating derivatives of F_4, however, we will require a different estimate which will turn out to be better in the region $r\leq \log^N(t)$, and which will require a more complicated argument than before. This refinement will also lead to slightly better estimates on $\partial_{tr}v_3$ and $r^2\partial_{rr}v_3$ near the origin.

LEMMA 3.17 (Near origin refinements of various v_3 related estimates).

$$|\partial_r v_3(t,r)| \leq \frac{C}{t^2 \log^{b+1}(t)} \left(\log(\log(t)) + \log(6+6r)\right) \tag{3.182}$$

$$|\partial_{tr} v_3(t,r)| \leq \frac{C \log(\log(t))}{t^3 \log^{b+1}(t)}, \quad r \leq \log^N(t)$$

$$r^2 |\partial_{rr} v_3| \leq \frac{Cr \log(\log(t))}{t^2 \log^{b+1}(t)}, \quad r \leq \log^N(t) \tag{3.183}$$

PROOF. We first use the same decomposition on $\partial_r v_3$ used before, and only need to use a different argument for one term, namely:

$$-\int_t^\infty \frac{1}{(s-t)} \int_0^{s-t} \rho \frac{\lambda''(s)}{r} \partial_r \left(\frac{-1-\rho^2+r^2}{\sqrt{(-1-\rho^2+r^2)^2+4r^2}} + F_3(r,\rho,\lambda(s))\right) d\rho ds$$

We treat different regions of the variable $s-t$. First, from our previous $\partial_r v_3$ estimate, we have

$$|\int_0^{s-t} \rho \frac{\lambda''(s)}{r} \partial_r \left(\frac{-1-\rho^2+r^2}{\sqrt{(-1-\rho^2+r^2)^2+4r^2}}\right) d\rho| \leq C(s-t)^2 |\lambda''(s)|, \quad s-t \leq \frac{1}{2}$$

which gives

$$\begin{aligned}&|-\int_t^{t+\frac{1}{2}} \frac{1}{(s-t)} \int_0^{s-t} \rho \frac{\lambda''(s)}{r} \partial_r \left(\frac{-1-\rho^2+r^2}{\sqrt{(-1-\rho^2+r^2)^2+4r^2}}\right) d\rho ds| \\ &\leq \frac{C}{t^2 \log^{b+1}(t)}\end{aligned}$$

Next, we have

$$\begin{aligned}&|-\int_{t+\frac{1}{2}}^{t+6} \frac{1}{(s-t)} \int_0^{s-t} \rho \frac{\lambda''(s)}{r} \partial_r \left(\frac{-1-\rho^2+r^2}{\sqrt{(-1-\rho^2+r^2)^2+4r^2}}\right) d\rho ds| \\ &\leq \int_{t+\frac{1}{2}}^{t+6} \frac{1}{(s-t)} \int_0^\infty \rho \frac{|\lambda''(s)|}{r} |\partial_r \left(\frac{-1-\rho^2+r^2}{\sqrt{(-1-\rho^2+r^2)^2+4r^2}}\right) | d\rho ds \\ &\leq C \int_{t+\frac{1}{2}}^{t+6} \frac{|\lambda''(s)|}{(s-t)} ds \\ &\leq C \frac{1}{t^2 \log^{b+1}(t)}\end{aligned}$$

For convenience, we recall the definition of F_3:

$$F_3(r,\rho,\lambda(s)) = \frac{1-(r^2-\rho^2)\lambda(s)^{2\alpha-2}}{\sqrt{4r^2\lambda(s)^{2\alpha-2} + (1-(r^2-\rho^2)\lambda(s)^{2\alpha-2})^2}}$$

Similarly, we have

$$|\partial_r F_3(r,\rho,\lambda(s))| \leq \frac{Cr\lambda(s)^{2+2\alpha}((\rho^2+r^2)\lambda(s)^{2\alpha-2}+1)}{\lambda(s)^4(1+2(\rho^2+r^2)\lambda(s)^{2\alpha-2} + (\rho^2-r^2)^2\lambda(s)^{4\alpha-4})^{3/2}}$$

For the region $s-t \leq \lambda(t)^{1-\alpha}$, we continue this estimate, and get

$$|\partial_r F_3(r,\rho,\lambda(s))| \leq Cr\lambda(s)^{2\alpha-2}$$

So,

$$\begin{aligned}
&|\int_t^{t+\lambda(t)^{1-\alpha}} \frac{1}{(s-t)} \int_0^{s-t} \frac{\rho\lambda''(s)}{r} \partial_r F_3(r,\rho,\lambda(s)) d\rho ds| \\
&\leq C \int_t^{t+\lambda(t)^{1-\alpha}} \frac{1}{(s-t)} \int_0^{s-t} \frac{\rho}{r} |\lambda''(s)| r \lambda(s)^{2\alpha-2} d\rho ds \\
&\leq \frac{C}{t^2 \log^{b+1}(t)}
\end{aligned}$$

$$\begin{aligned}
&|\int_{t+\lambda(t)^{1-\alpha}}^{t+6} \frac{1}{(s-t)} \int_0^{s-t} \frac{\rho\lambda''(s)}{r} \partial_r F_3(r,\rho,\lambda(s)) d\rho ds| \\
&\leq C \int_{t+\lambda(t)^{1-\alpha}}^{t+6} \frac{1}{(s-t)} \int_0^{\infty} \frac{\rho|\lambda''(s)|}{r} |\partial_r F_3(r,\rho,\lambda(s))| d\rho ds \\
&\leq \frac{C}{r} \int_{t+\lambda(t)^{1-\alpha}}^{t+6} \frac{|\lambda''(s)|}{(s-t)} r ds \leq \frac{C\log(\log(t))}{t^2 \log^{b+1}(t)}
\end{aligned}$$

Then, we treat the region $6 \leq s-t \leq 6+6r$ in a similar fashion:

$$\begin{aligned}
&|-\int_{t+6}^{t+6+6r} \frac{1}{(s-t)} \int_0^{s-t} \rho\frac{\lambda''(s)}{r} \partial_r \left(\frac{-1-\rho^2+r^2}{\sqrt{(-1-\rho^2+r^2)^2+4r^2}} + F_3(r,\rho,\lambda(s))\right) d\rho ds| \\
&\leq \int_{t+6}^{t+6+6r} \frac{1}{(s-t)} \int_0^{\infty} \rho\frac{|\lambda''(s)|}{r} \left(|\partial_r \left(\frac{-1-\rho^2+r^2}{\sqrt{(-1-\rho^2+r^2)^2+4r^2}}\right)| + |\partial_r F_3(r,\rho,\lambda(s))|\right) d\rho ds \\
&\leq C \int_{t+6}^{t+6+6r} \frac{|\lambda''(s)|}{(s-t)} ds \\
&\leq \frac{C\log(6+6r)}{t^2 \log^{b+1}(t)}
\end{aligned}$$

Now, we study the region $s-t \geq 6+6r$:
We first note that

$$\int_0^{\infty} \partial_r \left(\frac{-1-\rho^2+r^2}{\sqrt{(1+\rho^2-r^2)^2+4r^2}} + F_3(r,\rho,\lambda(s))\right) \rho d\rho = 0$$

which follows from the fact that the integrand of the expression above is equal to

$$\begin{aligned}
&\partial_\rho \left(r\left(\frac{-1+\rho^2-r^2}{\sqrt{\rho^4-2\rho^2(-1+r^2)+(1+r^2)^2}}\right.\right. \\
&\qquad\left.\left. + \frac{1+(r^2-\rho^2)\lambda(s)^{2\alpha-2}}{\sqrt{1+2(\rho^2+r^2)\lambda(s)^{2\alpha-2}+(\rho^2-r^2)^2\lambda(s)^{4\alpha-4}}}\right)\right)
\end{aligned}$$

Then, we use

$$r\left(\frac{-1+(s-t)^2-r^2}{\sqrt{(s-t)^4-2(s-t)^2(-1+r^2)+(1+r^2)^2}}\right) = r\left(1+E_{\partial_r v_3,1}\right)$$

with

$$|E_{\partial_r v_3,1}| \leq C\frac{(1+r)^2}{(s-t)^2}, \quad s-t \geq 6+6r$$

and

$$r\left(\frac{1+(r^2-(s-t)^2)\lambda(s)^{2\alpha-2}}{\sqrt{1+2((s-t)^2+r^2)\lambda(s)^{2\alpha-2}+((s-t)^2-r^2)^2\lambda(s)^{4\alpha-4}}}\right)=r\left(-1+E_{\partial_r v_3,2}\right)$$

with

$$|E_{\partial_r v_3,2}|\leq C\frac{(1+r)^2}{(s-t)^2},\quad s-t\geq 6+6r$$

which gives

$$\begin{aligned}
&|-\int_{t+6+6r}^{\infty}\frac{1}{(s-t)}\int_0^{s-t}\rho\frac{\lambda''(s)}{r}\partial_r\left(\frac{-1-\rho^2+r^2}{\sqrt{(-1-\rho^2+r^2)^2+4r^2}}+F_3(r,\rho,\lambda(s))\right)d\rho ds|\\
&=|\int_{t+6+6r}^{\infty}\frac{1}{(s-t)}\int_{s-t}^{\infty}\rho\frac{\lambda''(s)}{r}\partial_r\left(\frac{-1-\rho^2+r^2}{\sqrt{(-1-\rho^2+r^2)^2+4r^2}}+F_3(r,\rho,\lambda(s))\right)d\rho ds|\\
&=|\int_{t+6+6r}^{\infty}\frac{1}{(s-t)}\frac{\lambda''(s)}{r}r\left(E_{\partial_r v_3,1}+E_{\partial_r v_3,2}\right)ds|\\
&\leq C\int_{t+6+6r}^{\infty}\frac{|\lambda''(s)|}{(s-t)}\frac{(1+r)^2}{(s-t)^2}ds\\
&\leq\frac{C}{t^2\log^{b+1}(t)}
\end{aligned}$$

In total, we get

$$|\partial_r v_3(t,r)|\leq\frac{C}{t^2\log^{b+1}(t)}\left(\log(\log(t))+\log(6+6r)\right)$$

Note that this estimate is better than (3.79), in the region $r\leq\log^N(t)$. For $\partial_{tr}v_3$, we recall that the partial r derivative of the second line of (3.147) was bounded above in absolute value by $\frac{C}{t^3\log^{b+1}(t)}$, and an estimate of the r derivative of the first line of (3.147) was inferred from an estimate on $\partial_r v_3$. Using the above near origin refinement of the $\partial_r v_3$ estimate, instead of the previous one, gives

$$|\partial_{tr}v_3(t,r)|\leq\frac{C\log(\log(t))}{t^3\log^{b+1}(t)},\quad r\leq\log^N(t)$$

Now, we can prove a different estimate for $r^2\partial_r^2 v_3$ than what follows from previous work.

$$r^2\partial_{rr}v_3=r^2F_{0,1}(t,r)+r^2\partial_{tt}v_3-r\partial_r v_3+v_3$$

Using (3.182), as well as our previous estimates on v_3, we get

$$r^2|\partial_{rr}v_3|\leq\frac{Cr\log(\log(t))}{t^2\log^{b+1}(t)},\quad r\leq\log^N(t)$$ □

Finally, the estimate for $\partial_t v_5$ used when estimating λ''' was based on estimates for $N_2(f)$ and $\partial_r N_2(f)$ which only used at most two derivatives of λ. For future use, we will prove a stronger estimate on $\partial_t v_5$, which uses the final estimate on λ'''. (Note that this is the reason why no estimates on $\partial_t v_5$ were presented in the proposition statement in the λ''' section). As part of the process, we will also obtain an estimate on $\partial_{tr}v_4$.

LEMMA 3.18 (Improved $\partial_t v_5$, $\partial_{tr}v_4$ estimates).

$$|\partial_t v_5(t,r)|\leq\frac{Cr}{t^{9/2}\log^{\frac{3N}{2}+3b-2}(t)},\quad r\leq\frac{t}{2}\tag{3.184}$$

(3.185)
$$|\partial_{tr}v_4(t,r)| \leq \begin{cases} \frac{C}{t^3\log^{3b+2N-2}(t)}, & r\leq \frac{t}{2} \\ \frac{C}{t^{35/12}\log^{2b-1}(t)}, & r\geq \frac{t}{2} \end{cases}$$

PROOF. For this estimate, we start with

$$\begin{aligned}&\partial_t v_5(t,r)\\ &=\frac{-r}{2\pi}\int_0^1\int_t^\infty\int_{B_{s-t}(0)}\frac{1}{\sqrt{(s-t)^2-|y|^2}}\left(\frac{\partial_{12}N_2(f_{v_5})(s,|\beta x+y|)\left((\beta x+y)\cdot\hat{x}\right)^2}{|\beta x+y|^2}\right.\\ &-\frac{\partial_1 N_2(f_{v_5})(s,|\beta x+y|)\left((\beta x+y)\cdot\hat{x}\right)^2}{|\beta x+y|^3}\\ &\left.+\frac{\partial_1 N_2(f_{v_5})(s,|\beta x+y|)}{|\beta x+y|}\right)dA(y)dsd\beta\end{aligned}$$

This gives

(3.186)
$$\begin{aligned}&|\partial_t v_5(t,r)|\\ &\leq Cr\int_0^1\int_t^\infty\int_{B_{s-t}(0)}\frac{1}{\sqrt{(s-t)^2-|y|^2}}\left(|\partial_{12}N_2(f_{v_5})|(s,|\beta x+y|)+\frac{|\partial_1 N_2(f_{v_5})|(s,|\beta x+y|)}{|\beta x+y|}\right)dA(y)dsd\beta\end{aligned}$$

In order to proceed, we will have to estimate $\partial_{tr}v_4$, which was not done previously. We start by noting that, the procedure used to obtain (3.149) was to write

$$\partial_t v_4(t,r) = r\int_0^1 \partial_{tr}v_4(t,r\beta)d\beta$$

and then, to estimate $\partial_{tr}v_4(t,r\beta)$ uniformly for $0\leq\beta\leq 1$. So, in the region $r\leq\frac{t}{2}$, we have

$$|\partial_{tr}v_4(t,r)|\leq\frac{C}{t^3\log^{3b+2N-2}(t)},\quad r\leq\frac{t}{2}$$

For the region $r\geq\frac{t}{2}$, we start by introducing the vector field

$$V = t\partial_t + r\partial_r$$

and recalling that

$$-\partial_{tt}v_4+\partial_{rr}v_4+\frac{1}{r}\partial_r v_4-\frac{v_4}{r^2}=v_{4,c}$$

So, we have

$$\left(-\partial_{tt}+\partial_{rr}+\frac{1}{r}\partial_r-\frac{1}{r^2}\right)\left(V(\partial_t v_4)\right)=V\left(\partial_t v_{4,c}\right)+2\partial_t v_{4,c}$$

We use (3.148), (3.173), and

$$|\partial_t^2 v_{4,c}(t,r)|+|\partial_{tr}v_{4,c}(t,r)|\leq\frac{C}{t^{9/2}\log^{2b}(t)},\quad t-t^{1/6}\leq r\leq t+t^{1/6}$$

to get

$$|V(\partial_t v_{4,c})|(t,r)\leq C\begin{cases}\frac{\mathbb{1}_{\{r\geq\frac{\log^N(t)}{2}\}}}{r^3t^3\log^{3b}(t)}, & r\leq\frac{t}{2}\\ \frac{\log(r)}{r^3\log^{2b}(t)|t-r|^3}+\frac{\log(r)}{r^3t\log^{2b}(t)(t-r)^2}+\frac{\log(r)}{\log^{2b+1}(t)r^4t|t-r|} & \\ +\frac{\log(r)}{r^3\log^{3b+1}(t)t^3}, & \frac{t}{2}\leq r\leq t-t^{1/6},\text{ or } r\geq t+t^{1/6}\\ \frac{1}{t^{7/2}\log^{2b}(t)}, & t-t^{1/6}\leq r\leq t+t^{1/6}\end{cases}$$

and we get

$$||V(\partial_t v_{4,c})||_{L^2(rdr)}\leq\frac{C}{t^{35/12}\log^{2b-1}(t)}$$

Next, we recall (3.143), and get

$$||\partial_t v_{4,c}||_{L^2(rdr)} \leq \frac{C}{t^3 \log^{2N+3b}(t)}$$

Then, we apply the same procedure used before to estimate (e.g.) $\partial_t v_4$ in the region $r \geq \frac{t}{2}$, to the equation for $V(\partial_t v_4)$, and get

$$|V(\partial_t v_4)|(t,r) \leq \frac{C}{t^{23/12} \log^{2b-1}(t)}$$

But, then,

$$\begin{aligned}\partial_{tr} v_4(t,r) &= \frac{(t\partial_t^2 v_4 + r\partial_{tr} v_4)}{r} - \frac{t}{r}\partial_t^2 v_4 \\ &= \frac{V(\partial_t v_4)}{r} - \frac{t}{r}\partial_t^2 v_4\end{aligned}$$

and we have already estimated $\partial_t^2 v_4$, during our study of λ''''. So, we get

$$|\partial_{tr} v_4(t,r)| \leq \frac{C}{t^{35/12} \log^{2b-1}(t)}, \quad r \geq \frac{t}{2}$$

Now, we can estimate $\partial_{tr} N_2(f)$, and get

$$\begin{aligned}&|\partial_{tr} N_2(f)|(t,r) \\ &\leq \begin{cases} \frac{C}{t^5 \log^{3b}(t)(r^2+\lambda(t)^2)}, & r \leq \frac{t}{2} \\ \frac{C\log^3(r)}{r^2|t-r|^5} + \frac{C\log^3(r)}{r^{5/2}\sqrt{t}(t-r)^4} + \frac{C\log^2(r)}{r^2 t^{3/2}|t-r|^3 \log^{3b-1+\frac{5N}{2}}(t)}, & t > r > \frac{t}{2}\end{cases}\end{aligned}$$

(Note that we only need to estimate $\partial_{tr} N_2(f_{v_5})$ in the region $r < t$, which is why we used $\frac{1}{|t-r|} \geq \frac{1}{t}$ to simply the above estimate. In previous estimates, we did not proceed analogously because we eventually used a single estimate for the entire region $\frac{t}{2} \leq r \leq t - t^\gamma$, or $r \geq t + t^\gamma$ for some $\gamma > 0$).

Using the estimates on $\partial_t^j v_k, \quad j = 0,1, \quad k = 1,2,3,4$, we get

$$|\partial_t N_2(f)(t,r)| \leq \begin{cases} \frac{Cr}{t^5 \log^{3b}(t)(r^2+\lambda(t)^2)}, & r \leq \frac{t}{2} \\ \frac{C}{r^3 t^{7/2} \log^{\frac{3N}{2}+5b}(t)} + \frac{C\log^3(r)}{r^2(t-r)^4}, & t > r > \frac{t}{2}\end{cases}$$

Now, we return to (3.186), and use the same procedure used to estimate v_5 in the region $r \leq \frac{t}{2}$ earlier. (In particular, we handle the contribution to $\partial_t v_5$ coming from the term $\frac{\log^3(r)}{r^2|t-r|^5}$, which arises in the estimate for $\partial_{tr} N_2(f_{v_5})$, using the same procedure as in (3.96)). This results in

$$|\partial_t v_5(t,r)| \leq \frac{Cr}{t^{9/2} \log^{\frac{3N}{2}+3b-2}(t)}, \quad r \leq \frac{t}{2}$$

□

Now, we proceed to estimate F_4, and various of its derivatives. We recall

$$\begin{aligned}F_4(t,r) &= \left(1 - \chi_{\geq 1}\left(\frac{2r}{\log^N(t)}\right)\right)\left(F_{0,2}(t,r) + \left(\frac{\cos(2Q_{\frac{1}{\lambda(t)}}(r)) - 1}{r^2}\right)(v_1+v_2+v_3)\right) \\ &+ \left(1 - \chi_{\geq 1}\left(\frac{4r}{t}\right)\right)\left(\frac{\cos(2Q_{\frac{1}{\lambda(t)}}(r)) - 1}{r^2}\right)(v_4+v_5)\end{aligned}$$

Then, using (3.77), (3.83), (3.91), and (3.176), we get

$$|F_4(t,r)| \leq C\frac{\left(1-\chi_{\geq 1}(\frac{2r}{\log^N(t)})\right)}{(r^2+\lambda(t)^2)^2}\left(\frac{r}{t^2\log^{3b+1-2b\alpha}(t)}\right) + C\frac{\left(1-\chi_{\geq 1}(\frac{4r}{t})\right)}{(r^2+\lambda(t)^2)^2}\frac{r}{t^2\log^{5b+2N-1}(t)} \tag{3.187}$$

Next, we will need to record estimates on $\partial_t F_4$, and $r\partial_r F_4$. By combining (3.122), and (3.184), we get

$$|\partial_t v_4 + \partial_t v_5|(t,r) \leq \frac{Cr}{t^3\log^{3b+2N-2}(t)}, \quad r \leq \frac{t}{2}$$

This, combined with the explicit formula for $F_{0,2}$, (3.178), and (3.119), gives

$$|\partial_t F_4(t,r)| \leq \frac{C\mathbb{1}_{\{r\leq \log^N(t)\}}r}{t^3\log^{3b+1-2b\alpha}(t)(r^2+\lambda(t)^2)^2} + \frac{C\mathbb{1}_{\{r\leq\frac{t}{2}\}}}{(r^2+\lambda(t)^2)^2}\frac{r}{t^3\log^{5b+2N-2}(t)}$$

where we estimate derivatives on the cutoff functions by (e.g.)

$$|\chi_{\geq 1}'(\frac{2r}{\log^N(t)})|\frac{Nr}{t\log^{N+1}(t)} \leq \frac{C\mathbb{1}_{\{r\leq\log^N(t)\}}}{t\log(t)}$$

Similarly, for the partial r derivative of F_4, we use the explicit formula for $F_{0,2}$, (3.182), (3.177), (3.84), and (3.92), to get

$$r|\partial_r F_4(t,r)| \leq \frac{Cr\mathbb{1}_{\{r\leq\log^N(t)\}}}{t^2\log^{3b+1-2b\alpha}(t)(r^2+\lambda(t)^2)^2} + C\frac{\mathbb{1}_{\{r\leq\frac{t}{2}\}}}{(r^2+\lambda(t)^2)^2}\frac{r}{t^2\log^{5b+2N-1}(t)} \tag{3.188}$$

Treating the terms involving $\chi_{\geq 1}$ and $F_{0,2}$ as before, and using (3.179), (3.151), and (3.152), we get

$$|\partial_t^2 F_4(t,r)| \leq \frac{C\mathbb{1}_{\{r\leq\log^N(t)\}}r}{t^4(r^2+\lambda(t)^2)^2\log^{3b+1-2b\alpha}(t)} + \frac{C\mathbb{1}_{\{r\leq\frac{t}{2}\}}r}{t^4\log^{5b+2N-2}(t)(r^2+\lambda(t)^2)^2} + \frac{C\mathbb{1}_{\{r\leq\frac{t}{2}\}}}{t^4\log^{5b+N-2}(t)(r^2+\lambda(t)^2)^2}$$

Next, we note that, exactly as was the case with $\partial_{tr}v_4$, we can infer the following estimate on $\partial_{tr}v_5$ by inspecting (3.184), and the procedure used to obtain it:

$$|\partial_{tr}v_5(t,r)| \leq \frac{C}{t^{9/2}\log^{\frac{3N}{2}+3b-2}(t)}, \quad r \leq \frac{t}{2} \tag{3.189}$$

We use the same procedure, (3.181), (3.189), and (3.185), to get

$$r|\partial_{rt}F_4(t,r)| \leq C\frac{\mathbb{1}_{\{r\leq\log^N(t)\}}r}{t^3\log^{3b+1-2b\alpha}(t)(r^2+\lambda(t)^2)^2} + C\frac{\mathbb{1}_{\{r\leq\frac{t}{2}\}}r}{t^3\log^{5b+2N-2}(t)(r^2+\lambda(t)^2)^2}$$

Next, we use the same procedure used for v_1+v_2, v_3 to estimate $r^2\partial_r^2 v_4, r^2\partial_r^2 v_5$:

$$r^2\partial_r^2 v_4 = r^2 v_{4,c}(t,r) + r^2\partial_{tt}v_4 - r\partial_r v_4 + v_4$$

So,

(3.190)
$$\begin{aligned}&|r^2\partial_r^2 v_4(t,r)|\\ &\leq \frac{Cr}{t^2\log^{3b+2N}(t)}+\frac{Cr^2}{t^4\log^{3b-2+N}(t)}+\frac{Cr}{t^2\log^{3b+2N-1}(t)}+\frac{Cr}{t^2\log^{3b+2N-1}(t)}\\ &\leq \frac{Cr}{t^2\log^{3b+2N-1}(t)}, \quad r\leq \frac{t}{2}\end{aligned}$$

and

(3.191)
$$\begin{aligned}&|r^2\partial_r^2 v_5(t,r)|\\ &\leq \frac{Cr}{t^4\log^{3b}(t)}+\frac{Cr^2}{t^4\log^{3N+b-2}(t)}+\frac{Cr}{t^{7/2}\log^{\frac{5N}{2}+3b-3}(t)}+\frac{Cr}{t^{7/2}\log^{\frac{5N}{2}+3b-3}(t)}\\ &\leq \frac{Cr}{t^3\log^{3N+b-2}(t)}, \quad r\leq \frac{t}{2}\end{aligned}$$

By using (3.180), (3.183), (3.190), and (3.191), we get

$$r^2|\partial_{rr}F_4(t,r)|\leq C\frac{\mathbb{1}_{\{r\leq \log^N(t)\}}r}{t^2\log^{3b+1-2b\alpha}(t)(r^2+\lambda(t)^2)^2}+\frac{C\mathbb{1}_{\{r\leq\frac{t}{2}\}}r}{t^2\log^{5b+2N-1}(t)(r^2+\lambda(t)^2)^2}$$
□

3.0.10. Estimates on F_5. Recall that F_5 was one of the v_6-independent error terms on the right-hand side of (3.26) which was not included in the modulation equation, and is therefore, not necessarily orthogonal to $\phi_0(\frac{\cdot}{\lambda(t)})$. Hence, we must prove that it decays sufficiently quickly in sufficiently many norms. This is the result of the following lemma.

LEMMA 3.19. *We have the following estimates on F_5:*

$$\frac{1}{\lambda(x)^2}||F_5(x,r)||_{L^2(rdr)}\leq \frac{C\log^{6+2b}(x)}{x^{17/4}}$$

$$\frac{||F_5(x,\cdot\lambda(x))||_{\dot{H}^1_e}}{\lambda(x)}\leq\frac{C\log^{6+b}(x)}{x^{35/8}}$$

PROOF. We recall (3.28):

$$\begin{aligned}F_5(t,r)&=N_2(v_5)(t,r)\\ &+\frac{\sin(2(v_1+v_2+v_3+v_4))}{2r^2}\left(\cos(2Q_{\frac{1}{\lambda(t)}}+2v_5)-\cos(2Q_{\frac{1}{\lambda(t)}})\right)\\ &+\left(\frac{\cos(2(v_1+v_2+v_3+v_4))-1}{2r^2}\right)\left(\sin(2Q_{\frac{1}{\lambda(t)}}+2v_5)-\sin(2Q_{\frac{1}{\lambda(t)}})\right)\end{aligned}$$

Then, we start with

$$|N_2(v_5)|(t,r)\leq\begin{cases}\frac{Cr}{(r^2+\lambda(t)^2)t^7\log^{7b-6+5N}(t)}, & r\leq\frac{t}{2}\\ \frac{C\log^8(t)}{r^{7/2}\log^b(t)t^{7/2}}, & r\geq\frac{t}{2}\end{cases}$$

Next, we have

$$\begin{aligned}&|\frac{\sin(2(v_1+v_2+v_3+v_4))}{2r^2}\left(\cos(2Q_{\frac{1}{\lambda(t)}})\left(\cos(2v_5)-1\right)-\sin(2Q_{\frac{1}{\lambda(t)}})\sin(2v_5)\right)|\\&\leq C\frac{|v_1+v_2+v_3+v_4|}{r^2}\left(v_5(t,r)^2+\frac{r\lambda(t)}{r^2+\lambda(t)^2}|v_5(t,r)|\right)\end{aligned}$$

$$\begin{aligned}&|\left(\frac{\cos(2(v_1+v_2+v_3+v_4))-1}{2r^2}\right)\left(\sin(2Q_{\frac{1}{\lambda(t)}})\left(\cos(2v_5)-1\right)+\cos(2Q_{\frac{1}{\lambda(t)}})\sin(2v_5)\right)|\\&\leq\frac{C(v_1+v_2+v_3+v_4)^2}{r^2}\left(\frac{r\lambda(t)v_5^2}{r^2+\lambda(t)^2}+|v_5|\right)\end{aligned}$$

Using our previous pointwise estimates on v_1, v_2, v_3, v_4, v_5, we get

$$|F_5(t,r)|\leq\begin{cases}\frac{Cr}{(r^2+\lambda(t)^2)t^{11/2}\log^{5b-3+\frac{5N}{2}}(t)}, & r\leq\frac{t}{2}\\ \frac{C\log(r)\log^4(t)}{r^3t^2\log^b(t)|t-r|}+\frac{C\log^4(t)\log^2(r)}{(t-r)^2r^{5/2}t^{3/2}}\\ +\frac{C\log^4(t)}{r^{7/2}t^3\log^{\frac{3N}{2}+4b-1}(t)}, & \frac{t}{2}\leq r\leq t-\sqrt{t},\text{ or } r>t+\sqrt{t}\\ \frac{\log^4(t)}{r^{7/2}t^{3/2}}, & t-\sqrt{t}\leq r\leq t+\sqrt{t}\end{cases}$$

This gives

$$\frac{1}{\lambda(x)^2}||F_5(x,r)||_{L^2(rdr)}\leq\frac{C\log^{6+2b}(x)}{x^{17/4}}$$

We then proceed to estimate $||F_5(t,\cdot\lambda(t))||_{\dot{H}^1_e}$
We start with

$$\begin{aligned}|\partial_rN_2(v_5)(t,r)|\leq&\frac{C\lambda(t)}{r^2(r^2+\lambda(t)^2)}v_5(t,r)^2+\frac{C\lambda(t)|v_5(t,r)\partial_rv_5(t,r)|}{r(r^2+\lambda(t)^2)}\\&+\frac{C|\partial_rv_5(t,r)|v_5(t,r)^2}{r^2}+\frac{C|v_5(t,r)|^3}{r^3}\end{aligned}\tag{3.192}$$

For the two terms on the right-hand side of the above equation which involve $\partial_r v_5$, we estimate as follows:

$$\begin{aligned}&||\frac{\lambda(t)|v_5(t,r)\partial_rv_5(t,r)|}{r(r^2+\lambda(t)^2)}|_{r=R\lambda(t)}||_{L^2(RdR)}\\&\leq C\lambda(t)||\frac{v_5(t,r)}{r(r^2+\lambda(t)^2)}||_{L^\infty}\cdot||(\partial_2v_5)(t,\cdot\lambda(t))||_{L^2(RdR)}\leq\frac{C}{t^{21/4}\log^{b-6+\frac{5N}{2}}(t)}\end{aligned}$$

$$\begin{aligned}||\frac{|\partial_rv_5(t,r)|v_5(t,r)^2}{r^2}|_{r=R\lambda(t)}||_{L^2(RdR)}&\leq||(\partial_2v_5)(t,\cdot\lambda(t))||_{L^2(RdR)}\cdot||\frac{v_5^2}{r^2}||_{L^\infty}\\&\leq\frac{C\log^{11+b}(t)}{t^{31/4}}\end{aligned}$$

For the other terms in (3.192), we use the v_5 pointwise estimates. In total, we get

$$||(\partial_2N_2(v_5))(t,\cdot\lambda(t))||_{L^2(RdR)}\leq\frac{C}{t^{21/4}\log^{b-6+\frac{5N}{2}}(t)}$$

Next, we have

$$\partial_r \left(\frac{\sin(2(v_1+v_2+v_3+v_4))}{2r^2} \left(\cos(2Q_{\frac{1}{\lambda(t)}}+2v_5) - \cos(2Q_{\frac{1}{\lambda(t)}}) \right) \right)$$
$$= \frac{\cos(2(v_1+v_2+v_3+v_4))\partial_r(v_1+v_2+v_3+v_4)}{r^2} \left(\cos(2Q_{\frac{1}{\lambda(t)}}+2v_5) - \cos(2Q_{\frac{1}{\lambda(t)}}) \right)$$
$$- \frac{\sin(2(v_1+v_2+v_3+v_4))}{r^3} \left(\cos(2Q_{\frac{1}{\lambda(t)}}+2v_5) - \cos(2Q_{\frac{1}{\lambda(t)}}) \right)$$
$$+ \frac{\sin(2(v_1+v_2+v_3+v_4))}{r^2} \left(-\sin(2Q_{\frac{1}{\lambda(t)}}+2v_5)\partial_r(Q_{\frac{1}{\lambda(t)}}+v_5) + \sin(2Q_{\frac{1}{\lambda(t)}})\partial_r Q_{\frac{1}{\lambda(t)}} \right)$$

For $\partial_r v_k$, $\quad k=1,2,4$, we use the pointwise estimates from the previous subsections. On the other hand, for $\partial_r v_3$, we use the energy estimate procedure, previously used for $\partial_r v_5$, to get

$$||\partial_r v_3(t)||_{L^2(rdr)} \leq C\int_t^\infty ||F_{0,1}(s)||_{L^2(rdr)} ds \leq C\int_t^\infty \frac{\sqrt{\log(\log(s))}}{s^2 \log^{b+1}(s)} ds$$
$$\leq \frac{C\sqrt{\log(\log(t))}}{t\log^{b+1}(t)}$$

Then, we have

$$||\frac{\cos(2(v_1+v_2+v_3+v_4))\partial_r(v_1+v_2+v_3+v_4)}{r^2} \left(\cos(2Q_{\frac{1}{\lambda(t)}}+2v_5) - \cos(2Q_{\frac{1}{\lambda(t)}}) \right) |_{r=R\lambda(t)}||_{L^2(RdR)}$$
$$\leq ||\frac{\partial_r(v_1+v_2+v_4)}{r^2} \left(\cos(2Q_{\frac{1}{\lambda(t)}}+2v_5) - \cos(2Q_{\frac{1}{\lambda(t)}}) \right) |_{r=R\lambda(t)}||_{L^2(RdR)}$$
$$+ ||(\partial_2 v_3)(t,R\lambda(t))||_{L^2(RdR)} || \frac{\left(\cos(2Q_{\frac{1}{\lambda(t)}}+2v_5) - \cos(2Q_{\frac{1}{\lambda(t)}}) \right)}{r^2} ||_{L^\infty}$$
$$\leq \frac{C\sqrt{\log(\log(t))}}{t^{9/2}\log^{2b-2+\frac{5N}{2}}(t)}$$

Using the pointwise estimates on v_5, we get

$$|| - \frac{\sin(2(v_1+v_2+v_3+v_4))}{r^3} \left(\cos(2Q_{\frac{1}{\lambda(t)}}+2v_5) - \cos(2Q_{\frac{1}{\lambda(t)}}) \right) |_{r=R\lambda(t)}||_{L^2(RdR)}$$
$$\leq \frac{C}{t^{11/2}\log^{3b-3+\frac{5N}{2}}(t)}$$

We use

(3.193)
$$|\frac{\sin(2(v_1+v_2+v_3+v_4))}{r^2} \left(-\sin(2Q_{\frac{1}{\lambda(t)}}+2v_5)\partial_r(Q_{\frac{1}{\lambda(t)}}+v_5) + \sin(2Q_{\frac{1}{\lambda(t)}})\partial_r Q_{\frac{1}{\lambda(t)}} \right)|$$
$$\leq \frac{C|v_1+v_2+v_3+v_4|}{r^2} \left(\frac{r\lambda(t)^2 v_5^2}{(r^2+\lambda(t)^2)^2} + \frac{\lambda(t)|v_5(t,r)|}{r^2+\lambda(t)^2} + \frac{|\partial_r v_5| r\lambda(t)}{r^2+\lambda(t)^2} + |v_5 \partial_r v_5| \right)$$

The first term in (3.193) which involves $\partial_r v_5$ is estimated as follows:

$$||\frac{|v_1+v_2+v_3+v_4|}{r^2} \left(\frac{|\partial_r v_5| r\lambda(t)}{r^2+\lambda(t)^2} \right) |_{r=R\lambda(t)}||_{L^2(RdR)}$$
$$\leq C||\frac{|v_1+v_2+v_3+v_4|}{r^2} \left(\frac{|\partial_r v_5| r\lambda(t)}{r^2+\lambda(t)^2} \right) |_{r=R\lambda(t)}||_{L^2((0,\frac{t}{2\lambda(t)}),RdR)} \tag{3.194}$$
$$+ C||(\partial_2 v_5)(t,R\lambda(t))||_{L^2(RdR)} \cdot ||\frac{(v_1+v_2+v_3+v_4)}{r(r^2+\lambda(t)^2)}\lambda(t)||_{L^\infty(r\geq\frac{t}{2})}$$

where the second line of (3.194) is estimated using the pointwise estimates on $\partial_r v_5$ in the region $r \leq \frac{t}{2}$. In total, we get

$$||\frac{|v_1+v_2+v_3+v_4|}{r^2}\left(\frac{|\partial_r v_5| r\lambda(t)}{r^2+\lambda(t)^2}\right)|_{r=R\lambda(t)}||_{L^2(RdR)} \leq \frac{C\log^4(t)}{t^{21/4}}$$

The second term of (3.193) which involves $\partial_r v_5$ is estimated as follows:

$$||\frac{v_1+v_2+v_3+v_4}{r}||_{L^\infty} \cdot ||\frac{v_5}{r}||_{L^\infty} \cdot ||(\partial_2 v_5)(t,\cdot\lambda(t))||_{L^2(RdR)} \leq \frac{C\log^8(t)}{t^{\frac{25}{4}}\lambda(t)}$$

The other terms of (3.193) are estimated using the v_k pointwise estimates for $1 \leq k \leq 5$. In total, we get

$$||\frac{\sin(2(v_1+v_2+v_3+v_4))}{r^2}\left(-\sin(2Q_{\frac{1}{\lambda(t)}}+2v_5)\partial_r(Q_{\frac{1}{\lambda(t)}}+v_5)+\sin(2Q_{\frac{1}{\lambda(t)}})\partial_r Q_{\frac{1}{\lambda(t)}}\right)|_{r=R\lambda(t)}||_{L^2(RdR)}$$
$$\leq \frac{C\log^4(t)}{t^{21/4}}$$

Finally, we have

(3.195)

$$\begin{aligned}
&\partial_r\left(\left(\frac{\cos(2(v_1+v_2+v_3+v_4))-1}{2r^2}\right)\left(\sin(2Q_{\frac{1}{\lambda(t)}}+2v_5)-\sin(2Q_{\frac{1}{\lambda(t)}})\right)\right)\\
&=\frac{-\sin(2(v_1+v_2+v_3+v_4))}{r^2}\partial_r(v_1+v_2+v_3+v_4)\left(\sin(2Q_{\frac{1}{\lambda(t)}}+2v_5)-\sin(2Q_{\frac{1}{\lambda(t)}})\right)\\
&-\frac{(\cos(2(v_1+v_2+v_3+v_4))-1)}{r^3}\left(\sin(2Q_{\frac{1}{\lambda(t)}}+2v_5)-\sin(2Q_{\frac{1}{\lambda(t)}})\right)\\
&+\left(\frac{\cos(2(v_1+v_2+v_3+v_4))-1}{r^2}\right)\left(\cos(2Q_{\frac{1}{\lambda(t)}}+2v_5)(\partial_r Q_{\frac{1}{\lambda(t)}}+\partial_r v_5)-\cos(2Q_{\frac{1}{\lambda(t)}})\partial_r Q_{\frac{1}{\lambda(t)}}\right)
\end{aligned}$$

The second line of (3.195) is estimated by

$$\begin{aligned}
&||\frac{-\sin(2(v_1+v_2+v_3+v_4))}{r^2}\partial_r(v_1+v_2+v_3+v_4)\left(\sin(2Q_{\frac{1}{\lambda(t)}}+2v_5)-\sin(2Q_{\frac{1}{\lambda(t)}})\right)|_{r=R\lambda(t)}||_{L^2(RdR)}\\
&\leq C||\frac{v_1+v_2+v_3+v_4}{r}||_{L^\infty}\left(||\frac{v_5\partial_r(v_1+v_2+v_4)}{r}|_{r=R\lambda(t)}||_{L^2(RdR)}+||\frac{v_5}{r}||_{L^\infty}||(\partial_2 v_3)(t,\cdot\lambda(t))||_{L^2(RdR)}\right)\\
&\leq \frac{C\log^{6+b}(t)}{t^{35/8}}
\end{aligned}$$

The third line of (3.195) is estimated by

$$\begin{aligned}
&||-\frac{(\cos(2(v_1+v_2+v_3+v_4))-1)}{r^3}\left(\sin(2Q_{\frac{1}{\lambda(t)}}+2v_5)-\sin(2Q_{\frac{1}{\lambda(t)}})\right)|_{r=R\lambda(t)}||_{L^2(RdR)}\\
&\leq C||\frac{v_1+v_2+v_3+v_4}{r}||_{L^\infty}^2||\frac{v_5}{r}|_{r=R\lambda(t)}||_{L^2(RdR)}\\
&\leq \frac{C\log^{6+b}(t)}{t^5}
\end{aligned}$$

Finally, the fourth line of (3.195) is estimated by

$$\begin{aligned}
&||\left(\frac{\cos(2(v_1+v_2+v_3+v_4))-1}{r^2}\right)\left(\cos(2Q_{\frac{1}{\lambda(t)}}+2v_5)(\partial_r Q_{\frac{1}{\lambda(t)}}+\partial_r v_5)-\cos(2Q_{\frac{1}{\lambda(t)}})\partial_r Q_{\frac{1}{\lambda(t)}}\right)|_{r=R\lambda(t)}||_{L^2(RdR)}\\
&\leq ||\frac{v_1+v_2+v_3+v_4}{r}||_{L^\infty}^2\left(||\left(\lambda(t)\frac{v_5^2}{\lambda(t)^2+r^2}+\frac{r\lambda(t)^2 v_5}{(r^2+\lambda(t)^2)^2}\right)|_{r=R\lambda(t)}||_{L^2(RdR)}+||(\partial_2 v_5)(t,\cdot\lambda(t))||_{L^2(RdR)}\right)\\
&\leq \frac{C\log^{5+b}(t)}{t^{19/4}}
\end{aligned}$$

Finally, we use our pointwise estimates on all v_k to get

$$\left(\frac{1}{\lambda(t)^2}\int_0^\infty \frac{(F_5(t,R\lambda(t)))^2}{R^2}RdR\right)^{1/2} \leq \frac{C\log^{6+b}(t)}{t^{21/4}}$$

Combining all of these estimates, we get

$$\frac{1}{\lambda(t)}||F_5(t,\cdot\lambda(t))||_{\dot{H}^1_e} \leq \frac{C\log^{6+b}(t)}{t^{35/8}}$$ □

3.0.11. Estimates on F_6. We prove the analogous estimates on F_6:

LEMMA 3.20.

$$\frac{1}{\lambda(t)^2}||F_6(t,r)||_{L^2(rdr)} \leq \frac{C}{t^4\log^{3b+2N-1}(t)} \tag{3.196}$$

$$\frac{1}{\lambda(t)}||F_6(t,r)||_{\dot{H}^1_e} \leq \frac{C}{t^{9/2}\log^{4b-1+\frac{5N}{2}}(t)}$$

PROOF. We start with

$$\begin{aligned}||F_6(t,r)||^2_{L^2(rdr)} &\leq C\int_{\frac{t}{4}}^{\frac{t}{2}}\frac{\lambda(t)^4 r(v_4(t,r)^2+v_5(t,r)^2)}{r^8}dr + C\int_{\frac{t}{2}}^{\infty}\frac{\lambda(t)^4(v_4^2+v_5^2)rdr}{r^8}\\ &\leq \frac{C}{t^8\log^{10b+4N-2}(t)}\end{aligned}$$

where we used (3.83), (3.91). This concludes the proof of (3.196).

Next, we have

$$\begin{aligned}\partial_r F_6(t,r) &= \chi'_{\geq 1}(\frac{4r}{t})\left(\frac{4}{t}\right)\left(\frac{\cos(2Q_{\frac{1}{\lambda(t)}}(r))-1}{r^2}\right)(v_4+v_5)\\ &+\chi_{\geq 1}(\frac{4r}{t})(v_4+v_5)\,\partial_r\left(\frac{\cos(2Q_{\frac{1}{\lambda(t)}}(r))-1}{r^2}\right)\\ &+\chi_{\geq 1}(\frac{4r}{t})\left(\frac{\cos(2Q_{\frac{1}{\lambda(t)}}(r))-1}{r^2}\right)\partial_r(v_4+v_5)\end{aligned}$$

We estimate the L^2 norm as follows.

$$\begin{aligned}||\partial_r F_6(t,r)||_{L^2(rdr)} &\leq \frac{C}{t^{9/2}\log^{5b-1+\frac{5N}{2}}(t)} + ||\partial_r v_5||_{L^2(rdr)}\cdot||\frac{\chi_{\geq 1}(\frac{4r}{t})\lambda(t)^2}{(r^2+\lambda(t)^2)^2}||_{L^\infty_r}\\ &\leq \frac{C}{t^{9/2}\log^{5b-1+\frac{5N}{2}}(t)}\end{aligned}$$

where we used (3.83), (3.84), (3.91), and (3.93). The last term to estimate is

$$||\frac{F_6(t,r)}{r}||_{L^2(rdr)} \leq \frac{C}{t^5\log^{5b+2N-1}(t)}$$

where we used (3.83) and (3.91). Combining these, we get

$$\frac{1}{\lambda(t)}||F_6(t,r)||_{\dot{H}^1_e} \leq \frac{C}{t^{9/2}\log^{4b-1+\frac{5N}{2}}(t)}$$ □

3.0.12. Estimates on v_{corr}-dependent quantities. Finally, we define $v_{corr} := v_1 + v_2 + v_3 + v_4 + v_5$ and record some estimates on v_{corr}-dependent quantities which will appear as coefficients of various error terms involving the final correction, which is to be constructed in the next section.

LEMMA 3.21. *We have the following estimates*

$$||\frac{v_{corr}(x, R\lambda(x))}{R\lambda(x)}||_{L^\infty}^2 + ||\frac{v_{corr}(x, R\lambda(x))}{R\lambda(x)^2(1+R^2)}||_{L^\infty} \leq \frac{C\log(\log(x))}{x^2\log(x)}$$

$$1 + ||\frac{v_{corr}(x, R\lambda(x))}{R}||_{L^\infty} + ||\partial_R(v_{corr}(x, R\lambda(x)))||_{L^\infty} \leq C$$

$$\begin{aligned}&||\frac{v_{corr}(x, R\lambda(x))\partial_R(v_{corr}(x, R\lambda(x)))}{R\lambda(x)^2}||_{L^\infty_R((0,1))}\\&+ ||\frac{v_{corr}(x, R\lambda(x))\partial_R(v_{corr}(x, R\lambda(x)))}{R^2\lambda(x)^2}||_{L^\infty_R((1,\infty))} + ||\frac{\partial_R(v_{corr}(x, R\lambda(x)))}{(1+R^2)\lambda(x)^2}||_{L^\infty}\\&\leq \frac{C\log(\log(x))}{x^2\log(x)}\end{aligned}$$

PROOF. From (3.176), (3.77), (3.78), (3.83), and (3.91), we get

$$|v_{corr}(t,r)| \leq \begin{cases}\frac{Cr\log(\log(t))}{t^2\log^{b+1}(t)}, & r \leq \log^N(t)\\ \frac{Cr}{t^2\log^b(t)}, & \log^N(t) \leq r \leq \frac{t}{2}\\ \frac{C}{\sqrt{r}}, & \frac{t}{2} < r\end{cases}$$

This gives

$$||\frac{v_{corr}(x, R\lambda(x))}{R\lambda(x)}||_{L^\infty}^2 + ||\frac{v_{corr}(x, R\lambda(x))}{R\lambda(x)^2(1+R^2)}||_{L^\infty} \leq \frac{C\log(\log(x))}{x^2\log(x)}$$

Then, we use (3.177),(3.182), (3.79), (3.84), and (3.92) to get

$$|\partial_R(v_{corr}(x, R\lambda(x)))| \leq C\lambda(x)\begin{cases}\frac{\log(\log(x))}{x^2\log^{b+1}(x)}, & R\lambda(x) \leq \log^N(x)\\ \frac{1}{x^2\log^b(x)}, & \log^N(x) \leq R\lambda(x) \leq \frac{x}{2}\\ \frac{1}{\sqrt{x}}, & R\lambda(x) > \frac{x}{2}\end{cases}$$

This implies

$$1 + ||\frac{v_{corr}(x, R\lambda(x))}{R}||_{L^\infty} + ||\partial_R(v_{corr}(x, R\lambda(x)))||_{L^\infty} \leq C$$

Next, we have

$$|\frac{v_{corr}(x, R\lambda(x))}{R\lambda(x)}| \cdot |\frac{\partial_R(v_{corr}(x, R\lambda(x)))}{\lambda(x)}| \leq C\frac{(\log(\log(x)))^2}{x^4\log^{2b+2}(x)}, \quad R \leq 1$$

$$|\frac{v_{corr}(x, R\lambda(x))}{R\lambda(x)}| \cdot |\frac{\partial_R(v_{corr}(x, R\lambda(x)))}{R\lambda(x)}| \leq \begin{cases}\frac{C(\log(\log(x)))^2}{Rx^4\log^{2b+2}(x)}, & 1 \leq R \leq \frac{\log^N(x)}{\lambda(x)}\\ \frac{C}{Rx^4\log^{2b}(x)}, & \log^N(x) \leq R\lambda(x) \leq \frac{x}{2}\\ \frac{C}{R^{5/2}\lambda(x)^{3/2}\sqrt{x}}, & R\lambda(x) > \frac{x}{2}\end{cases}$$

and

$$\frac{|\partial_R(v_{corr}(x,R\lambda(x)))|}{(1+R^2)\lambda(x)^2} \leq \begin{cases} \frac{C\log(\log(x))}{x^2\log(x)(1+R^2)}, & R\lambda(x) \leq \log^N(x) \\ \frac{1}{x^2(1+R^2)}, & \log^N(x) \leq R\lambda(x) \leq \frac{x}{2} \\ \frac{\log^b(x)}{\sqrt{x}} \cdot \frac{1}{1+R^2}, & R\lambda(x) > \frac{x}{2} \end{cases}$$

which imply

$$\begin{aligned} &||\frac{v_{corr}(x,R\lambda(x))\partial_R(v_{corr}(x,R\lambda(x)))}{R\lambda(x)^2}||_{L^\infty_R((0,1))} \\ &+||\frac{v_{corr}(x,R\lambda(x))\partial_R(v_{corr}(x,R\lambda(x)))}{R^2\lambda(x)^2}||_{L^\infty_R((1,\infty))} + ||\frac{\partial_R(v_{corr}(x,R\lambda(x)))}{(1+R^2)\lambda(x)^2}||_{L^\infty} \\ &\leq \frac{C\log(\log(x))}{x^2\log(x)} \end{aligned}$$

□

CHAPTER 4

Solving the final equation

The full equation to solve is (3.26) with 0 Cauchy data at infinity. For ease of notation, let us set $u = v_6$, and re-write (3.26) as

$$-\partial_{tt}u + \partial_{rr}u + \frac{1}{r}\partial_r u - \frac{\cos(2Q_{\frac{1}{\lambda(t)}}(r))}{r^2}u = F(t,r) + F_3(t,r)$$

where

$$F(t,r) = F_4(t,r) + F_5(t,r) + F_6(t,r)$$

and we recall that F_4, F_5, and F_6 are defined in (3.27), (3.28), and (3.29), and are estimated in theorem 3.1. and

$$F_3 = N(u) + L_1(u) \tag{4.1}$$

with

$$N(f) = \left(\frac{\sin(2f) - 2f}{2r^2}\right)\cos(2Q_{\frac{1}{\lambda(t)}}) + \left(\frac{\cos(2f) - 1}{2r^2}\right)\sin(2(Q_{\frac{1}{\lambda(t)}} + v_{corr}))$$

$$L_1(f) = \frac{\sin(2f)}{2r^2}\cos(2Q_{\frac{1}{\lambda(t)}})(\cos(2v_{corr}) - 1) - \frac{\sin(2f)}{2r^2}\sin(2Q_{\frac{1}{\lambda(t)}})\sin(2v_{corr})$$

$$v_{corr} = v_1 + v_2 + v_3 + v_4 + v_5$$

Note that we will utilize the crucial fact that

$$\langle F_4(t,\cdot), \phi_0(\frac{\cdot}{\lambda(t)})\rangle = 0$$

4.0.1. The equation for $\mathcal{F}(u)$. We will make appropriate changes of variables in order to (formally) derive the equation for the distorted Fourier transform, discussed in section 4 of [**12**], of u. (Note, however, that we will not renormalize the time variable, unlike in [**12**].) We will denote the distorted Fourier transform of a function f by $\mathcal{F}(f)$. Let

$$u(t,r) = v(t, \frac{r}{\lambda(t)})$$

Then, if we evaluate the equation for u at the point $(t, R\lambda(t))$, we obtain

$$\begin{aligned}
&-\partial_{11}v(t,R) + 2\frac{\lambda'(t)}{\lambda(t)}R\partial_{12}v(t,R) + \left(\frac{\lambda''(t)}{\lambda(t)} - 2\frac{\lambda'(t)^2}{\lambda(t)^2}\right)R\partial_2 v(t,R)\\
&-\frac{\lambda'(t)^2}{\lambda(t)^2}R^2\partial_{22}v(t,R) + \frac{1}{\lambda(t)^2}\left(\partial_{22}v(t,R) + \frac{1}{R}\partial_2 v(t,R) - \frac{\cos(2Q_1(R))}{R^2}v(t,R)\right)\\
&= F(t, R\lambda(t)) + F_3(t, R\lambda(t))
\end{aligned}$$

Now, let

$$v(t,R) = \frac{w(t,R)}{\sqrt{R}}$$

to get

$$
\begin{aligned}
&-\partial_{11}w(t,R)-\frac{\lambda'(t)}{\lambda(t)}\partial_1 w(t,R)+2\frac{\lambda'(t)}{\lambda(t)}\partial_1(R\partial_2 w)(t,R)\\
&+\left(\frac{-\lambda''(t)}{2\lambda(t)}+\frac{1}{4}\frac{\lambda'(t)^2}{\lambda(t)^2}\right)w(t,R)+\left(\frac{\lambda''(t)}{\lambda(t)}-\frac{\lambda'(t)^2}{\lambda(t)^2}\right)R\partial_2 w(t,R)\\
&-\frac{\lambda'(t)^2}{\lambda(t)^2}R^2\partial_{22}w(t,R)+\frac{1}{\lambda(t)^2}\left(\partial_{22}w(t,R)-\left(\frac{3}{4R^2}-\frac{8}{(1+R^2)^2}\right)w(t,R)\right)\\
&=\sqrt{R}F(t,R\lambda(t))+\sqrt{R}F_3(t,R\lambda(t))
\end{aligned} \tag{4.2}
$$

Next, from (5.1) of [**12**], we have

$$\mathcal{F}(R\partial_R w)=-2\xi\partial_\xi\mathcal{F}(w)+K(\mathcal{F}(w))$$

where we will use various estimates on K, proven in [**12**], later on. Making the final change of variable

$$y(t,\xi)=\mathcal{F}(w)(t,\xi\lambda(t)^2)$$

and evaluating the distorted Fourier transform of (4.2) at the point $(t,\omega\lambda(t)^2)$, we get

(4.3)

$$
\begin{aligned}
&-\partial_{tt}y(t,\omega)-\omega y(t,\omega)-\frac{\lambda'(t)}{\lambda(t)}\partial_t y(t,\omega)+\frac{2\lambda'(t)}{\lambda(t)}K\left(\partial_1 y(t,\frac{\cdot}{\lambda(t)^2})\right)(\omega\lambda(t)^2)\\
&+\left(\frac{-\lambda''(t)}{2\lambda(t)}+\frac{\lambda'(t)^2}{4\lambda(t)^2}\right)y(t,\omega)+\frac{\lambda''(t)}{\lambda(t)}K\left(y(t,\frac{\cdot}{\lambda(t)^2})\right)(\omega\lambda(t)^2)\\
&+2\frac{\lambda'(t)^2}{\lambda(t)^2}\left([\xi\partial_\xi,K](y(t,\frac{\cdot}{\lambda(t)^2}))\right)(\omega\lambda(t)^2)-\frac{\lambda'(t)^2}{\lambda(t)^2}K\left(K(y(t,\frac{\cdot}{\lambda(t)^2}))\right)(\omega\lambda(t)^2)\\
&=\mathcal{F}(\sqrt{\cdot}F(t,\cdot\lambda(t)))(\omega\lambda(t)^2)+\mathcal{F}(\sqrt{\cdot}F_3(u(y))(t,\cdot\lambda(t)))(\omega\lambda(t)^2)
\end{aligned}
$$

where we write $F_3(u(y))$ to emphasize the dependence of F_3 on y, which is related to u via

$$y(t,\xi)=\mathcal{F}(\sqrt{\cdot}u(t,\cdot\lambda(t)))(\xi\lambda(t)^2)$$

4.0.2. Estimates on F_2. Let

$$
\begin{aligned}
F_2(y)(t,\omega)=&-\frac{\lambda'(t)}{\lambda(t)}\partial_t y(t,\omega)+\frac{2\lambda'(t)}{\lambda(t)}K\left(\partial_1 y(t,\frac{\cdot}{\lambda(t)^2})\right)(\omega\lambda(t)^2)\\
&+\left(\frac{-\lambda''(t)}{2\lambda(t)}+\frac{\lambda'(t)^2}{4\lambda(t)^2}\right)y(t,\omega)+\frac{\lambda''(t)}{\lambda(t)}K\left(y(t,\frac{\cdot}{\lambda(t)^2})\right)(\omega\lambda(t)^2)\\
&+2\frac{\lambda'(t)^2}{\lambda(t)^2}\left([\xi\partial_\xi,K](y(t,\frac{\cdot}{\lambda(t)^2}))\right)(\omega\lambda(t)^2)\\
&-\frac{\lambda'(t)^2}{\lambda(t)^2}K\left(K(y(t,\frac{\cdot}{\lambda(t)^2}))\right)(\omega\lambda(t)^2)
\end{aligned}
$$

Note that (4.3) becomes

$$
\begin{aligned}
&\partial_{tt}y+\omega y\\
&\quad=-\mathcal{F}(\sqrt{\cdot}F(t,\cdot\lambda(t)))(\omega\lambda(t)^2)+F_2(y)(t,\omega)-\mathcal{F}(\sqrt{\cdot}F_3(u(y))(t,\cdot\lambda(t)))(\omega\lambda(t)^2)
\end{aligned}
$$

Our goal is to prove the following proposition:

PROPOSITION 4.1. *There exists $C_1 > 0$ such that, for all y satisfying*

$$y(t,\omega)\sqrt{\rho(\omega\lambda(t)^2)}\langle\omega\lambda(t)^2\rangle \in C^0_t([T_0,\infty), L^2(d\omega))$$

and

$$\partial_t y(t,\omega)\langle\sqrt{\omega}\lambda(t)\rangle\sqrt{\rho(\omega\lambda(t)^2)} \in C^0_t([T_0,\infty), L^2(d\omega))$$

we have the following inequalities, for $x \geq T_0$:

$$\begin{aligned}||F_2(y)(x)||_{L^2(\rho(\omega\lambda(x)^2)d\omega)} \leq \frac{C_1}{x\log(x)}\Big(&||\partial_t y(x)||_{L^2(\rho(\omega\lambda(x)^2)d\omega)}\\ &+\frac{1}{x}||y(x)||_{L^2(\rho(\omega\lambda(x)^2)d\omega)}\Big)\end{aligned} \tag{4.4}$$

and

$$\begin{aligned}&||\sqrt{\omega}\lambda(x)F_2(y)(x)||_{L^2(\rho(\omega\lambda(x)^2)d\omega)}\\ &\leq \frac{C_1}{x\log(x)}\left(||\sqrt{\omega}\lambda(x)\partial_t y(x)||_{L^2(\rho(\omega\lambda(x)^2)d\omega)} + ||\partial_t y(x)||_{L^2(\rho(\omega\lambda(x)^2)d\omega)}\right)\\ &+\frac{C_1}{x^2\log(x)}\left(||\sqrt{\omega}\lambda(x)y(x)||_{L^2(\rho(\omega\lambda(x)^2)d\omega)} + ||y(x)||_{L^2(\rho(\omega\lambda(x)^2)d\omega)}\right)\end{aligned} \tag{4.5}$$

PROOF. By the symbol type bounds on a from Proposition 4.7 of [**12**], we have

$$|\frac{\xi\rho'(\xi)}{\rho(\xi)}| = |\frac{\xi(a\overline{a}' + \overline{a}a')}{|a|^2}| \leq C$$

Then, by (5.3) of [**12**], there exists a constant C such that, for $f \in C^\infty_c((0,\infty))$ and $\alpha = 0, \frac{1}{2}$, we have

$$||\langle\xi\rangle^\alpha(Kf)||_{L^2(\rho d\xi)} \leq C\left(||\langle\xi\rangle^\alpha f||_{L^2(\rho d\xi)} + ||\langle\xi\rangle^\alpha(K_0(f))||_{L^2(\rho d\xi)}\right)$$

In addition, we use Proposition 5.2 of [**12**] to get, for $\alpha = 0, \frac{1}{2}$,

$$||\langle\xi\rangle^{\alpha+1/2}K_0 f||_{L^2(\rho d\xi)} \leq C||\langle\xi\rangle^\alpha f||_{L^2(\rho d\xi)}$$

$$||\langle\xi\rangle^\alpha[\xi\partial_\xi, K_0]f||_{L^2(\rho d\xi)} \leq C||\langle\xi\rangle^\alpha f||_{L^2(\rho d\xi)}$$

So, for $\alpha = 0, \frac{1}{2}$

$$||\langle\xi\rangle^\alpha Kf||_{L^2(\rho d\xi)} \leq C||\langle\xi\rangle^\alpha f||_{L^2(\rho d\xi)}$$

Also, since

$$[\xi\partial_\xi, K]f = [\xi\partial_\xi, K_0]f - \xi\partial_\xi\left(\frac{\xi\rho'(\xi)}{\rho(\xi)}\right)f$$

the symbol type bounds on a imply that for $\alpha = 0, \frac{1}{2}$,

$$||\langle\xi\rangle^\alpha[\xi\partial_\xi, K]f||_{L^2(\rho d\xi)} \leq C||\langle\xi\rangle^\alpha f||_{L^2(\rho d\xi)}$$

For later convenience, we will record estimates on some terms appearing in (4.3), treating homogeneous components of a norm with weight $\langle \omega\lambda(x)^2\rangle^{1/2}$ separately:

$$\begin{aligned}
&||K(\partial_1 y(x,\frac{\cdot}{\lambda(x)^2}))(\omega\lambda(x)^2)||_{L^2(\rho(\omega\lambda(x)^2)d\omega)} \\
&= \left(\int_0^\infty \rho(\xi)\left(K(\partial_1 y(x,\frac{\cdot}{\lambda(x)^2}))(\xi)\right)^2 \frac{d\xi}{\lambda(x)^2}\right)^{1/2} \\
&\leq \frac{C}{\lambda(x)}\left(\int_0^\infty \rho(\xi)\left(\partial_1 y(x,\frac{\xi}{\lambda(x)^2})\right)^2 d\xi\right)^{1/2} \\
&\leq C||\partial_1 y(x)||_{L^2(\rho(\omega\lambda(x)^2)d\omega)}
\end{aligned}$$

$$\begin{aligned}
&||\sqrt{\omega}\lambda(x)K(\partial_1 y(x,\frac{\cdot}{\lambda(x)^2}))(\omega\lambda(x)^2)||^2_{L^2(\rho(\omega\lambda(x)^2)d\omega)} \\
&\leq C\int_0^\infty \left(K(\partial_1 y(x,\frac{\cdot}{\lambda(x)^2}))(\omega\lambda(x)^2)\right)^2 \left(1+\lambda(x)^4\omega^2\right)^{1/2}\rho(\omega\lambda(x)^2)d\omega \\
&\leq C\int_0^\infty \left(K(\partial_1 y(x,\frac{\cdot}{\lambda(x)^2}))(\xi)\right)^2 \sqrt{1+\xi^2}\frac{\rho(\xi)d\xi}{\lambda(x)^2} \\
&\leq \frac{C}{\lambda(x)^2}\int_0^\infty \left(\partial_1 y(x,\frac{\xi}{\lambda(x)^2})\right)^2 \sqrt{1+\xi^2}\rho(\xi)d\xi \\
&\leq C\int_0^\infty (\partial_1 y(x,\omega))^2\left(1+\omega\lambda(x)^2\right)\rho(\omega\lambda(x)^2)d\omega
\end{aligned}$$

So,

$$\begin{aligned}
&||\sqrt{\omega}\lambda(x)K(\partial_1 y(x,\frac{\cdot}{\lambda(x)^2}))(\omega\lambda(x)^2)||_{L^2(\rho(\omega\lambda(x)^2)d\omega)} \\
&\leq C\left(||\partial_1 y(x)||_{L^2(\rho(\omega\lambda(x)^2)d\omega)} + ||\sqrt{\omega}\lambda(x)\partial_1 y(x)||_{L^2(\rho(\omega\lambda(x)^2)d\omega)}\right)
\end{aligned}$$

Similarly,

$$||\left([\xi\partial_\xi, K](y(x,\frac{\cdot}{\lambda(x)^2}))\right)(\omega\lambda(x)^2)||_{L^2(\rho(\omega\lambda(x)^2)d\omega)} \leq C||y(x)||_{L^2(\rho(\omega\lambda(x)^2)d\omega)}$$

$$||K\left(K(y(x,\frac{\cdot}{\lambda(x)^2}))\right)(\omega\lambda(x)^2)||_{L^2(\rho(\omega\lambda(x)^2)d\omega)} \leq C||y(x)||_{L^2(\rho(\omega\lambda(x)^2)d\omega)}$$

$$||K\left(y(x,\frac{\cdot}{\lambda(x)^2})\right)(\omega\lambda(x)^2)||_{L^2(\rho(\omega\lambda(x)^2)d\omega)} \leq C||y(x)||_{L^2(\rho(\omega\lambda(x)^2)d\omega)}$$

$$\begin{aligned}
&||\sqrt{\omega}\lambda(x)\left([\xi\partial_\xi, K](y(x,\frac{\cdot}{\lambda(x)^2}))\right)(\omega\lambda(x)^2)||_{L^2(\rho(\omega\lambda(x)^2)d\omega)} \\
&\leq C\left(||y(x)||_{L^2(\rho(\omega\lambda(x)^2)d\omega)} + ||\sqrt{\omega}\lambda(x)y(x)||_{L^2(\rho(\omega\lambda(x)^2)d\omega)}\right)
\end{aligned}$$

$$\begin{aligned}
&||\sqrt{\omega}\lambda(x)\left(K\left(y(x,\frac{\cdot}{\lambda(x)^2})\right)(\omega\lambda(x)^2)\right)||_{L^2(\rho(\omega\lambda(x)^2)d\omega)} \\
&\leq C\left(||y(x)||_{L^2(\rho(\omega\lambda(x)^2)d\omega)} + ||\sqrt{\omega}\lambda(x)y(x)||_{L^2(\rho(\omega\lambda(x)^2)d\omega)}\right)
\end{aligned}$$

$$\begin{aligned}
&||\sqrt{\omega}\lambda(x)\left(K\left(K(y(x,\frac{\cdot}{\lambda(x)^2}))\right)(\omega\lambda(x)^2)\right)||_{L^2(\rho(\omega\lambda(x)^2)d\omega)} \\
&\leq C\left(||y(x)||_{L^2(\rho(\omega\lambda(x)^2)d\omega)} + ||\sqrt{\omega}\lambda(x)y(x)||_{L^2(\rho(\omega\lambda(x)^2)d\omega)}\right)
\end{aligned}$$

We thus conclude the inequalities (4.4) and (4.5). □

4.0.3. F_3 Estimates. Our goal in this subsection is to prove the following proposition.

PROPOSITION 4.2. *For y satisfying*

$$y(t,\omega)\sqrt{\rho(\omega\lambda(t)^2)}\langle\omega\lambda(t)^2\rangle \in C^0_t([T_0,\infty),L^2(d\omega))$$

let $F_3(u(y))$ be given by the expression (4.1), *where*

$$u(t,r) = \sqrt{\frac{\lambda(t)}{r}}\mathcal{F}^{-1}\left(y(t,\frac{\cdot}{\lambda(t)^2})\right)(\frac{r}{\lambda(t)}), \quad r>0$$

Then, there exists an absolute constant $C>0$ independent *of y, such that*

$$\begin{aligned}
&||\mathcal{F}(\sqrt{\cdot}F_3(u(y))(t,\cdot\lambda(t)))(\omega\lambda(t)^2)||_{L^2(\rho(\omega\lambda(t)^2)d\omega)}\\
&\leq C||y(t)||_{L^2(\rho(\omega\lambda(t)^2)d\omega)}\left(||\frac{v_{corr}(t,R\lambda(t))}{R\lambda(t)}||^2_{L^\infty}+||\frac{v_{corr}(t,R\lambda(t))}{R\lambda(t)^2(1+R^2)}||_{L^\infty}\right)\\
&+C||\langle\omega\lambda(t)^2\rangle y(t)||^3_{L^2(\rho(\omega\lambda(t)^2)d\omega)}\\
&+\frac{C}{\lambda(t)}||\langle\sqrt{\omega}\lambda(t)\rangle y(t)||^2_{L^2(\rho(\omega\lambda(t)^2)d\omega)}\left(1+||\frac{v_{corr}(t,R\lambda(t))}{R}||_{L^\infty}\right)
\end{aligned} \tag{4.6}$$

and

$$\begin{aligned}
&||\sqrt{\omega}\lambda(t)\mathcal{F}(\sqrt{\cdot}F_3(u(y))(t,\cdot\lambda(t)))(\omega\lambda(t)^2)||_{L^2(\rho(\omega\lambda(t)^2)d\omega)}\\
&\leq C||\langle\sqrt{\omega}\lambda(t)\rangle y(t)||_{L^2(\rho(\omega\lambda(t)^2)d\omega)}\cdot\left(||\frac{v_{corr}(t,R\lambda(t))}{R\lambda(t)}||^2_{L^\infty}+||\frac{v_{corr}(t,R\lambda(t))}{R\lambda(t)^2(1+R^2)}||_{L^\infty}\right.\\
&+||\frac{\partial_R(v_{corr}(t,R\lambda(t)))}{(1+R^2)\lambda(t)^2}||_{L^\infty}+||\frac{v_{corr}(t,R\lambda(t))\partial_R(v_{corr}(t,R\lambda(t)))}{R\lambda(t)^2}||_{L^\infty_R((0,1))}\\
&\left.+||\frac{v_{corr}(t,R\lambda(t))\partial_R(v_{corr}(t,R\lambda(t)))}{R^2\lambda(t)^2}||_{L^\infty_R((1,\infty))}\right)\\
&+C||\langle\omega\lambda(t)^2\rangle y(t)||^3_{L^2(\rho(\omega\lambda(t)^2)d\omega)}\\
&+\frac{C}{\lambda(t)}||\langle\omega\lambda(t)^2\rangle y(t)||^2_{L^2(\rho(\omega\lambda(t)^2)d\omega)}\\
&\cdot\left(1+||\frac{v_{corr}(t,R\lambda(t))}{R}||_{L^\infty}+||\partial_R(v_{corr}(t,R\lambda(t)))||_{L^\infty}\right)
\end{aligned} \tag{4.7}$$

where

$$v_{corr} = v_1+v_2+v_3+v_4+v_5$$

PROOF. To prove these estimates, it will be convenient to use the distorted Fourier transforms of [**1**], associated to the operators L^*L and LL^*, where

$$L = \partial_r - \frac{\cos(Q_1)}{r}, \quad L^* = -\partial_r - \frac{(\cos(Q_1)+1)}{r}$$

Let us use $\tilde{\phi}_\xi(r)$ to denote the eigenfunctions denoted by $\phi_\xi(r)$ in [**1**], and $\phi(r,\xi)$ to denote the eigenfunctions of [**12**]. By the definitions of these eigenfunctions, there exists f such that

$$\sqrt{r}\tilde{\phi}_{\sqrt{\xi}}(r) = f(\sqrt{\xi})\phi(r,\xi) \tag{4.8}$$

To determine the expression for f, let $g \in \mathcal{S}$, with $g(0) = 0$. Using the definitions of the Fourier transforms from [**1**] and [**12**], and (4.8), we have

$$\mathcal{F}(g)(\xi) = \int_0^\infty \phi(r,\xi)g(r)dr = \int_0^\infty \frac{\sqrt{r}\tilde{\phi}_{\sqrt{\xi}}(r)}{f(\sqrt{\xi})}g(r)dr = \frac{1}{f(\sqrt{\xi})}\mathcal{F}_H(\frac{g(\cdot)}{\sqrt{\cdot}})(\sqrt{\xi})$$

where $\mathcal{F}_H$ is the operator defined in section 3.1 of [**1**] Using the inversion formula from [**12**] we get

$$g(r) = \int_0^\infty \phi(r,\xi)\mathcal{F}(g)(\xi)\rho(\xi)d\xi = \int_0^\infty \frac{\sqrt{r}}{f(\sqrt{\xi})^2}\mathcal{F}_H(\frac{g(\cdot)}{\sqrt{\cdot}})(\sqrt{\xi})\rho(\xi)\tilde{\phi}_{\sqrt{\xi}}(r)d\xi$$

So, for $g \in \mathcal{S}$, with $g(0) = 0$, we have, for all $r \neq 0$,

$$\frac{g(r)}{\sqrt{r}} = \int_0^\infty \frac{2}{f(u)^2}\mathcal{F}_H(\frac{g(\cdot)}{\sqrt{\cdot}})(u)\rho(u^2)\tilde{\phi}_u(r)udu$$

Comparing this to the inversion formula from [**1**], we get

$$f(u) = \sqrt{2u\rho(u^2)}$$

In order to estimate the F_3 terms, we will also use the following lemma

LEMMA 4.1. *There exists $C > 0$ such that, for all $\overline{y}$ with $\overline{y}(\xi)\langle\xi\rangle \in L^2((0,\infty),\rho(\xi)d\xi)$, if $\overline{v}$ is given by*

$$\overline{v}(r) = \frac{1}{\sqrt{r}}\int_0^\infty \overline{y}(\xi)\phi(r,\xi)\rho(\xi)d\xi, \quad r > 0$$

then, $\overline{v} \in C^0(0,\infty)$, and $\frac{\overline{v}(r)}{r\langle\log(r)\rangle}$ admits a continuous extension to $[0,\infty)$ with $\frac{\overline{v}(r)}{r\langle\log(r)\rangle}(0) = \lim_{r\to\infty}\overline{v}(r) = 0$.

$$\overline{v}, L(\overline{v}), L^*L(\overline{v}) \in L^2((0,\infty), rdr)$$

with

$$||\overline{v}(r)||_{L^2(rdr)} = ||\mathcal{F}^{-1}(\overline{y})(r)||_{L^2(dr)} = ||\overline{y}||_{L^2(\rho(\xi)d\xi)} \tag{4.9}$$

$$||\sqrt{\xi}\overline{y}(\xi)||^2_{L^2(\rho(\xi)d\xi)} = ||L\overline{v}||^2_{L^2(rdr)} \tag{4.10}$$

and

$$||\xi\overline{y}(\xi)||^2_{L^2(\rho(\xi)d\xi)} = ||L^*L\overline{v}||^2_{L^2(rdr)} \tag{4.11}$$

Moreover, $\overline{v} \in \dot{H}^1_e$, with

$$||\overline{v}||_{\dot{H}^1_e} \leq C\left(||\overline{v}||_{L^2(rdr)} + ||L(\overline{v})||_{L^2(rdr)}\right) \tag{4.12}$$

$$|\overline{v}(r)| \leq C||\overline{v}||_{\dot{H}^1_e}, \quad r \geq 0 \tag{4.13}$$

$$|\frac{\overline{v}(r)}{r\langle\log(r)\rangle}| \leq C\left(||\overline{v}||_{L^2(rdr)} + ||L(\overline{v})||_{L^2(rdr)} + ||L^*L\overline{v}||_{L^2(rdr)}\right), \quad r > 0 \tag{4.14}$$

$$\int_0^\infty \left(\frac{(L(\overline{v})(r))^2}{r^2(1+r^2)}\right) rdr \leq C||L^*L\overline{v}||^2_{L^2(rdr)} \tag{4.15}$$

PROOF. If $f \in C^1([0,\infty)) \cap L^2((0,\infty), rdr)$, $Lf \in L^2((0,\infty), rdr)$, and $f(0) = \lim_{r\to\infty} f(r) = 0$, then, since

$$|\frac{\cos(Q_1(r))-1}{r}| \leq 1, \quad r>0$$

we have

$$\partial_r f - \frac{f}{r} = Lf + \left(\frac{\cos(Q_1(r))-1}{r}\right) f \in L^2((0,\infty), rdr)$$

Then, if $M>1$,

$$\begin{aligned}\int_{\frac{1}{M}}^M (\partial_r f - \frac{f}{r})^2 rdr &= \int_{\frac{1}{M}}^M \left((\partial_r f)^2 - \frac{\partial_r(f^2)}{r} + \frac{f^2}{r^2}\right) rdr \\ &= \int_0^\infty \mathbb{1}_{[\frac{1}{M},M]}(r) \left((\partial_r f)^2 + \frac{f^2}{r^2}\right) rdr - (f(M))^2 + (f(\frac{1}{M}))^2\end{aligned}$$

Letting $M\to\infty$, and using the monotone convergence theorem, we have

$$||f||^2_{\dot{H}^1_e} = \int_0^\infty \left((\partial_r f)^2 + \frac{f^2}{r^2}\right) rdr \leq C\left(||Lf||^2_{L^2(rdr)} + ||f||^2_{L^2(rdr)}\right) \tag{4.16}$$

Next, for $f \in C^1([0,\infty)) \cap \dot{H}^1_e$ satisfying $f(0)=0$, we have

$$f^2(r) = \int_0^r (2f(s)f'(s))\, ds = 2\int_0^r \frac{f(s)}{\sqrt{s}}(f'(s)\sqrt{s})ds \leq 2||\frac{f}{r}||_{L^2(rdr)}||f'||_{L^2(rdr)}$$

So,

$$||f||_\infty \leq C||\frac{f}{r}||^{1/2}_{L^2(rdr)}||f'||^{1/2}_{L^2(rdr)} \leq C||f||_{\dot{H}^1_e} \tag{4.17}$$

Next, for any $g \in C^1((0,\infty)) \cap C^0([0,\infty))$ such that $g(0) = \lim_{r\to\infty} g(r) = 0$, $L^*g \in L^2((0,\infty), rdr)$, and for any $M \geq 1$,

$$\begin{aligned}\int_{\frac{1}{M}}^M (L^*g)^2 rdr &= \int_{\frac{1}{M}}^M (g'(r))^2 rdr + \int_{\frac{1}{M}}^M (\cos(Q_1(r))+1)\frac{d}{dr}\left(g(r)^2\right) dr \\ &+ \int_{\frac{1}{M}}^M \left(\frac{\cos(Q_1(r))+1}{r}\right)^2 g(r)^2 rdr \\ &= (\cos(Q_1(M))+1)g(M)^2 - \left(\cos(Q_1(\frac{1}{M}))+1\right) g(\frac{1}{M})^2 \\ &+ \int_0^\infty \mathbb{1}_{[\frac{1}{M},M]}(r)\left((g'(r))^2 + \frac{4g(r)^2}{r^2(1+r^2)}\right) rdr\end{aligned}$$

By the monotone convergence theorem,

$$||L^*g||^2_{L^2(rdr)} = \int_0^\infty \left((g'(r))^2 + \frac{4g(r)^2}{r^2(1+r^2)}\right) rdr \tag{4.18}$$

If $\overline{y}$ is as in the lemma statement, for $M\geq 4$, define

$$\overline{v}_M(r) := \frac{1}{\sqrt{r}}\int_0^\infty \overline{y}(\xi)\phi(r,\xi)\chi_{\leq 1}(\frac{\xi}{M})\rho(\xi)d\xi, \quad r>0$$

We will now record some estimates on $\partial_r^k \phi(r,\xi), \quad k \leq 1$, which will allow us to prove a certain regularity of $\overline{v}$ and $\overline{v_M}$. From [**12**], we have

$$\frac{1}{\sqrt{r}}\phi(r,\xi) = \frac{1}{2}\phi_0(r) + \frac{1}{r}\sum_{j=1}^{\infty}(r^2\xi)^j \phi_j(r^2), \quad r^2\xi \leq 4$$

and

$$\frac{1}{\sqrt{r}}\phi(r,\xi) = \frac{2\text{Re}\left(a(\xi)\psi^+(r,\xi)\right)}{\sqrt{r}}, \quad r^2\xi > 4$$

Therefore,

$$|\frac{1}{r^{3/2}\langle\log(r)\rangle}\phi(r,\xi)| \leq \begin{cases} \frac{C}{\langle\log(r)\rangle}\left(\frac{\phi_0(r)}{r} + \frac{\log(1+r^2)}{r^2}\right), & r^2\xi \leq 4 \\ \frac{C|a(\xi)|}{\xi^{1/4}r^{1/2}\cdot r\langle\log(r)\rangle}, & r^2\xi > 4 \end{cases} \tag{4.19}$$

We use

$$\frac{\langle\log(\xi)\rangle^2}{\langle\log(r)\rangle^2} \leq C, \quad r^2\xi \leq 4, \quad \xi \geq 1$$

and the fact that

$$r \mapsto \frac{1}{\sqrt{r}\langle\log(r)\rangle} \text{ is decreasing on } (0,\infty)$$

which gives

$$\frac{1}{\sqrt{r}\langle\log(r)\rangle} \leq \frac{C\xi^{1/4}}{\langle\log(\xi)\rangle}, \quad r^2\xi > 4$$

in (4.19) to get, for all $r \geq 0$

$$|\frac{\phi(r,\xi)}{r^{3/2}\langle\log(r)\rangle}| \leq C\frac{|a(\xi)|\sqrt{\xi}}{\langle\log(\xi)\rangle} + C\begin{cases} \frac{1}{\langle\log(\xi)\rangle}, & \xi \geq 1 \\ 1, & \xi \leq 1 \end{cases}$$

Then, if

$$g(\xi) = \left(\frac{|a(\xi)|^2\xi}{\langle\log(\xi)\rangle^2} + \begin{cases} \frac{1}{\langle\log(\xi)\rangle^2}, & \xi \geq 1 \\ 1, & \xi \leq 1 \end{cases}\right)\frac{\rho(\xi)}{\langle\xi\rangle^2}$$

we have, for all $r \geq 0$

$$\begin{aligned} |\overline{y}(\xi)\frac{\phi(r,\xi)}{r^{3/2}\langle\log(r)\rangle}\rho(\xi)| &\leq |\overline{y}(\xi)|\langle\xi\rangle\sqrt{\rho(\xi)}\cdot\left(|\frac{\phi(r,\xi)}{r^{3/2}\langle\log(r)\rangle}|\frac{\sqrt{\rho(\xi)}}{\langle\xi\rangle}\right) \\ &\leq C|\overline{y}(\xi)|\langle\xi\rangle\sqrt{\rho(\xi)}\cdot\sqrt{g(\xi)} \end{aligned}$$

But,

$$|\overline{y}(\xi)|\langle\xi\rangle\sqrt{\rho(\xi)}\cdot\sqrt{g(\xi)} \in L^1((0,\infty),d\xi)$$

by Cauchy-Schwartz, due to the assumptions on $\overline{y}$ and the fact that

$$\int_0^\infty g(\xi)d\xi \leq C$$

Therefore, by the dominated convergence theorem, $\overline{v}$, defined in the lemma statement satisfies $\frac{\overline{v}(r)}{r\langle\log(r)\rangle}$ is continuous on $(0,\infty)$, and we have

$$|\frac{\overline{v}(r)}{r\langle\log(r)\rangle}| \leq \int_0^\infty |\overline{y}(\xi)\frac{\phi(r,\xi)}{r^{3/2}\langle\log(r)\rangle}\rho(\xi)|d\xi \leq C||\overline{y}(\xi)\langle\xi\rangle||_{L^2(\rho(\xi)d\xi)}, \quad r > 0$$

and

$$\lim_{r\to 0}\frac{\overline{v}(r)}{r\langle\log(r)\rangle}=\int_0^\infty \overline{y}(\xi)\lim_{r\to 0}\left(\frac{\phi(r,\xi)}{r^{3/2}\langle\log(r)\rangle}\right)\rho(\xi)d\xi=0$$

Similarly,

$$|\frac{\phi(r,\xi)}{\sqrt{r}}\frac{\sqrt{\rho(\xi)}}{\langle\xi\rangle}|\le C\sqrt{g_2(\xi)}$$

with

$$g_2(\xi)=\left(|a(\xi)|^2+\begin{cases}1, & \xi\le 1\\ \frac{1}{\xi}, & \xi>1\end{cases}\right)\frac{\rho(\xi)}{\langle\xi\rangle^2}\in L^1((0,\infty),d\xi)$$

So, again by the dominated convergence theorem,

$$\lim_{r\to\infty}\overline{v}(r)=\int_0^\infty \overline{y}(\xi)\lim_{r\to\infty}\left(\frac{\phi(r,\xi)}{\sqrt{r}}\right)\rho(\xi)d\xi=0$$

The same argument shows that

$$\lim_{r\to\infty}\overline{v_M}(r)=0$$

From (4.19), we also have

$$|\frac{\phi(r,\xi)}{r^{3/2}}|\le\begin{cases}C, & r^2\xi\le 4\\ C|a(\xi)|\sqrt{\xi}, & r^2\xi>4\end{cases}$$

whence,

$$|\frac{\phi(r,\xi)}{r^{3/2}}|\le C+C|a(\xi)|\sqrt{\xi},\quad r\ge 0$$

Then, the dominated convergence theorem shows that $\overline{v_M}$ satisfies that $\frac{\overline{v_M}(r)}{r}$ admits a continuous extension to a function defined on $[0,\infty)$. Similarly,

$$|\partial_r\left(\frac{\phi(r,\xi)}{\sqrt{r}}\right)|\le\begin{cases}C\left(|\phi_0'(r)|+\xi\log(1+r^2)\right), & r^2\xi\le 4\\ \frac{C|a(\xi)|\xi^{1/4}}{r^{1/2}}, & r^2\xi>4\end{cases}$$

which shows

$$|\partial_r\left(\frac{\phi(r,\xi)}{\sqrt{r}}\right)|\le C+C|a(\xi)|\sqrt{\xi},\quad r\ge 0$$

Therefore, $\overline{v_M}\in C^1([0,\infty))$. Moreover,

$$L(\overline{v_M})(0)=\int_0^\infty \overline{y}(\xi)\lim_{r\to 0}L\left(\frac{\phi(r,\xi)}{\sqrt{r}}\right)\chi_{\le 1}\left(\frac{\xi}{M}\right)\rho(\xi)d\xi=0$$

where we used $L(\phi_0)=0$. Again, by the dominated convergence theorem,

$$\lim_{r\to\infty}\partial_r\overline{v_M}(r)=\int_0^\infty \overline{y}(\xi)\lim_{r\to\infty}\partial_r\left(\frac{\phi(r,\xi)}{\sqrt{r}}\right)\chi_{\le 1}\left(\frac{\xi}{M}\right)\rho(\xi)d\xi=0$$

Finally, we have

$$|\partial_r^2\left(\frac{\phi(r,\xi)}{\sqrt{r}}\right)|\le C\left(1+\sqrt{\xi}+\xi|a(\xi)|\right),\quad r\ge 0$$

and the same dominated convergence theorem based procedure used above shows that $\overline{v_M}\in C^2((0,\infty))$.

Next, we have

$$||\overline{v}(r)||_{L^2(rdr)}=||\mathcal{F}^{-1}(\overline{y})(r)||_{L^2(dr)}=||\overline{y}||_{L^2(\rho(\xi)d\xi)}$$

$$||\sqrt{\xi}\overline{y}(\xi)||^2_{L^2(\rho(\xi)d\xi)} = \int_0^\infty \xi|\mathcal{F}(\sqrt{\cdot}\overline{v}(\cdot))(\xi)|^2\rho(\xi)d\xi = \int_0^\infty \xi\frac{|\mathcal{F}_H(\overline{v})(\sqrt{\xi})|^2}{2\sqrt{\xi}}d\xi$$
$$= \int_0^\infty \eta^2|\mathcal{F}_H(\overline{v})(\eta)|^2 d\eta = ||\eta\mathcal{F}_H(\overline{v})(\eta)||^2_{L^2(d\eta)}$$
$$= ||L\overline{v}||^2_{L^2(rdr)}$$

where we used the L^2 isometry property of $\mathcal{F}_H$ from [**1**]. Finally,

$$||\xi\overline{y}(\xi)||^2_{L^2(\rho(\xi)d\xi)} = \int_0^\infty \xi^2\frac{|\mathcal{F}_H(\overline{v})(\sqrt{\xi})|^2}{2\sqrt{\xi}}d\xi = \int_0^\infty \eta^4|\mathcal{F}_H(\overline{v})(\eta)|^2 d\eta$$
$$= \int_0^\infty |\mathcal{F}_H(L^*L\overline{v})(\eta)|^2 d\eta = ||L^*L\overline{v}||^2_{L^2(rdr)}$$

This shows that $\overline{v}, \overline{v_M}, L(\overline{v}), L(\overline{v_M}), L^*L(\overline{v}), L^*L(\overline{v_M}) \in L^2((0,\infty), rdr)$. Combining these facts with our estimates on $\frac{\phi(r,\xi)}{\sqrt{r}}\frac{\sqrt{\rho(\xi)}}{\langle\xi\rangle}$ and

$$|\overline{v}(r) - \overline{v_M}(r)| \leq \int_0^\infty |\overline{y}(t,\xi)\frac{\phi(r,\xi)}{\sqrt{r}}| \cdot |\chi_{\leq 1}(\frac{\xi}{M}) - 1|\rho(\xi)d\xi$$

the Dominated convergence theorem gives

$$\overline{v_M} \to \overline{v}, \text{ pointwise, and in } L^2((0,\infty), rdr), \quad M \to \infty$$
$$L(\overline{v_M}) \to L(\overline{v}), \text{in } L^2((0,\infty), rdr), \quad M \to \infty$$
$$L^*L(\overline{v_M}) \to L^*L(\overline{v}), \text{in } L^2((0,\infty), rdr), \quad M \to \infty$$

We conclude the proof of the lemma by noting that (4.16) and (4.17) hold for $\overline{v_M}$, and (4.18) holds for $g = L(\overline{v_M})$. So, by approximation, we have (4.12), (4.13), and (4.15). □

Now, we can estimate the $F_3(u(y))$ terms, for y such that $y(t,\omega)\sqrt{\rho(\omega\lambda(t)^2)}\langle\omega\lambda(t)^2\rangle \in C^0_t([T_0,\infty), L^2(d\omega))$. This is sufficient for our purposes, since all y in the space Z (which is the space in which we will construct a solution to (4.3), and is defined later on) satisfy this condition. Recall that

$$F_3(u)(t,r) = N(u)(t,r) + L_1(u)(t,r)$$

where u and y are related by

$$y(t,\xi) = \mathcal{F}(\sqrt{\cdot}u(t,\cdot\lambda(t)))(\xi\lambda(t)^2) \tag{4.20}$$

and

(4.21)

$$L_1(u)(t,r)$$
$$= \frac{\sin(2u(t,r))}{2r^2}\left(\cos(2Q_{\frac{1}{\lambda(t)}}(r))\left(\cos(2v_{corr}) - 1\right) - \sin(2Q_{\frac{1}{\lambda(t)}})\sin(2v_{corr})\right)$$
$$N(u)(t,r) = \left(\frac{\sin(2u(t,r)) - 2u(t,r)}{2r^2}\right)\cos(2Q_{\frac{1}{\lambda(t)}}(r))$$
$$+ \left(\frac{\cos(2u(t,r)) - 1}{2r^2}\right)\sin(2Q_{\frac{1}{\lambda(t)}}(r) + 2v_{corr})$$

We start with the L^2 estimate on the $L_1(u)$ term. Using the same procedure as in (4.9), we get

$$||\mathcal{F}(\sqrt{\cdot}L_1(u)(t,\cdot\lambda(t)))(\omega\lambda(t)^2)||^2_{L^2(\rho(\omega\lambda(t)^2)d\omega)} = \frac{1}{\lambda(t)^2}\int_0^\infty R(L_1(u)(t,R\lambda(t)))^2 dR$$

Then,

$$\begin{aligned}&\frac{1}{\lambda(t)^2}\int_0^\infty R(L_1(u)(t,R\lambda(t)))^2 dR\\ &\leq \frac{C}{\lambda(t)^2}\int_0^\infty R\frac{(u(t,R\lambda(t)))^2}{R^4\lambda(t)^4}\left(v_{corr}(t,R\lambda(t))\right)^4 dR\\ &+\frac{C}{\lambda(t)^2}\int_0^\infty R\frac{(u(t,R\lambda(t)))^2}{R^4\lambda(t)^4}\frac{R^2}{(1+R^2)^2}\left(v_{corr}(t,R\lambda(t))\right)^2 dR\end{aligned}$$

Using the functions v and w (introduced when deriving the equation for y) defined by

$$u(t,r) = v(t,\frac{r}{\lambda(t)})$$

$$v(t,R) = \frac{w(t,R)}{\sqrt{R}}$$

we have

$$\begin{aligned}||u(t,\cdot\lambda(t))||^2_{L^2(RdR)} &= ||v(t,\cdot)||^2_{L^2(RdR)} = ||w(t)||^2_{L^2(dR)} = ||\mathcal{F}(w)(t)||^2_{L^2(\rho(\xi)d\xi)}\\ &= \lambda(t)^2||y(t)||^2_{L^2(\rho(\omega\lambda(t)^2)d\omega)}\end{aligned}$$

So, we end up with

$$\begin{aligned}&||\mathcal{F}(\sqrt{\cdot}L_1(u)(t,\cdot\lambda(t)))(\omega\lambda(t)^2)||_{L^2(\rho(\omega\lambda(t)^2)d\omega)}\\ &\leq C||y(t)||_{L^2(\rho(\omega\lambda(t)^2)d\omega)}\left(||\frac{v_{corr}(t,R\lambda(t))}{R\lambda(t)}||^2_{L^\infty_R} + ||\frac{v_{corr}(t,R\lambda(t))}{R\lambda(t)^2(1+R^2)}||_{L^\infty_R}\right)\end{aligned}$$

Next, we apply (4.10) to our current setting to get

$$\begin{aligned}&||\sqrt{\omega}\lambda(t)\mathcal{F}(\sqrt{\cdot}L_1(u)(t,\cdot\lambda(t)))(\omega\lambda(t)^2)||^2_{L^2(\rho(\omega\lambda(t)^2)d\omega)}\\ &= \frac{1}{\lambda(t)^2}||L(L_1(u)(t,\cdot\lambda(t)))||^2_{L^2(RdR)}\end{aligned}$$

Using (4.21), we get

$$\begin{aligned}&|\partial_R(L_1(u)(t,R\lambda(t)))|\\ &\leq C\left(|Lv(t,R)| + \frac{|v(t,R)|}{R}\right)\left(\frac{(v_{corr}(t,R\lambda(t)))^2}{R^2\lambda(t)^2} + \frac{|v_{corr}(t,R\lambda(t))|}{R\lambda(t)^2(1+R^2)}\right)\\ &+C\frac{|v(t,R)|}{R^2\lambda(t)^2}\left(\frac{R|\partial_R(v_{corr}(t,R\lambda(t)))|}{1+R^2} + |\partial_R(v_{corr}(t,R\lambda(t)))v_{corr}(t,R\lambda(t))|\right)\end{aligned}$$

On the other hand,

$$|\frac{L_1(u)(t,R\lambda(t))}{R}| \leq \frac{C|v(t,R)|}{R^3\lambda(t)^2}\left(v_{corr}(t,R\lambda(t))^2 + \frac{|v_{corr}(t,R\lambda(t))|R}{(1+R^2)}\right)$$

So, we get

$$
\begin{aligned}
&||L(L_1(u)(t,\cdot\lambda(t)))||_{L^2(RdR)}\\
&\leq C\left(||Lv(t)||_{L^2(RdR)}+||v(t)||_{\dot{H}^1_e}\right)\left(||\frac{v_{corr}(t,R\lambda(t))}{R\lambda(t)}||^2_{L^\infty_R}+||\frac{v_{corr}(t,R\lambda(t))}{R\lambda(t)^2(1+R^2)}||_{L^\infty_R}\right)\\
&+C||v(t)||_{\dot{H}^1_e}\cdot||\frac{\partial_R(v_{corr}(t,R\lambda(t)))}{(1+R^2)\lambda(t)^2}||_{L^\infty_R}\\
&+C||v(t)||_{\dot{H}^1_e}\cdot||\frac{v_{corr}(t,R\lambda(t))\partial_R(v_{corr}(t,R\lambda(t)))}{R\lambda(t)^2}||_{L^\infty_R((0,1))}\\
&+C||v(t)||_{L^2((1,\infty),RdR)}\cdot||\frac{\partial_R(v_{corr}(t,R\lambda(t)))\cdot v_{corr}(t,R\lambda(t))}{R^2\lambda(t)^2}||_{L^\infty_R((1,\infty))}
\end{aligned}
$$

Now, we use (4.9) and (4.10) to translate the right-hand side in terms of y:

$$
\begin{aligned}
||Lv(t)||_{L^2(RdR)}&=\lambda(t)||\sqrt{\omega}\lambda(t)y(t)||_{L^2(\rho(\omega\lambda(t)^2)d\omega)}\\
||v(t)||_{L^2(RdR)}&=\lambda(t)||y(t)||_{L^2(\rho(\omega\lambda(t)^2)d\omega)}
\end{aligned}
$$

Then, we get

$$
\begin{aligned}
&||\sqrt{\omega}\lambda(t)\mathcal{F}(\sqrt{\cdot}L_1(u)(t,\cdot\lambda(t)))(\omega\lambda(t)^2)||_{L^2(\rho(\omega\lambda(t)^2)d\omega)}\\
&\leq C\left(||\sqrt{\omega}\lambda(t)y(t)||_{L^2(\rho(\omega\lambda(t)^2)d\omega)}+||y(t)||_{L^2(\rho(\omega\lambda(t)^2)d\omega)}\right)\\
&\cdot\left(||\frac{v_{corr}(t,R\lambda(t))}{R\lambda(t)}||^2_{L^\infty_R}+||\frac{v_{corr}(t,R\lambda(t))}{R\lambda(t)^2(1+R^2)}||_{L^\infty_R}+||\frac{\partial_R(v_{corr}(t,R\lambda(t)))}{(1+R^2)\lambda(t)^2}||_{L^\infty_R}\right.\\
&+||\frac{v_{corr}(t,R\lambda(t))\partial_R(v_{corr}(t,R\lambda(t)))}{R\lambda(t)^2}||_{L^\infty_R((0,1))}\\
&\left.+||\frac{v_{corr}(t,R\lambda(t))\partial_R(v_{corr}(t,R\lambda(t)))}{R^2\lambda(t)^2}||_{L^\infty_R((1,\infty))}\right)
\end{aligned}
$$

Now, we treat the N terms. Recall that

$$
\begin{aligned}
N(u)(t,R\lambda(t))=&\left(\frac{\sin(2v(t,R))-2v(t,R)}{2R^2\lambda(t)^2}\right)\cos(2Q_1(R))\\
&+\left(\frac{\cos(2v(t,R))-1}{2R^2\lambda(t)^2}\right)\sin(2Q_1(R)+2v_{corr}(t,R\lambda(t)))
\end{aligned}
$$

We then use (4.14) and the previous estimates to conclude that, if $R<1$, then,

$$
\begin{aligned}
|N(u)(t,R\lambda(t))|\leq& C\left(||v(t)||^2_{\dot{H}^1_e}+||L^*Lv(t)||^2_{L^2(RdR)}\right)\left(\log^2(\frac{1}{R})+1\right)\cdot\frac{|v(t,R)|}{\lambda(t)^2}\\
&+C||v(t)||_{\dot{H}^1_e}\frac{|v(t,R)|}{R\lambda(t)^2}\left(1+\frac{|v_{corr}(t,R\lambda(t))|}{R}\right)
\end{aligned}
$$

On the other hand, if $R>1$, then,

$$
|N(u)(t,R\lambda(t))|\leq C\frac{|v(t,R)|^3}{\lambda(t)^2}+C\frac{(v(t,R))^2}{\lambda(t)^2}\left(1+\frac{|v_{corr}(t,R\lambda(t))|}{R}\right)
$$

Then, we get

$$
\begin{aligned}
&\frac{1}{\lambda(t)^2}||N(u)(t,\cdot\lambda(t))||^2_{L^2(RdR)}\\
&\leq\frac{C}{\lambda(t)^6}\left(||v(t)||^6_{L^2(RdR)}+||Lv(t)||^6_{L^2(RdR)}+||L^*Lv(t)||^6_{L^2(RdR)}\right)\\
&+\frac{C}{\lambda(t)^6}\left(||v(t)||^4_{L^2(RdR)}+||Lv(t)||^4_{L^2(RdR)}\right)\left(1+||\frac{v_{corr}(t,R\lambda(t))}{R}||^2_{L^\infty_R}\right)
\end{aligned}
\tag{4.22}
$$

Finally, we apply (4.11) in our current setting:

$$\int_0^\infty (L^* Lv(t,R))^2 RdR = \lambda(t)^2 ||\omega\lambda(t)^2 y(t)||^2_{L^2(\rho(\omega\lambda(t)^2)d\omega)}$$

Using the same procedure as used for L_1 above, we then translate the rest of the right hand side, and the left hand side of (4.22) in terms of y, and $\mathcal{F}(\sqrt{\cdot}N(u)(t,\cdot\lambda(t)))(\omega\lambda(t)^2)$, respectively, to get

$$\begin{aligned}
&||\mathcal{F}(\sqrt{\cdot}N(u)(t,\cdot\lambda(t)))(\omega\lambda(t)^2)||_{L^2(\rho(\omega\lambda(t)^2)d\omega)}\\
&\leq C\left(||y(t)||^3_{L^2(\rho(\omega\lambda(t)^2)d\omega)} + ||\sqrt{\omega}\lambda(t)y(t)||^3_{L^2(\rho(\omega\lambda(t)^2)d\omega)} + ||\omega\lambda(t)^2 y(t)||^3_{L^2(\rho(\omega\lambda(t)^2)d\omega)}\right)\\
&+\frac{C}{\lambda(t)}\left(||y(t)||^2_{L^2(\rho(\omega\lambda(t)^2)d\omega)} + ||\sqrt{\omega}\lambda(t)y(t)||^2_{L^2(\rho(\omega\lambda(t)^2)d\omega)}\right)\left(1+||\frac{v_{corr}(t,R\lambda(t))}{R}||_{L^\infty}\right)
\end{aligned}$$

We will now estimate $||\sqrt{\omega}\lambda(t)\mathcal{F}(\sqrt{\cdot}N(u)(t,\cdot\lambda(t)))(\omega\lambda(t)^2)||_{L^2(\rho(\omega\lambda(t)^2)d\omega)}$, starting with

$$\begin{aligned}
&|\partial_R(N(u)(t,R\lambda(t)))|\\
&\leq \frac{C|v(t,R)|^3}{R^3\lambda(t)^2} + C\frac{|\partial_R v(t,R)|(v(t,R))^2}{R^2\lambda(t)^2}\\
&+C\frac{|Q_1'(R)+\partial_R(v_{corr}(t,R\lambda(t)))|}{R^2\lambda(t)^2}(v(t,R))^2\\
&+C\frac{|Q_1(R)|+|v_{corr}(t,R\lambda(t))|}{R^2\lambda(t)^2}|v(t,R)|\left(\frac{|v(t,R)|}{R}+|\partial_R v(t,R)|\right)
\end{aligned}$$

Then, we use our previous estimates to get, for $R \leq 1$:

$$\begin{aligned}
&|\partial_R(N(u)(t,R\lambda(t)))|\\
&\leq \frac{C}{\lambda(t)^2}\left(||v(t)||^3_{\dot{H}^1_e} + ||L^*Lv(t)||^3_{L^2(RdR)}\right)\left(1+\log^2(\frac{1}{R})\right)^{3/2}\\
&+C\frac{\langle\log(R)\rangle^2}{\lambda(t)^2}\left(||v(t)||^2_{L^2(RdR)} + ||Lv(t)||^2_{L^2(RdR)} + ||L^*Lv(t)||^2_{L^2(RdR)}\right)\\
&\cdot\left(|Lv(t,R)|+\frac{|v(t,R)|}{R}\right)\\
&+C\frac{(1+||\partial_R(v_{corr}(t,R\lambda(t)))||_{L^\infty})}{\lambda(t)^2}\left(||v(t)||^2_{L^2(RdR)} + ||Lv(t)||^2_{L^2(RdR)} + ||L^*Lv(t)||^2_{L^2(RdR)}\right)\\
&\cdot\left(\log^2(\frac{1}{R})+1\right)\\
&+C\frac{\left(||v(t)||_{L^2(RdR)} + ||Lv(t)||_{L^2(RdR)} + ||L^*Lv(t)||_{L^2(RdR)}\right)\langle\log(R)\rangle|\partial_R v|}{\lambda(t)^2}\\
&\cdot\left(1+||\frac{v_{corr}(t,R\lambda(t))}{R}||_{L^\infty}\right)\\
&+\frac{C}{\lambda(t)^2}\left(1+||\frac{v_{corr}(t,R\lambda(t))}{R}||_{L^\infty}\right)\left(||v(t)||^2_{\dot{H}^1_e} + ||L^*Lv(t)||^2_{L^2(RdR)}\right)\\
&\cdot\left(\log^2(\frac{1}{R})+1\right)
\end{aligned}$$

For $R \geq 1$, we have

$$
\begin{aligned}
&|\partial_R(N(u)(t,R\lambda(t)))| \\
&\leq \frac{C}{\lambda(t)^2}||v(t)||_{\dot{H}^1_e}^2|v(t,R)| + \frac{C||v(t)||_{\dot{H}^1_e}^2}{\lambda(t)^2}\left(|Lv(t,R)| + \frac{|v(t,R)|}{R}\right) \\
&+ C\frac{\left(1 + ||\partial_R(v_{corr}(t,R\lambda(t)))||_{L^\infty} + ||\frac{v_{corr}(t,R\lambda(t))}{R}||_{L^\infty}\right)}{\lambda(t)^2}||v(t)||_{\dot{H}^1_e}|v(t,R)| \\
&+ \frac{C}{\lambda(t)^2}\left(1 + ||\frac{v_{corr}(t,R\lambda(t))}{R}||_{L^\infty}\right)\left(|Lv(t,R)| + \frac{|v(t,R)|}{R}\right)||v(t)||_{\dot{H}^1_e}
\end{aligned}
$$

Then, we get

$$
\begin{aligned}
&||\partial_R(N(u)(t,R\lambda(t)))||_{L^2(RdR)} \\
&\leq \frac{C}{\lambda(t)^2}\left(||v(t)||_{L^2(RdR)}^3 + ||Lv(t)||_{L^2(RdR)}^3 + ||L^*Lv(t)||_{L^2(RdR)}^3\right) \\
&+ \frac{C}{\lambda(t)^2}\left(1 + ||\partial_R(v_{corr}(t,R\lambda(t)))||_{L^\infty} + ||\frac{v_{corr}(t,R\lambda(t))}{R}||_{L^\infty}\right) \\
&\cdot\left(||v(t)||_{L^2(RdR)}^2 + ||Lv(t)||_{L^2(RdR)}^2 + ||L^*Lv(t)||_{L^2(RdR)}^2\right)
\end{aligned}
$$

Finally, we consider, for $R \leq 1$:

$$
\begin{aligned}
|\frac{N(u)(t,R\lambda(t))}{R}| \leq &\frac{C}{\lambda(t)^2}\left(||v(t)||_{\dot{H}^1_e}^3 + ||L^*Lv(t)||_{L^2(RdR)}^3\right)\left(1 + \log^2(\frac{1}{R})\right)^{3/2} \\
&+ \frac{C}{\lambda(t)^2}\left(||v(t)||_{\dot{H}^1_e}^2 + ||L^*Lv(t)||_{L^2(RdR)}^2\right)\left(1 + \log^2(\frac{1}{R})\right) \\
&\cdot\left(1 + ||\frac{v_{corr}(t,R\lambda(t))}{R}||_{L^\infty}\right)
\end{aligned}
$$

and for $R > 1$, we have

$$
\begin{aligned}
|\frac{N(u)(t,R\lambda(t))}{R}| \leq &C\frac{||v(t)||_{\dot{H}^1_e}^2}{\lambda(t)^2}|v(t,R)| \\
&+ C\frac{||v(t)||_{\dot{H}^1_e}|v(t,R)|}{\lambda(t)^2}\left(1 + ||\frac{v_{corr}(t,R\lambda(t))}{R}||_{L^\infty}\right)
\end{aligned}
$$

Then, we get

$$
\begin{aligned}
&||L(N(u)(t,\cdot\lambda(t)))||_{L^2(RdR)} \\
&\leq \frac{C}{\lambda(t)^2}\left(||v(t)||_{L^2(RdR)}^3 + ||Lv(t)||_{L^2(RdR)}^3 + ||L^*Lv(t)||_{L^2(RdR)}^3\right) \\
&+ \frac{C}{\lambda(t)^2}\left(1 + ||\partial_R(v_{corr}(t,R\lambda(t)))||_{L^\infty} + ||\frac{v_{corr}(t,R\lambda(t))}{R}||_{L^\infty}\right) \\
&\cdot\left(||v(t)||_{L^2(RdR)}^2 + ||Lv(t)||_{L^2(RdR)}^2 + ||L^*Lv(t)||_{L^2(RdR)}^2\right)
\end{aligned}
$$

We use the same procedure as in the previous estimates, to translate the left and right hand sides of the above estimate, and combine everything, to get (4.6) and (4.7). □

Before we proceed, we will need one more estimate. First, because λ is decreasing, if $x \geq t$, then, $\lambda(x) \leq \lambda(t)$. Next, we use Proposition 4.7 b of [**12**] to conclude that there exists a constant $C_1 > 0$ such that

$$\frac{1}{C_1 \xi \log^2(\xi)} \leq \rho(\xi) \leq \frac{C_1}{\xi \log^2(\xi)}, \quad 0 < \xi < \frac{1}{2e^2}$$
$$\frac{\xi}{C_1} \leq \rho(\xi) \leq C_1 \xi, \quad \xi > \frac{1}{2e^2}$$

Then, if $x \geq t$, if $\omega\lambda(t)^2 \leq \frac{1}{2e^2}$, then, $\omega\lambda(x)^2 \leq \frac{1}{2e^2}$, and

$$\begin{aligned}\frac{\rho(\omega\lambda(t)^2)}{\rho(\omega\lambda(x)^2)} &\leq C\frac{\omega\lambda(x)^2 \log^2(\omega\lambda(x)^2)}{\omega\lambda(t)^2 \log^2(\omega\lambda(t)^2)}\\ &\leq C\end{aligned}$$

where we used the fact that $x \mapsto x\log^2(x)$ is increasing on $(0, \frac{1}{2e^2})$.

If $\omega\lambda(t)^2 \geq \frac{1}{2e^2}$, but $\omega\lambda(x)^2 \leq \frac{1}{2e^2}$, then,

$$\begin{aligned}\frac{\rho(\omega\lambda(t)^2)}{\rho(\omega\lambda(x)^2)} &\leq C\omega\lambda(t)^2\omega\lambda(x)^2 \log^2(\omega\lambda(x)^2)\\ &\leq C\frac{\lambda(t)^2}{\lambda(x)^2}\left(\omega\lambda(x)^2\right)^2 \log^2(\omega\lambda(x)^2)\\ &\leq \frac{C\lambda(t)^2}{\lambda(x)^2}\end{aligned}$$

Finally, if $\omega\lambda(t)^2 \geq \frac{1}{2e^2}$, and $\omega\lambda(x)^2 \geq \frac{1}{2e^2}$, then,

$$\frac{\rho(\omega\lambda(t)^2)}{\rho(\omega\lambda(x)^2)} \leq C\frac{\lambda(t)^2}{\lambda(x)^2}$$

In all cases, we have: there exists $C_\rho > 0$ such that, if $x \geq t$, then,

$$\frac{\rho(\omega\lambda(t)^2)}{\rho(\omega\lambda(x)^2)} \leq C_\rho \frac{\lambda(t)^2}{\lambda(x)^2} \tag{4.23}$$

Before setting up the final iteration, we will need to estimate various oscillatory integrals involving F_4:

4.0.4. Estimates on F_4-related oscillatory integrals.

LEMMA 4.2. *We have the following estimates*

$$\begin{aligned}&||\int_t^\infty \frac{\sin((t-x)\sqrt{\omega})}{\sqrt{\omega}} \mathcal{F}(\sqrt{\cdot}F_4(x,\cdot\lambda(x)))(\omega\lambda(x)^2)dx||_{L^2(\rho(\omega\lambda(t)^2)d\omega)}\\ &\leq \frac{C(\log(\log(t))^2}{t^2 \log^{b+1-2\alpha b}(t)}\end{aligned}$$

$$\begin{aligned}&||\int_t^\infty \cos((t-x)\sqrt{\omega})\mathcal{F}(\sqrt{\cdot}F_4(x,\cdot\lambda(x)))(\omega\lambda(x)^2)dx||_{L^2(\rho(\omega\lambda(t)^2)d\omega)}\\ &\leq \frac{C(\log(\log(t)))^2}{t^3 \log^{b+1-2\alpha b}(t)}\end{aligned}$$

$$\begin{aligned}&||\sqrt{\omega}\lambda(t)\int_t^\infty \frac{\sin((t-x)\sqrt{\omega})}{\sqrt{\omega}}\mathcal{F}(\sqrt{\cdot}F_4(x,\cdot\lambda(x)))(\omega\lambda(x)^2)dx||_{L^2(\rho(\omega\lambda(t)^2)d\omega)}\\&\leq \frac{C(\log(\log(t)))^2}{t^2\log^{1+b-2\alpha b}(t)}\end{aligned}$$

$$\begin{aligned}&||\sqrt{\omega}\lambda(t)\partial_t\left(\int_t^\infty \frac{\sin((t-x)\sqrt{\omega})}{\sqrt{\omega}}\mathcal{F}(\sqrt{\cdot}F_4(x,\cdot\lambda(x)))(\omega\lambda(x)^2)dx\right)||_{L^2(\rho(\omega\lambda(t)^2)d\omega)}\\&\leq \frac{C(\log(\log(t)))^2}{t^3\log^{1+b-2\alpha b}(t)}\end{aligned}$$

$$\begin{aligned}&||\omega\lambda(t)^2\int_t^\infty \frac{\sin((t-x)\sqrt{\omega})}{\sqrt{\omega}}\mathcal{F}(\sqrt{\cdot}F_4(x,\cdot\lambda(x)))(\omega\lambda(x)^2)dx||_{L^2(\rho(\omega\lambda(t)^2)d\omega)}\\&\leq \frac{C}{t^2\log^{1+b-2b\alpha}(t)}\end{aligned}$$

Proof. We start with

$$\begin{aligned}&\int_t^\infty \frac{\sin((t-x)\sqrt{\omega})}{\sqrt{\omega}}\mathcal{F}(\sqrt{\cdot}F_4(x,\cdot\lambda(x)))(\omega\lambda(x)^2)dx\\&=-\frac{\mathcal{F}(\sqrt{\cdot}F_4(t,\cdot\lambda(t)))(\omega\lambda(t)^2)}{\omega}\\&-\int_t^\infty \frac{\cos((t-x)\sqrt{\omega})}{\omega}\left(\mathcal{F}(\sqrt{\cdot}F_4(x,\cdot\lambda(x)))'(\omega\lambda(x)^2)\cdot 2\lambda(x)\lambda'(x)\omega+\mathcal{F}(\sqrt{\cdot}\partial_x\left(F_4(x,\cdot\lambda(x))\right))(\omega\lambda(x)^2)\right)dx\end{aligned}$$

Then,

$$\begin{aligned}-\frac{\mathcal{F}(\sqrt{\cdot}F_4(t,\cdot\lambda(t)))(\omega\lambda(t)^2)}{\omega}&=\frac{-1}{\omega}\int_0^{\frac{2}{\sqrt{\omega\lambda(t)}}}\phi(r,\omega\lambda(t)^2)\sqrt{r}F_4(t,r\lambda(t))dr\\&-\frac{1}{\omega}\int_{\frac{2}{\sqrt{\omega\lambda(t)}}}^\infty \phi(r,\omega\lambda(t)^2)\sqrt{r}F_4(t,r\lambda(t))dr\end{aligned}$$

In the region $r\sqrt{\omega\lambda(t)^2}\leq 2$, we use proposition 4.4 of [**12**] to make the decomposition

$$\phi(r,\omega\lambda(t)^2)=\widetilde{\phi_0}(r)+\frac{1}{\sqrt{r}}\sum_{j=1}^\infty (r^2\omega\lambda(t)^2)^j\phi_j(r^2)$$

where we denote, by $\widetilde{\phi_0}$, what was denoted by ϕ_0 in [**12**]. In our notation,

$$\widetilde{\phi_0}(r)=\frac{\sqrt{r}}{2}\phi_0(r)$$

The first term to consider is then

$$-\frac{1}{\omega}\int_0^{\frac{2}{\sqrt{\omega\lambda(t)}}}\widetilde{\phi_0}(r)\sqrt{r}F_4(t,r\lambda(t))dr=\frac{-1}{2\omega}\int_0^{\frac{2}{\sqrt{\omega\lambda(t)}}}r\phi_0(r)F_4(t,r\lambda(t))dr$$

We will then consider several cases of ω.

Case 1: $1\leq \frac{2}{\sqrt{\omega\lambda(t)}}\leq \frac{\log^N(t)}{\lambda(t)}$. In this region, we use the orthogonality of $F_4(t,\cdot\lambda(t))$ to ϕ_0, to get

$$\frac{-1}{2\omega}\int_0^{\frac{2}{\sqrt{\omega\lambda(t)}}}r\phi_0(r)F_4(t,r\lambda(t))dr=\frac{1}{2\omega}\int_{\frac{2}{\sqrt{\omega\lambda(t)}}}^\infty \phi_0(r)F_4(t,r\lambda(t))rdr$$

which gives

$$\begin{aligned}
&|\frac{-1}{2\omega}\int_0^{\frac{2}{\sqrt{\omega}\lambda(t)}} r\phi_0(r)F_4(t,r\lambda(t))dr| \\
&\leq \frac{C}{\omega}\int_{\frac{2}{\sqrt{\omega}\lambda(t)}}^{\frac{\log^N(t)}{\lambda(t)}} \frac{\phi_0(r)}{\lambda(t)^4 r^4}\left(\frac{r\lambda(t)}{t^2\log^{3b+1-2\alpha b}(t)}\right) rdr \\
&+\frac{C}{\omega}\int_{\frac{2}{\sqrt{\omega}\lambda(t)}}^{\frac{t}{2\lambda(t)}} \frac{\phi_0(r)}{r^4\lambda(t)^4\log^{5b+2N-2}(t)}\frac{r\lambda(t)}{t^2} rdr \\
&\leq \frac{C}{t^2\log^{2b+1-2\alpha b}(t)}, \quad 1\leq \frac{2}{\sqrt{\omega}\lambda(t)}\leq \frac{\log^N(t)}{\lambda(t)}
\end{aligned}$$

Case 2: $\frac{2}{\sqrt{\omega}\lambda(t)} \leq 1$. Here, we have

$$\begin{aligned}
&|\frac{-1}{2\omega}\int_0^{\frac{2}{\sqrt{\omega}\lambda(t)}} r\phi_0(r)F_4(t,r\lambda(t))dr| \\
&\leq \frac{C}{\omega}\int_0^{\frac{2}{\sqrt{\omega}\lambda(t)}} \phi_0(r)\left(\frac{r\lambda(t)}{\lambda(t)^4(r^2+1)^2t^2\log^{3b+1-2\alpha b}(t)}\right) rdr \\
&+\frac{C}{\omega}\int_0^{\frac{2}{\sqrt{\omega}\lambda(t)}} \frac{\phi_0(r)r\lambda(t)rdr}{\lambda(t)^4(r^2+1)^2\log^{5b+2N-2}(t)t^2} \\
&\leq \frac{C}{\omega\left(\sqrt{\omega}\lambda(t)\right)^4 t^2\log^{1-2\alpha b}(t)}, \quad \frac{2}{\sqrt{\omega}\lambda(t)}\leq 1
\end{aligned}$$

Case 3: $\frac{\log^N(t)}{\lambda(t)} \leq \frac{2}{\sqrt{\omega}\lambda(t)} \leq \frac{t}{2\lambda(t)}$. We use the orthogonality condition here, and recall that the $v_1+v_2+v_3$ and $F_{0,2}$ terms are supported in the region $r\leq \log^N(t)$. Then, in the integral below, only the v_4+v_5 term contributes:

$$\begin{aligned}
&|\frac{-1}{2\omega}\int_0^{\frac{2}{\sqrt{\omega}\lambda(t)}} r\phi_0(r)F_4(t,r\lambda(t))dr| = |\frac{1}{2\omega}\int_{\frac{2}{\sqrt{\omega}\lambda(t)}}^{\infty} \phi_0(r)F_4(t,r\lambda(t))rdr| \\
&\leq \frac{C}{\omega}\int_{\frac{2}{\sqrt{\omega}\lambda(t)}}^{\frac{t}{2\lambda(t)}} \frac{rdr}{t^2\log^{2b+2N-2}(t)(r^2+1)^2} \\
&\leq \frac{C}{t^2\log^{4b+2N-2}(t)}, \quad \frac{\log^N(t)}{\lambda(t)}\leq \frac{2}{\sqrt{\omega}\lambda(t)}\leq \frac{t}{2\lambda(t)}
\end{aligned}$$

Case 4: $\frac{2}{\sqrt{\omega}\lambda(t)} \geq \frac{t}{2\lambda(t)}$. We use the orthogonality condition, and note that $F_4(t,r)=0, \quad r\geq \frac{t}{2}$. This gives

$$|\frac{-1}{2\omega}\int_0^{\frac{2}{\sqrt{\omega}\lambda(t)}} r\phi_0(r)F_4(t,r\lambda(t))dr| = |\frac{1}{2\omega}\int_{\frac{2}{\sqrt{\omega}\lambda(t)}}^{\infty} \phi_0(r)F_4(t,r\lambda(t))rdr| = 0$$

Combining these estimates, we get

$$|\frac{-1}{2\omega}\int_0^{\frac{2}{\sqrt{\omega}\lambda(t)}} r\phi_0(r)F_4(t,r\lambda(t))dr| \leq \begin{cases} \frac{C}{\omega^3\lambda(t)^4t^2\log^{1-2\alpha b}(t)}, & \frac{2}{\sqrt{\omega}\lambda(t)} \leq 1 \\ \frac{C}{t^2\log^{2b+1-2\alpha b}(t)}, & 1 \leq \frac{2}{\sqrt{\omega}\lambda(t)} \leq \frac{\log^N(t)}{\lambda(t)} \\ \frac{C}{t^2\log^{4b+2N-2}(t)}, & \frac{\log^N(t)}{\lambda(t)} \leq \frac{2}{\sqrt{\omega}\lambda(t)} \leq \frac{t}{2\lambda(t)} \\ 0, \quad \frac{t}{2\lambda(t)} \leq \frac{2}{\sqrt{\omega}\lambda(t)} \end{cases}$$

Using proposition 4.7 b of [**12**], we have (for example)

$$\rho(x) \leq \begin{cases} \frac{C}{x\log^2(x)}, & x \leq \frac{1}{4} \\ Cx, & x \geq \frac{1}{4} \end{cases}$$

This leads to

$$||\frac{-1}{2\omega}\int_0^{\frac{2}{\sqrt{\omega}\lambda(t)}} r\phi_0(r)F_4(t,r\lambda(t))dr||_{L^2(\rho(\omega\lambda(t)^2)d\omega)} \leq \frac{C}{t^2\log^{b+1-2\alpha b}(t)} \tag{4.24}$$

The next integral to consider is

$$-\frac{1}{\omega}\sum_{j=1}^{\infty}\int_0^{\frac{2}{\sqrt{\omega}\lambda(t)}} r^{2j}\omega^j\lambda(t)^{2j}\phi_j(r^2)F_4(t,r\lambda(t))dr$$

For this integral, there is of course no orthogonality to exploit. In all cases of ω, we will use the estimate (from proposition 4.4 of [**12**])

$$|\phi_j(r^2)| \leq \frac{3C_1^j}{(j-1)!}\log(1+r^2)$$

We again treat various cases of ω.

Case 1: $1 \leq \frac{2}{\sqrt{\omega}\lambda(t)} \leq \frac{\log^N(t)}{\lambda(t)}$. Then, we have

$$\begin{aligned}
&|-\frac{1}{\omega}\sum_{j=1}^{\infty}\int_0^{\frac{2}{\sqrt{\omega}\lambda(t)}} r^{2j}\omega^j\lambda(t)^{2j}\phi_j(r^2)F_4(t,r\lambda(t))dr| \\
&\leq \frac{C}{\omega}\sum_{j=1}^{\infty}\int_0^{\frac{2}{\sqrt{\omega}\lambda(t)}} r^{2j}\omega^j\lambda(t)^{2j}\frac{C_1^j}{(j-1)!}\left(\frac{\log(\log(t))r}{\lambda(t)^3(r^2+1)^2t^2\log^{3b+1-2b\alpha}(t)} + \frac{\log(t)r\lambda(t)}{t^2\lambda(t)^4(r^2+1)^2\log^{5b+2N-2}(t)}\right)dr \\
&\leq \frac{C}{\omega}\sum_{j=1}^{\infty}\left(\int_0^1 \frac{r^{2j+1}\omega^j\lambda(t)^{2j}C_1^j}{(j-1)!}\frac{\log(\log(t))}{t^2\log^{1-2b\alpha}(t)}dr + \int_1^{\frac{2}{\sqrt{\omega}\lambda(t)}}\frac{\omega^j\lambda(t)^{2j}C_1^j}{(j-1)!}\frac{\log(\log(t))}{t^2\log^{1-2b\alpha}(t)}r^{2j-3}dr\right) \\
&\leq \frac{C(\log(\log(t)))^2}{t^2\log^{1+2b-2b\alpha}(t)}
\end{aligned}$$

where we used the support properties of the various terms in F_4, and the largeness of N.

Case 2: $\frac{2}{\sqrt{\omega}\lambda(t)} \leq 1$. Here, we have

$$\begin{aligned}
&|-\frac{1}{\omega}\sum_{j=1}^{\infty}\int_0^{\frac{2}{\sqrt{\omega}\lambda(t)}} r^{2j}\omega^j\lambda(t)^{2j}\phi_j(r^2)F_4(t,r\lambda(t))dr| \\
&\leq \frac{C}{\omega}\sum_{j=1}^{\infty}\int_0^{\frac{2}{\sqrt{\omega}\lambda(t)}} r^{2j}\omega^j\lambda(t)^{2j}\frac{C_1^j}{(j-1)!}\log(1+r^2) \\
&\qquad\left(\frac{r\lambda(t)}{\lambda(t)^4(r^2+1)^2}\frac{1}{t^2\log^{3b+1-2b\alpha}(t)} + \frac{r\lambda(t)}{\lambda(t)^4(r^2+1)^2}\frac{1}{t^2\log^{5b+2N-2}(t)}\right)dr \\
&\leq \frac{C}{\omega}\sum_{j=1}^{\infty}\left(\int_0^{\frac{2}{\sqrt{\omega}\lambda(t)}} r^{2j+3}dr\right)\frac{\omega^j\lambda(t)^{2j}C_1^j}{(j-1)!t^2\log^{1-2b\alpha}(t)} \\
&\leq \frac{C}{\omega^3\lambda(t)^4t^2\log^{1-2b\alpha}(t)}, \quad \frac{2}{\sqrt{\omega}\lambda(t)} \leq 1
\end{aligned}$$

Case 3: $\frac{\log^N(t)}{\lambda(t)} \leq \frac{2}{\sqrt{\omega}\lambda(t)} \leq \frac{t}{2\lambda(t)}$. Again, we use the support properties of the various terms in F_4, to get

$$\begin{aligned}
&|-\frac{1}{\omega}\sum_{j=1}^{\infty}\int_0^{\frac{2}{\sqrt{\omega}\lambda(t)}} r^{2j}\omega^j\lambda(t)^{2j}\phi_j(r^2)F_4(t,r\lambda(t))dr| \\
&\leq \frac{C}{\omega}\sum_{j=1}^{\infty}\left(\int_0^{\frac{\log^N(t)}{\lambda(t)}}\frac{r^{2j+1}dr}{(r^2+1)^2}\right)\frac{\omega^j\lambda(t)^{2j}C_1^j}{(j-1)!}\frac{\log(\log(t))}{t^2\log^{1-2b\alpha}(t)} \\
&+\frac{C}{\omega}\sum_{j=1}^{\infty}\left(\int_0^{\frac{2}{\sqrt{\omega}\lambda(t)}}\frac{r^{2j+1}dr}{(r^2+1)^2}\right)\frac{\omega^j\lambda(t)^{2j}C_1^j}{(j-1)!t^2\log^{2b+2N-3}(t)} \\
&\leq \frac{C(\log(\log(t)))^2}{t^2\log^{1+2b-2b\alpha}(t)}, \quad \frac{\log^N(t)}{\lambda(t)} \leq \frac{2}{\sqrt{\omega}\lambda(t)} \leq \frac{t}{2\lambda(t)}
\end{aligned}$$

Case 4: $\frac{2}{\sqrt{\omega}\lambda(t)} \geq \frac{t}{2\lambda(t)}$.

$$\begin{aligned}
&|-\frac{1}{\omega}\sum_{j=1}^{\infty}\int_0^{\frac{2}{\sqrt{\omega}\lambda(t)}} r^{2j}\omega^j\lambda(t)^{2j}\phi_j(r^2)F_4(t,r\lambda(t))dr| \\
&\leq \frac{C}{\omega}\sum_{j=1}^{\infty}\left(\int_0^{\frac{\log^N(t)}{\lambda(t)}}\frac{r^{2j+1}dr}{(r^2+1)^2}\right)\frac{\omega^j\lambda(t)^{2j}C_1^j}{(j-1)!}\frac{\log(\log(t))}{t^2\log^{1-2b\alpha}(t)} \\
&+\frac{C}{\omega}\sum_{j=1}^{\infty}\left(\int_0^{\frac{t}{2\lambda(t)}}\frac{r^{2j+1}dr}{(r^2+1)^2}\right)\frac{\omega^j\lambda(t)^{2j}C_1^j}{(j-1)!t^2\log^{2b+2N-3}(t)} \\
&\leq \frac{C(\log(\log(t)))^2}{t^2\log^{1+2b-2b\alpha}(t)}, \quad \frac{2}{\sqrt{\omega}\lambda(t)} \geq \frac{t}{2\lambda(t)}
\end{aligned}$$

Then, we get

(4.25)
$$||-\frac{1}{\omega}\sum_{j=1}^{\infty}\int_0^{\frac{2}{\sqrt{\omega}\lambda(t)}} r^{2j}\omega^j\lambda(t)^{2j}\phi_j(r^2)F_4(t,r\lambda(t))dr||_{L^2(\rho(\omega\lambda(t)^2)d\omega)} \leq \frac{C(\log(\log(t)))^2}{t^2\log^{b+1-2b\alpha}(t)}$$

The third integral to estimate is

$$\begin{aligned}
&-\frac{1}{\omega}\int_{\frac{2}{\sqrt{\omega}\lambda(t)}}^{\infty}\phi(r,\omega\lambda(t)^2)\sqrt{r}F_4(t,r\lambda(t))dr \\
&= -\frac{2}{\omega}\text{Re}\left(\int_{\frac{2}{\sqrt{\omega}\lambda(t)}}^{\infty} a(\omega\lambda(t)^2)\psi^+(r,\omega\lambda(t)^2)\sqrt{r}F_4(t,r\lambda(t))dr\right)
\end{aligned}$$

where we used propositions 4.6 and 4.7 of [**12**]. The estimates from these propositions also imply

$$\begin{aligned}
&|-\frac{1}{\omega}\int_{\frac{2}{\sqrt{\omega}\lambda(t)}}^{\infty}\phi(r,\omega\lambda(t)^2)\sqrt{r}F_4(t,r\lambda(t))dr| \\
&\leq \frac{C|a(\omega\lambda(t)^2)|}{\omega^{5/4}\lambda(t)^{1/2}}\int_{\frac{2}{\sqrt{\omega}\lambda(t)}}^{\infty}\sqrt{r}|F_4(t,r\lambda(t))|dr
\end{aligned}$$

Case 1: $1 \leq \frac{2}{\sqrt{\omega}\lambda(t)} \leq \frac{\log^N(t)}{\lambda(t)}$.

$$
\begin{aligned}
&|-\frac{1}{\omega}\int_{\frac{2}{\sqrt{\omega}\lambda(t)}}^\infty \phi(r,\omega\lambda(t)^2)\sqrt{r}F_4(t,r\lambda(t))dr| \\
&\leq \frac{C|a(\omega\lambda(t)^2)|}{\omega^{5/4}\lambda(t)^{1/2}}\left(\int_{\frac{2}{\sqrt{\omega}\lambda(t)}}^{\frac{\log^N(t)}{\lambda(t)}} \frac{\sqrt{r}dr}{\lambda(t)^3 r^3 t^2 \log^{3b+1-2b\alpha}(t)} + \int_{\frac{2}{\sqrt{\omega}\lambda(t)}}^\infty \frac{r^{3/2}dr}{\lambda(t)^3 r^4 \log^{5b+2N-2}(t) t^2}\right) \\
&\leq C\frac{|a(\omega\lambda(t)^2)|}{\omega^{5/4}\lambda(t)^{1/2}} \cdot \frac{\omega^{3/4}\lambda(t)^{3/2}}{t^2\log^{3b+1-2\alpha b}(t)\lambda(t)^3} \\
&\leq \frac{C|a(\omega\lambda(t)^2)|}{\omega}\cdot\frac{\sqrt{\omega}\lambda(t)}{t^2\log^{1-2\alpha b}(t)}, \quad 1 \leq \frac{2}{\sqrt{\omega}\lambda(t)} \leq \frac{\log^N(t)}{\lambda(t)}
\end{aligned}
$$

Case 2: $\frac{2}{\sqrt{\omega}\lambda(t)} \leq 1$.

$$
\begin{aligned}
&|-\frac{1}{\omega}\int_{\frac{2}{\sqrt{\omega}\lambda(t)}}^\infty \phi(r,\omega\lambda(t)^2)\sqrt{r}F_4(t,r\lambda(t))dr| \\
&\leq \frac{C|a(\omega\lambda(t)^2)|}{\omega^{5/4}\lambda(t)^{1/2}}\left(\frac{1}{t^2\log^{1-2\alpha b}(t)}\int_0^{\frac{\log^N(t)}{\lambda(t)}} \frac{r^{3/2}dr}{(r^2+1)^2} + \frac{1}{t^2\log^{2b+2N-2}(t)}\int_0^\infty \frac{r^{3/2}dr}{(r^2+1)^2}\right) \\
&\leq \frac{C|a(\omega\lambda(t)^2)|}{\omega^{5/4}\lambda(t)^{1/2}t^2\log^{1-2\alpha b}(t)}, \quad \frac{2}{\sqrt{\omega}\lambda(t)} \leq 1
\end{aligned}
$$

Case 3: $\frac{\log^N(t)}{\lambda(t)} \leq \frac{2}{\sqrt{\omega}\lambda(t)} \leq \frac{t}{2\lambda(t)}$. In this region, the $F_{0,2}$ and $v_1+v_2+v_3$ terms do not contribute to the integral because of the support properties of $1-\chi_{\geq 1}(\frac{r}{\log^N(t)})$. We get

$$
\begin{aligned}
&|-\frac{1}{\omega}\int_{\frac{2}{\sqrt{\omega}\lambda(t)}}^\infty \phi(r,\omega\lambda(t)^2)\sqrt{r}F_4(t,r\lambda(t))dr| \\
&\leq \frac{C|a(\omega\lambda(t)^2)|}{\omega^{5/4}\lambda(t)^{1/2}}\int_{\frac{2}{\sqrt{\omega}\lambda(t)}}^\infty \frac{\sqrt{r}dr}{\lambda(t)^3 r^3\log^{5b+2N-2}(t)t^2} \\
&\leq \frac{C|a(\omega\lambda(t)^2)|}{\omega}\cdot\frac{\sqrt{\omega}\lambda(t)}{t^2\log^{2b+2N-2}(t)}
\end{aligned}
$$

Case 4: $\frac{2}{\sqrt{\omega}\lambda(t)} \geq \frac{t}{2\lambda(t)}$. In this case, the integral is zero, because of the support properties of F_4.

Combining the above estimates, we get

$$
\begin{aligned}
&|-\frac{1}{\omega}\int_{\frac{2}{\sqrt{\omega}\lambda(t)}}^\infty \phi(r,\omega\lambda(t)^2)\sqrt{r}F_4(t,r\lambda(t))dr| \\
&\leq C|a(\omega\lambda(t)^2)|\begin{cases}\frac{1}{\omega^{5/4}\lambda(t)^{1/2}t^2\log^{1-2\alpha b}(t)}, & \frac{2}{\sqrt{\omega}\lambda(t)} \leq 1\\ \frac{1}{\sqrt{\omega}t^2\log^{1+b-2\alpha b}(t)}, & 1 \leq \frac{2}{\sqrt{\omega}\lambda(t)} \leq \frac{\log^N(t)}{\lambda(t)}\\ \frac{1}{\sqrt{\omega}t^2\log^{3b+2N-2}(t)}, & \frac{\log^N(t)}{\lambda(t)} \leq \frac{2}{\sqrt{\omega}\lambda(t)} \leq \frac{t}{2\lambda(t)}\\ 0, & \frac{t}{2\lambda(t)} \leq \frac{2}{\sqrt{\omega}\lambda(t)}\end{cases}
\end{aligned}
$$

which gives

(4.26)
$$||-\frac{1}{\omega}\int_{\frac{2}{\sqrt{\omega}\lambda(t)}}^{\infty}\phi(r,\omega\lambda(t)^2)\sqrt{r}F_4(t,r\lambda(t))dr||_{L^2(\rho(\omega\lambda(t)^2)d\omega)}\leq\frac{C\sqrt{\log(\log(t))}}{t^2\log^{1+b-2\alpha b}(t)}$$

(Note that, by proposition 4.7 b of [**12**], $\rho(\xi)=\frac{1}{\pi|a(\xi)|^2}$). Now, we consider

$$-\int_t^{\infty}\frac{\cos((t-x)\sqrt{\omega})}{\omega}\mathcal{F}(\sqrt{\cdot}F_4(x,\cdot\lambda(x)))'(\omega\lambda(x)^2)\cdot 2\omega\lambda(x)\lambda'(x)dx$$

We start by noting that, in the region $r\leq\frac{2}{\sqrt{\xi}}$,

$$\partial_2\phi(r,\xi)=\partial_\xi\left(\widetilde{\phi_0}(r)+\frac{1}{\sqrt{r}}\sum_{j=1}^{\infty}(r^{2j}\xi^j\phi_j(r^2))\right)$$

So,

$$\begin{aligned}(4.27)\qquad &\mathcal{F}(\sqrt{\cdot}F_4(x,\cdot\lambda(x)))'(\omega\lambda(x)^2)\\ &=\int_0^{\frac{2}{\sqrt{\omega}\lambda(x)}}\sqrt{r}F_4(x,r\lambda(x))\cdot\frac{1}{\sqrt{r}}\sum_{j=1}^{\infty}(r^{2j}j(\omega\lambda(x)^2)^{j-1}\phi_j(r^2))dr\\ &+\int_{\frac{2}{\sqrt{\omega}\lambda(x)}}^{\infty}\sqrt{r}F_4(x,r\lambda(x))\partial_2\phi(r,\omega\lambda(x)^2)dr\end{aligned}$$

We start with the first term:

$$\begin{aligned}&|-\int_t^{\infty}\frac{\cos((t-x)\sqrt{\omega})}{\omega}\left(\int_0^{\frac{2}{\sqrt{\omega}\lambda(x)}}\sqrt{r}F_4(x,r\lambda(x))\cdot\frac{1}{\sqrt{r}}\sum_{j=1}^{\infty}(r^{2j}j(\omega\lambda(x)^2)^{j-1}\phi_j(r^2))dr\right)\cdot 2\omega\lambda(x)\lambda'(x)dx|\\ &\leq C\sum_{j=1}^{\infty}j\frac{\omega^j}{\omega}\int_t^{\infty}\lambda(x)^{2j-1}|\lambda'(x)|\int_0^{\frac{2}{\sqrt{\omega}\lambda(x)}}|F_4(x,r\lambda(x))|r^{2j}\phi_j(r^2)drdx\\ &\leq C\sum_{j=1}^{\infty}\frac{j\omega^jC_1^j}{(j-1)!}\int_t^{\infty}\frac{\lambda(x)^{2j-1}|\lambda'(x)|}{\omega}\int_0^{\frac{2}{\sqrt{\omega}\lambda(x)}}|F_4(x,r\lambda(x))|r^{2j}\log(1+r^2)drdx\end{aligned}$$

The r integral in the last line of the above expression was estimated before, and we get

$$\begin{aligned}&|-\int_t^{\infty}\frac{\cos((t-x)\sqrt{\omega})}{\omega}\left(\int_0^{\frac{2}{\sqrt{\omega}\lambda(x)}}\sqrt{r}F_4(x,r\lambda(x))\cdot\frac{1}{\sqrt{r}}\sum_{j=1}^{\infty}(r^{2j}j(\omega\lambda(x)^2)^{j-1}\phi_j(r^2))dr\right)\\ &\cdot 2\omega\lambda(x)\lambda'(x)dx|\\ &\leq\int_t^{\infty}\frac{1}{x\log(x)}\begin{cases}\frac{C}{\omega^3\lambda(x)^4x^2\log^{1-2b\alpha}(x)}, & \frac{2}{\sqrt{\omega}\lambda(x)}\leq 1\\ \frac{C(\log(\log(x)))^2}{x^2\log^{1+2b-2b\alpha}(x)}, & 1\leq\frac{2}{\sqrt{\omega}\lambda(x)}\end{cases}dx\end{aligned}$$

Finally,

$$
\begin{aligned}
&||-\int_t^\infty \frac{\cos((t-x)\sqrt{\omega})}{\omega}\left(\int_0^{\frac{2}{\sqrt{\omega\lambda(x)}}}\sqrt{r}F_4(x,r\lambda(x))\cdot\frac{1}{\sqrt{r}}\sum_{j=1}^\infty (r^{2j}j(\omega\lambda(x)^2)^{j-1}\phi_j(r^2))dr\right)\\
&\quad\cdot 2\omega\lambda(x)\lambda'(x)dx||_{L^2(\rho(\omega\lambda(t)^2)d\omega)}\\
&\leq C\int_t^\infty ||\frac{1}{\omega}\cdot\left(\int_0^{\frac{2}{\sqrt{\omega\lambda(x)}}}\sqrt{r}F_4(x,r\lambda(x))\cdot\frac{1}{\sqrt{r}}\sum_{j=1}^\infty (r^{2j}j(\omega\lambda(x)^2)^{j-1}\phi_j(r^2))dr\right)\\
&\quad\cdot 2\omega\lambda(x)\lambda'(x)||_{L^2(\rho(\omega\lambda(t)^2)d\omega)}dx\\
&\leq C\int_t^\infty \left(\frac{\lambda(t)}{\lambda(x)}\right)||\frac{1}{\omega}\cdot\left(\int_0^{\frac{2}{\sqrt{\omega\lambda(x)}}}\sqrt{r}F_4(x,r\lambda(x))\cdot\frac{1}{\sqrt{r}}\sum_{j=1}^\infty (r^{2j}j(\omega\lambda(x)^2)^{j-1}\phi_j(r^2))dr\right)\\
&\quad\cdot 2\omega\lambda(x)\lambda'(x)||_{L^2(\rho(\omega\lambda(x)^2)d\omega)}dx\\
&\leq \frac{C(\log(\log(t)))^2}{t^2\log^{2+b-2b\alpha}(t)}
\end{aligned}
\tag{4.28}
$$

where we used (4.23). The next integral to treat is

$$
\begin{aligned}
&|-2\omega\int_t^\infty \frac{\cos((t-x)\sqrt{\omega})}{\omega}\lambda(x)\lambda'(x)\int_{\frac{2}{\sqrt{\omega\lambda(x)}}}^\infty \sqrt{r}F_4(x,r\lambda(x))\partial_2\phi(r,\omega\lambda(x)^2)drdx|\\
&\leq C\int_t^\infty \frac{1}{x\log^{2b+1}(x)}\int_{\frac{2}{\sqrt{\omega\lambda(x)}}}^\infty \sqrt{r}|F_4(x,r\lambda(x))||\partial_2\phi(r,\omega\lambda(x)^2)|drdx
\end{aligned}
$$

We again use

$$\phi(r,\xi)=2\text{Re}(a(\xi)\psi^+(r,\xi)),\quad r\geq\frac{2}{\sqrt{\xi}}$$

and the following symbol type estimate on a from proposition 4.7 of [**12**]:

$$|a^{(k)}(\xi)|\leq\frac{C_k|a(\xi)|}{\xi^k}$$

as well as, from proposition 4.6 of [**12**]

$$\psi^+(r,\xi)=\frac{e^{ir\sqrt{\xi}}}{\xi^{1/4}}\sigma(r\sqrt{\xi},r),\quad r\sqrt{\xi}\geq 2$$

and the asymptotic series representation for σ, to get

$$|\partial_2\phi(r,\xi)|\leq\frac{C|a(\xi)|}{\xi^{3/4}}r,\quad r\sqrt{\xi}\geq 2$$

This gives

$$
\begin{aligned}
&|-2\omega\int_t^\infty \frac{\cos((t-x)\sqrt{\omega})}{\omega}\lambda(x)\lambda'(x)\int_{\frac{2}{\sqrt{\omega\lambda(x)}}}^\infty \sqrt{r}F_4(x,r\lambda(x))\partial_2\phi(r,\omega\lambda(x)^2)drdx|\\
&\leq\frac{C|a(\omega\lambda(x)^2)|}{\omega^{3/4}\lambda(x)^{3/2}}\int_t^\infty\frac{1}{x\log^{2b+1}(x)}\int_{\frac{2}{\sqrt{\omega\lambda(x)}}}^\infty |F_4(x,r\lambda(x))|r^{3/2}drdx
\end{aligned}
$$

Note that estimates on this integral can not quite be inferred from estimates on a previously treated integral. So, we start considering cases of ω:

Case 1: $1 \leq \frac{2}{\sqrt{\omega}\lambda(x)} \leq \frac{\log^N(x)}{\lambda(x)}$. Here, we get

$$\begin{aligned}
&\frac{|a(\omega\lambda(x)^2)|}{\omega^{3/4}\lambda(x)^{3/2}}\int_{\frac{2}{\sqrt{\omega}\lambda(x)}}^{\infty} |F_4(x,r\lambda(x))|r^{3/2}dr \\
&\leq C\frac{|a(\omega\lambda(x)^2)|}{\omega^{3/4}\lambda(x)^{3/2}}\left(\int_{\frac{2}{\sqrt{\omega}\lambda(x)}}^{\frac{\log^N(x)}{\lambda(x)}} \frac{r^{5/2}\lambda(x)}{\lambda(x)^4(r^2+1)^2x^2\log^{3b+1-2b\alpha}(x)}dr + \int_{\frac{2}{\sqrt{\omega}\lambda(x)}}^{\frac{x}{2\lambda(x)}} \frac{r^{3/2}dr}{\lambda(x)^3r^3x^2\log^{5b+2N-2}(x)}\right) \\
&\leq \frac{C|a(\omega\lambda(x)^2)|}{x^2\log^{1-2b\alpha}(x)\sqrt{\omega}\lambda(x)}, \quad 1 \leq \frac{2}{\sqrt{\omega}\lambda(x)} \leq \frac{\log^N(x)}{\lambda(x)}
\end{aligned}$$

Case 2: $\frac{2}{\sqrt{\omega}\lambda(x)} \leq 1$.

$$\begin{aligned}
&\frac{|a(\omega\lambda(x)^2)|}{\omega^{3/4}\lambda(x)^{3/2}}\int_{\frac{2}{\sqrt{\omega}\lambda(x)}}^{\infty} |F_4(x,r\lambda(x))|r^{3/2}dr \\
&\leq \frac{C|a(\omega\lambda(x)^2)|}{\omega^{3/4}\lambda(x)^{3/2}}\left(\int_0^{\frac{\log^N(x)}{\lambda(x)}} \frac{r^{5/2}dr}{(r^2+1)^2x^2\log^{1-2b\alpha}(x)} + \int_0^{\infty} \frac{r^{5/2}dr}{x^2(r^2+1)^2\log^{2b+2N-2}(x)}\right) \\
&\leq \frac{C|a(\omega\lambda(x)^2)|}{\omega^{3/4}\lambda(x)^{3/2}x^2\log^{1-2b\alpha}(x)}, \quad \frac{2}{\sqrt{\omega}\lambda(x)} \leq 1
\end{aligned}$$

Case 3: $\frac{x}{2\lambda(x)} \geq \frac{2}{\sqrt{\omega}\lambda(x)} \geq \frac{\log^N(x)}{\lambda(x)}$. In this region, only the v_4+v_5 term in F_4 contributes:

$$\begin{aligned}
&\frac{|a(\omega\lambda(x)^2)|}{\omega^{3/4}\lambda(x)^{3/2}}\int_{\frac{2}{\sqrt{\omega}\lambda(x)}}^{\infty} |F_4(x,r\lambda(x))|r^{3/2}dr \\
&\leq \frac{C|a(\omega\lambda(x)^2)|}{\omega^{3/4}\lambda(x)^{3/2}}\int_{\frac{2}{\sqrt{\omega}\lambda(x)}}^{\frac{x}{2\lambda(x)}} \frac{r^{3/2}dr}{\lambda(x)^3r^3x^2\log^{5b+2N-2}(x)} \\
&\leq \frac{C|a(\omega\lambda(x)^2)|}{\sqrt{\omega}\lambda(x)^{3/2}x^2\log^{\frac{5b}{2}+2N-2}(x)}, \quad \frac{x}{2\lambda(x)} \geq \frac{2}{\sqrt{\omega}\lambda(x)} \geq \frac{\log^N(x)}{\lambda(x)}
\end{aligned}$$

Case 4: $\frac{2}{\sqrt{\omega}\lambda(x)} \geq \frac{x}{2\lambda(x)}$. In this region, the integral to estimate is zero, because of the support properties of F_4.

Combining these estimates, we get

(4.29)
$$\begin{aligned}
&\frac{|a(\omega\lambda(x)^2)|}{\omega^{3/4}\lambda(x)^{3/2}}\int_{\frac{2}{\sqrt{\omega}\lambda(x)}}^{\infty} |F_4(x,r\lambda(x))|r^{3/2}dr \\
&\leq C|a(\omega\lambda(x)^2)| \cdot \begin{cases} \frac{1}{x^2\log^{1-2b\alpha}(x)\sqrt{\omega}\lambda(x)}, & 1 \leq \frac{2}{\sqrt{\omega}\lambda(x)} \leq \frac{\log^N(x)}{\lambda(x)} \\ \frac{1}{\omega^{3/4}\lambda(x)^{3/2}x^2\log^{1-2b\alpha}(x)}, & \frac{2}{\sqrt{\omega}\lambda(x)} \leq 1 \\ \frac{1}{\sqrt{\omega}\lambda(x)^{3/2}x^2\log^{\frac{5b}{2}+2N-2}(x)}, & \frac{x}{2\lambda(x)} \geq \frac{2}{\sqrt{\omega}\lambda(x)} \geq \frac{\log^N(x)}{\lambda(x)} \\ 0, \quad \frac{2}{\sqrt{\omega}\lambda(x)} \geq \frac{x}{2\lambda(x)} \end{cases}
\end{aligned}$$

Using the same procedure as in (4.28), we get

(4.30)
$$\begin{aligned}
&|| -2\omega\int_t^{\infty} \frac{\cos((t-x)\sqrt{\omega})}{\omega}\lambda(x)\lambda'(x)\int_{\frac{2}{\sqrt{\omega}\lambda(x)}}^{\infty} \sqrt{r}F_4(x,r\lambda(x))\partial_2\phi(r,\omega\lambda(x)^2)drdx||_{L^2(\rho(\omega\lambda(t)^2)d\omega)} \\
&\leq \frac{C\sqrt{\log(\log(t))}}{t^2\log^{2+b-2b\alpha}(t)}
\end{aligned}$$

To treat the next integral:

$$-\int_t^\infty \frac{\cos((t-x)\sqrt{\omega})}{\omega}\mathcal{F}(\sqrt{\cdot}\partial_x(F_4(x,\cdot\lambda(x))))(\omega\lambda(x)^2)dx$$

we first note that

$$\int_0^\infty \phi_0(r)F_4(x,r\lambda(x))rdr=0, \quad x\geq T_0$$

implies

$$\int_0^\infty \phi_0(r)\partial_x(F_4(x,r\lambda(x)))rdr=0, \quad x\geq T_0$$

So, $\partial_x(F_4(x,r\lambda(x)))$ is still orthogonal to $\phi_0(r)$. Also, by noting the symbol-type estimates, (3.174), and inspecting the procedure used to estimate $\frac{\mathcal{F}(\sqrt{\cdot}F_4(x,\cdot\lambda(x)))(\omega\lambda(x)^2)}{\omega}$, we get

$$||\frac{\mathcal{F}(\sqrt{\cdot}\partial_x(F_4(x,\cdot\lambda(x))))(\omega\lambda(x)^2)}{\omega}||_{L^2(\rho(\omega\lambda(t)^2)d\omega)}\leq \frac{C\lambda(t)}{\lambda(x)}\left(\frac{(\log(\log(x)))^2}{x^3\log^{b+1-2\alpha b}(x)}\right)$$

where we again use (4.23). Using Minkowski's inequality, we get

$$\begin{aligned}(4.31)\quad &||-\int_t^\infty \frac{\cos((t-x)\sqrt{\omega})}{\omega}\mathcal{F}(\sqrt{\cdot}\partial_x(F_4(x,\cdot\lambda(x))))(\omega\lambda(x)^2)dx||_{L^2(\rho(\omega\lambda(t)^2)d\omega)}\\ &\leq \frac{C(\log(\log(t))^2}{t^2\log^{b+1-2\alpha b}(t)}\end{aligned}$$

Combining this with our other estimates in this section, we finally get

$$\begin{aligned}&||\int_t^\infty \frac{\sin((t-x)\sqrt{\omega})}{\sqrt{\omega}}\mathcal{F}(\sqrt{\cdot}F_4(x,\cdot\lambda(x)))(\omega\lambda(x)^2)dx||_{L^2(\rho(\omega\lambda(t)^2)d\omega)}\\ &\leq \frac{C(\log(\log(t))^2}{t^2\log^{b+1-2\alpha b}(t)}\end{aligned}$$

Next, we estimate

$$\begin{aligned}&\partial_t\left(\int_t^\infty \frac{\sin((t-x)\sqrt{\omega})}{\sqrt{\omega}}\mathcal{F}(\sqrt{\cdot}F_4(x,\cdot\lambda(x)))(\omega\lambda(x)^2)dx\right)\\ &=\int_t^\infty \cos((t-x)\sqrt{\omega})\mathcal{F}(\sqrt{\cdot}F_4(x,\cdot\lambda(x)))(\omega\lambda(x)^2)dx\\ &=\int_t^\infty \frac{\sin((t-x)\sqrt{\omega})}{\sqrt{\omega}}\left(\mathcal{F}(\sqrt{\cdot}F_4(x,\cdot\lambda(x)))'(\omega\lambda(x)^2)\right.\\ &\qquad\left.\cdot 2\lambda(x)\lambda'(x)\omega+\mathcal{F}(\sqrt{\cdot}\partial_x(F_4(x,\cdot\lambda(x))))(\omega\lambda(x)^2)\right)dx\\ &=-\left(\mathcal{F}(\sqrt{\cdot}F_4(t,\cdot\lambda(t)))'(\omega\lambda(t)^2)\cdot 2\lambda(t)\lambda'(t)+\frac{\mathcal{F}(\sqrt{\cdot}\partial_t(F_4(t,\cdot\lambda(t))))(\omega\lambda(t)^2)}{\omega}\right)\\ &-\int_t^\infty \frac{\cos((t-x)\sqrt{\omega})}{\omega}\left(\mathcal{F}(\sqrt{\cdot}F_4(x,\cdot\lambda(x)))''(\omega\lambda(x)^2)\cdot 4\lambda(x)^2\lambda'(x)^2\omega^2\right.\\ &\qquad\qquad+2\mathcal{F}(\sqrt{\cdot}\partial_x(F_4(x,\cdot\lambda(x))))'(\omega\lambda(x)^2)\cdot 2\lambda(x)\lambda'(x)\omega\\ &+\mathcal{F}(\sqrt{\cdot}F_4(x,\cdot\lambda(x)))'(\omega\lambda(x)^2)\cdot 2\omega((\lambda'(x))^2+\lambda(x)\lambda''(x))\\ &\left.+\mathcal{F}(\sqrt{\cdot}\partial_x^2(F_4(x,\cdot\lambda(x))))(\omega\lambda(x)^2)\right)dx\end{aligned}$$

Note that, while obtaining (4.28) and (4.30), we showed that

$$||\frac{\mathcal{F}(\sqrt{\cdot}F_4(x,\cdot\lambda(x)))'(\omega\lambda(x)^2)}{\omega}\cdot 2\omega\lambda(x)\lambda'(x)||_{L^2(\rho(\omega\lambda(x)^2)d\omega)}\leq C\frac{(\log(\log(x)))^2}{x^3\log^{2+b-2b\alpha}(x)}$$

Similarly, from the procedure used to obtain (4.31), we infer

$$||\frac{\mathcal{F}(\sqrt{\cdot}\partial_t(F_4(t,\cdot\lambda(t))))(\omega\lambda(t)^2)}{\omega}||_{L^2(\rho(\omega\lambda(t)^2)d\omega)}\leq\frac{C(\log(\log(t)))^2}{t^3\log^{b+1-2\alpha b}(t)}$$

Next, we consider the term involving $\mathcal{F}(\sqrt{\cdot}F_4(x,\cdot\lambda(x)))''$. For this, we start by studying, with $f(x,r)=\sqrt{r}F_4(x,r\lambda(x))$,

$$\mathcal{F}(f(x))''(\xi)=\int_0^{\frac{2}{\sqrt{\xi}}}f(x,r)\partial_2^2\phi(r,\xi)dr+\int_{\frac{2}{\sqrt{\xi}}}^{\infty}f(x,r)\partial_2^2\phi(r,\xi)dr$$

In the region $r\sqrt{\xi}\leq 2$, we use

$$\partial_2^2\phi(r,\xi)=\frac{1}{\sqrt{r}}\sum_{j=2}^{\infty}j(j-1)\xi^{j-2}r^{2j}\phi_j(r^2),\quad r\sqrt{\xi}\leq 2$$

Then,

$$|-4\int_t^{\infty}\cos((t-x)\sqrt{\omega})\lambda(x)^2\lambda'(x)^2\omega\sum_{j=2}^{\infty}j(j-1)(\omega\lambda(x)^2)^{j-2}\int_0^{\frac{2}{\sqrt{\omega\lambda(x)}}}F_4(x,r\lambda(x))r^{2j}\phi_j(r^2)drdx|$$
$$\leq C\sum_{j=2}^{\infty}j(j-1)\omega^{j-1}\int_t^{\infty}\left(\frac{\lambda'(x)}{\lambda(x)}\right)^2\lambda(x)^{2j}\int_0^{\frac{2}{\sqrt{\omega\lambda(x)}}}|F_4(x,r\lambda(x))|r^{2j}|\phi_j(r^2)|drdx$$

we then apply the same procedure as in (4.28), and get

$$||-4\int_t^{\infty}\cos((t-x)\sqrt{\omega})\lambda(x)^2\lambda'(x)^2\omega\sum_{j=2}^{\infty}j(j-1)(\omega\lambda(x)^2)^{j-2}\int_0^{\frac{2}{\sqrt{\omega\lambda(x)}}}$$
$$F_4(x,r\lambda(x))r^{2j}\phi_j(r^2)drdx||_{L^2(\rho(\omega\lambda(t)^2)d\omega)}$$
$$\leq\frac{C(\log(\log(t)))^2}{t^3\log^{3+b-2b\alpha}(t)}$$

Next, we have

$$\partial_2^2\phi(r,\xi)=(\partial_2^2\phi(r,\xi))_0+(\partial_2^2\phi(r,\xi))_1$$

with

$$(\partial_2^2\phi(r,\xi))_0=2\mathrm{Re}\left(a''(\xi)\psi^+(r,\xi)+2a'(\xi)\partial_2\psi^+(r,\xi)\right)$$
$$+2\mathrm{Re}\left(a(\xi)\left(\partial_\xi^2\left(\frac{e^{ir\sqrt{\xi}}}{\xi^{1/4}}\sigma(r\sqrt{\xi},r)\right)+\frac{r^2}{4\xi^{5/4}}e^{ir\sqrt{\xi}}\sigma(r\sqrt{\xi},r)\right)\right)$$

and

$$(\partial_2^2\phi(r,\xi))_1=-2\mathrm{Re}\left(a(\xi)\frac{r^2}{4\xi^{5/4}}e^{ir\sqrt{\xi}}\sigma(r\sqrt{\xi},r)\right)$$

$$|(\partial_2^2\phi(r,\xi))_0|\leq\frac{Cr|a(\xi)|}{\xi^{7/4}}$$

Again, with

$$f(x,r)=\sqrt{r}F_4(x,r\lambda(x))$$

we have

$$\begin{aligned}
&\int_{\frac{2}{\sqrt{\xi}}}^{\infty} f(x,r)(\partial_2^2\phi(r,\xi))_1 dr = -2\mathrm{Re}\left(\frac{a(\xi)}{4\xi^{5/4}}\int_{\frac{2}{\sqrt{\xi}}}^{\infty} f(x,r)r^2 e^{ir\sqrt{\xi}}\sigma(r\sqrt{\xi},r)dr\right)\\
&= \frac{1}{2}\mathrm{Re}\left(\frac{a(\xi)}{\xi^{5/4}}\frac{4f(x,\frac{2}{\sqrt{\xi}})}{i\xi^{3/2}}\sigma(2,\frac{2}{\sqrt{\xi}})e^{2i}\right)\\
&+\frac{1}{2}\mathrm{Re}\left(\frac{a(\xi)}{\xi^{5/4}}\int_{\frac{2}{\sqrt{\xi}}}^{\infty}\frac{e^{ir\sqrt{\xi}}}{i\sqrt{\xi}}\left(\partial_r f(x,r)r^2\sigma(r\sqrt{\xi},r)+2f(x,r)r\sigma(r\sqrt{\xi},r)\right.\right.\\
&\qquad\left.\left.+\, r^2 f(x,r)\left(\sqrt{\xi}\partial_1\sigma(r\sqrt{\xi},r)+\partial_2\sigma(r\sqrt{\xi},r)\right)\right)dr\right)
\end{aligned}$$

Again, using Proposition 4.6 of [**12**] to estimate the σ terms, we get

$$\begin{aligned}
&|\int_{\frac{2}{\sqrt{\xi}}}^{\infty} f(x,r)(\partial_2^2\phi(r,\xi))_1 dr|\\
&\leq \frac{C|a(\xi)|}{\xi^{11/4}}|f(x,\frac{2}{\sqrt{\xi}})| + C\frac{|a(\xi)|}{\xi^{7/4}}\int_{\frac{2}{\sqrt{\xi}}}^{\infty}\left(r|\partial_r f(x,r)|+|f(x,r)|\right) r dr
\end{aligned}$$

Then,

$$\begin{aligned}
&|-\int_t^\infty \frac{\cos((t-x)\sqrt{\omega})}{\omega}\left(\int_{\frac{2}{\sqrt{\omega}\lambda(x)}}^{\infty}\sqrt{r}F_4(x,r\lambda(x))\left((\partial_2^2\phi(r,\xi))_0+(\partial_2^2\phi(r,\xi))_1\right)|_{\xi=\omega\lambda(x)^2}dr\right)\cdot 4\lambda(x)^2\lambda'(x)^2\omega^2 dx|\\
&\leq C\int_t^\infty\left(\int_{\frac{2}{\sqrt{\omega}\lambda(x)}}^{\infty}\left(|F_4(x,r\lambda(x))|r^{3/2}+r^{3/2}\cdot r\lambda(x)|\partial_2 F_4(x,r\lambda(x))|\right)dr\right)\frac{|a(\omega\lambda(x)^2)|}{\omega^{3/4}\lambda(x)^{3/2}}\frac{dx}{x^2\log^{2b+2}(x)}\\
&+C\int_t^\infty\frac{|a(\omega\lambda(x)^2)|}{\omega^2\lambda(x)^6}\frac{|F_4(x,\frac{2}{\sqrt{\omega}})|}{x^2\log^{4b+2}(x)}dx
\end{aligned}$$

For the last term, we have

$$\begin{aligned}
&\int_t^\infty ||\frac{|a(\omega\lambda(x)^2)|}{\omega^2\lambda(x)^6}\frac{|F_4(x,\frac{2}{\sqrt{\omega}})|}{x^2\log^{4b+2}(x)}||_{L^2(\rho(\omega\lambda(t)^2)d\omega)}dx\\
&\leq C\int_t^\infty\frac{\lambda(t)}{\lambda(x)}\frac{1}{x^2\log^{2b+2}(x)\lambda(x)^4}\left(\int_0^\infty|F_4(x,y)|^2y^8\frac{dy}{y^3}\right)^{1/2}dx\\
&\leq\frac{C\sqrt{\log(\log(t))}}{t^3\log^{b+3-2\alpha b}(t)}
\end{aligned}$$

So,

$$||\int_t^\infty\frac{|a(\omega\lambda(x)^2)|}{\omega^2\lambda(x)^6}\frac{|F_4(x,\frac{2}{\sqrt{\omega}})|}{x^2\log^{4b+2}(x)}dx||_{L^2(\rho(\omega\lambda(t)^2)d\omega)}\leq\frac{C\sqrt{\log(\log(t))}}{t^3\log^{b+3-2\alpha b}(t)}$$

On the other hand, by the same procedure used to obtain (4.29) and (4.30), we get

$$||\int_t^\infty \left(\int_{\frac{2}{\sqrt{\omega\lambda(x)}}}^\infty \left(|F_4(x,r\lambda(x))|r^{3/2} + r^{3/2}\cdot r\lambda(x)|\partial_2 F_4(x,r\lambda(x))|\right) dr\right) \frac{|a(\omega\lambda(x)^2)|}{\omega^{3/4}\lambda(x)^{3/2}} \frac{dx}{x^2\log^{2b+2}(x)}||_{L^2(\rho(\omega\lambda(t)^2)d\omega)}$$
$$\leq \frac{C\sqrt{\log(\log(t))}}{t^3\log^{3+b-2b\alpha}(t)}$$

Next, by noting the symbol-type character of the estimates (3.174), and inspecting (4.28) and (4.30), we deduce the following estimate

$$||-\int_t^\infty \frac{\cos((t-x)\sqrt{\omega})}{\omega}\cdot 2\mathcal{F}(\sqrt{\cdot}\partial_x(F_4(x,\cdot\lambda(x))))'(\omega\lambda(x)^2)\cdot 2\lambda(x)\lambda'(x)\omega dx||_{L^2(\rho(\omega\lambda(t)^2)d\omega)}$$
$$\leq \frac{C(\log(\log(t)))^2}{t^3\log^{2+b-2b\alpha}(t)}$$

The following term was already estimated via (4.28), (4.30) (except with different λ-dependent coefficients). Taking into account the estimates on λ', λ'', we get

$$||-\int_t^\infty \frac{\cos((t-x)\sqrt{\omega})}{\omega}\mathcal{F}(\sqrt{\cdot}F_4(x,\cdot\lambda(x)))'(\omega\lambda(x)^2)\cdot 2\omega\left(\lambda'(x)^2+\lambda(x)\lambda''(x)\right)dx||_{L^2(\rho(\omega\lambda(t)^2)d\omega)}$$
$$\leq \frac{C(\log(\log(t)))^2}{t^3\log^{2+b-2b\alpha}(t)}$$

Finally, we start with estimating $\frac{\mathcal{F}(\sqrt{\cdot}\partial_x^2(F_4(x,\cdot\lambda(x))))(\omega\lambda(x)^2)}{\omega}$. Firstly, we note that $\partial_x^2(F_4(x,r\lambda(x)))$ is still orthogonal to $\phi_0(r)$. Then, we repeat the same procedure used to estimate $\frac{\mathcal{F}(\sqrt{\cdot}F_4(x,\cdot\lambda(x)))(\omega\lambda(x)^2)}{\omega}$. By again noting the symbol-type behavior of (3.174), the only contributions to $\frac{\mathcal{F}(\sqrt{\cdot}\partial_x^2(F_4(x,\cdot\lambda(x))))(\omega\lambda(x)^2)}{\omega}$ which need to be checked are those due to the last term in (3.175), which is not directly comparable to terms arising in (3.174). To be clear, we still use the orthogonality of the full function $\partial_x^2(F_4(x,r\lambda(x)))$ to $\phi_0(r)$; but, after using the orthogonality as needed, we can then deduce estimates on all integrals which do not involve the last term in (3.175) just by comparison with an analogous term in (3.174), as described above. We start with

Case 1: $\frac{2}{\sqrt{\omega\lambda(x)}} \leq 1$.

$$|\frac{-1}{\omega}\int_0^{\frac{2}{\sqrt{\omega\lambda(x)}}} \widetilde{\phi_0}(r)\sqrt{r}\frac{\mathbb{1}_{\{r\lambda(x)\leq\frac{x}{2}\}}}{x^4\log^{b+N-2}(x)(r^2+1)^2}dr|$$
$$\leq \frac{C}{\omega}\int_0^{\frac{2}{\sqrt{\omega\lambda(x)}}} \frac{\phi_0(r)rdr}{x^4\log^{b+N-2}(x)(r^2+1)^2}$$
$$\leq \frac{C}{\omega^{5/2}\lambda(x)^3x^4\log^{b+N-2}(x)}, \quad \frac{2}{\sqrt{\omega\lambda(x)}}\leq 1$$

Case 2: $1 \leq \frac{2}{\sqrt{\omega\lambda(x)}} \leq \frac{x}{2\lambda(x)}$. Here, we first use the orthogonality of $\partial_x^2(F_4(x,r\lambda(x)))$ to $\phi_0(r)$. Using the procedure described above, we only need to estimate

$$
\begin{aligned}
&|\frac{1}{\omega}\int_{\frac{2}{\sqrt{\omega}\lambda(x)}}^{\infty}\widetilde{\phi_0}(r)\sqrt{r}\frac{\mathbb{1}_{\{r\lambda(x)\leq\frac{x}{2}\}}}{x^4\log^{b+N-2}(x)(r^2+1)^2}dr| \leq \frac{C}{\omega}\int_{\frac{2}{\sqrt{\omega}\lambda(x)}}^{\frac{x}{2\lambda(x)}}\frac{dr}{x^4\log^{b+N-2}(x)r^4}\\
&\leq \frac{C\sqrt{\omega}}{x^4\log^{4b+N-2}(x)}, \quad 1\leq\frac{2}{\sqrt{\omega}\lambda(x)}\leq\frac{x}{2\lambda(x)}
\end{aligned}
$$

Case 3: $\frac{x}{2\lambda(x)} \leq \frac{2}{\sqrt{\omega}\lambda(x)}$. In this case, after again using the orthogonality of $\partial_x^2(F_4(x, r\lambda(x)))$ to $\phi_0(r)$, the only integral to be checked is zero, due to the support properties of F_4. Next, we have

Case 1: $1 \leq \frac{2}{\sqrt{\omega}\lambda(x)} \leq \frac{x}{2\lambda(x)}$.

$$
\begin{aligned}
&\frac{1}{\omega}\sum_{j=1}^{\infty}\int_0^{\frac{2}{\sqrt{\omega}\lambda(x)}} r^{2j}\omega^j\lambda(x)^{2j}|\phi_j(r^2)|\frac{\mathbb{1}_{\{r\lambda(x)\leq\frac{x}{2}\}}}{x^4\log^{b+N-2}(x)(r^2+1)^2}dr\\
&\leq \frac{C}{\omega}\sum_{j=1}^{\infty}\frac{\omega^j\lambda(x)^{2j}C_1^j}{(j-1)!}\int_0^{\frac{2}{\sqrt{\omega}\lambda(x)}}\frac{r^{2j}\log(1+r^2)dr}{x^4\log^{b+N-2}(x)(r^2+1)^2}\\
&\leq \frac{C}{x^4\log^{3b+N-3}(x)}, \quad 1\leq\frac{2}{\sqrt{\omega}\lambda(x)}\leq\frac{x}{2\lambda(x)}
\end{aligned}
$$

where we treat the integral in exactly the same way we did in obtaining (4.24)

Case 2: $\frac{2}{\sqrt{\omega}\lambda(x)} \leq 1$.

$$
\begin{aligned}
&\frac{1}{\omega}\sum_{j=1}^{\infty}\int_0^{\frac{2}{\sqrt{\omega}\lambda(x)}} r^{2j}\omega^j\lambda(x)^{2j}|\phi_j(r^2)|\frac{\mathbb{1}_{\{r\lambda(x)\leq\frac{x}{2}\}}}{x^4\log^{b+N-2}(x)(r^2+1)^2}dr\\
&\leq \frac{C}{\omega}\sum_{j=1}^{\infty}\frac{\omega^j\lambda(x)^{2j}C_1^j}{(j-1)!x^4\log^{b+N-2}(x)}\int_0^{\frac{2}{\sqrt{\omega}\lambda(x)}} r^{2j+2}dr\\
&\leq \frac{C}{\omega^{5/2}\lambda(x)^3x^4\log^{b+N-2}(x)}, \quad \frac{2}{\sqrt{\omega}\lambda(x)}\leq 1
\end{aligned}
$$

Case 3: $\frac{2}{\sqrt{\omega}\lambda(x)} \geq \frac{x}{2\lambda(x)}$.

$$
\begin{aligned}
&\frac{1}{\omega}\sum_{j=1}^{\infty}\int_0^{\frac{2}{\sqrt{\omega}\lambda(x)}} r^{2j}\omega^j\lambda(x)^{2j}|\phi_j(r^2)|\frac{\mathbb{1}_{\{r\lambda(x)\leq\frac{x}{2}\}}}{x^4\log^{b+N-2}(x)(r^2+1)^2}dr\\
&\leq \frac{C}{\omega}\sum_{j=1}^{\infty}\frac{\omega^j\lambda(x)^{2j}C_1^j}{(j-1)!}\int_0^{\frac{x}{2\lambda(x)}}\frac{r^{2j}\log(1+r^2)dr}{x^4\log^{b+N-2}(x)(r^2+1)^2}\\
&\leq \frac{C}{x^4\log^{3b+N-3}(x)}, \quad \frac{2}{\sqrt{\omega}\lambda(x)}\geq\frac{x}{2\lambda(x)}
\end{aligned}
$$

Finally, we consider the following integral in multiple cases:

Case 1: $\frac{2}{\sqrt{\omega}\lambda(x)} \leq 1$.

$$\begin{aligned}&\frac{1}{\omega}\int_{\frac{2}{\sqrt{\omega}\lambda(x)}}^{\infty} \frac{2|\mathrm{Re}\left(a(\omega\lambda(x)^2)\psi^{+}(r,\omega\lambda(x)^2)\right)|\sqrt{r}\mathbb{1}_{\{r\lambda(x)\leq\frac{x}{2}\}}}{x^4\log^{b+N-2}(x)(r^2+1)^2}dr\\ &\leq \frac{C}{\omega}\frac{|a(\omega\lambda(x)^2)|}{\omega^{1/4}\lambda(x)^{1/2}}\int_{\frac{2}{\sqrt{\omega}\lambda(x)}}^{\frac{x}{2\lambda(x)}} \frac{\sqrt{r}dr}{x^4\log^{b+N-2}(x)(r^2+1)^2}\\ &\leq \frac{C|a(\omega\lambda(x)^2)|}{\omega^{5/4}\sqrt{\lambda(x)}x^4\log^{b+N-2}(x)}, \quad \frac{2}{\sqrt{\omega}\lambda(x)}\leq 1\end{aligned}$$

Case 2: $1\leq \frac{2}{\sqrt{\omega}\lambda(x)} \leq \frac{x}{2\lambda(x)}$

$$\begin{aligned}&\frac{1}{\omega}\int_{\frac{2}{\sqrt{\omega}\lambda(x)}}^{\infty} 2|\mathrm{Re}\left(a(\omega\lambda(x)^2)\psi^{+}(r,\omega\lambda(x)^2)\right)|\sqrt{r}\frac{\mathbb{1}_{\{r\lambda(x)\leq\frac{x}{2}\}}}{x^4\log^{b+N-2}(x)(r^2+1)^2}dr\\ &\leq \frac{C|a(\omega\lambda(x)^2)|}{\omega^{5/4}\lambda(x)^{1/2}}\int_{\frac{2}{\sqrt{\omega}\lambda(x)}}^{\frac{x}{2\lambda(x)}} \frac{dr}{r^{7/2}x^4\log^{b+N-2}(x)}\\ &\leq \frac{C|a(\omega\lambda(x)^2)|}{x^4\log^{3b+N-2}(x)}, \quad 1\leq\frac{2}{\sqrt{\omega}\lambda(x)}\leq\frac{x}{2\lambda(x)}\end{aligned}$$

Case 3: $\frac{2}{\sqrt{\omega}\lambda(x)} > \frac{x}{2\lambda(x)}$. In this case, the integral to estimate is zero. Combining all of the above estimates, and using the procedure described above, we get

$$\begin{aligned}&||-\int_t^{\infty}\frac{\cos((t-x)\sqrt{\omega})}{\omega}\mathcal{F}(\sqrt{\cdot}\partial_x^2(F_4(x,\cdot\lambda(x))))(\omega\lambda(x)^2)dx||_{L^2(\rho(\omega\lambda(t)^2)d\omega)}\\ &\leq \frac{C(\log(\log(t)))^2}{t^3\log^{b+1-2\alpha b}(t)}\end{aligned}$$

In total, we finally get

$$\begin{aligned}||\int_t^{\infty}\cos((t-x)\sqrt{\omega})\mathcal{F}(\sqrt{\cdot}F_4(x,\cdot\lambda(x)))(\omega\lambda(x)^2)dx||_{L^2(\rho(\omega\lambda(t)^2)d\omega)}\\ \leq \frac{C(\log(\log(t)))^2}{t^3\log^{b+1-2\alpha b}(t)}\end{aligned}\tag{4.32}$$

The next integral to estimate is

$$\begin{aligned}&\lambda(t)\int_t^{\infty}\sin((t-x)\sqrt{\omega})\mathcal{F}(\sqrt{\cdot}F_4(x,\cdot\lambda(x)))(\omega\lambda(x)^2)dx\\ &=\lambda(t)\left(-\frac{\mathcal{F}(\sqrt{\cdot}F_4(t,\cdot\lambda(t)))(\omega\lambda(t)^2)}{\sqrt{\omega}}\right.\\ &\qquad\left.-\int_t^{\infty}\frac{\cos((t-x)\sqrt{\omega})}{\sqrt{\omega}}\partial_x\left(\mathcal{F}(\sqrt{\cdot}F_4(x,\cdot\lambda(x)))(\omega\lambda(x)^2)\right)dx\right)\\ &=\frac{-\lambda(t)\mathcal{F}(\sqrt{\cdot}F_4(t,\cdot\lambda(t)))(\omega\lambda(t)^2)}{\sqrt{\omega}}\\ &\quad -\lambda(t)\int_t^{\infty}\frac{\sin((t-x)\sqrt{\omega})}{\omega}\partial_x^2\left(\mathcal{F}(\sqrt{\cdot}F_4(x,\cdot\lambda(x)))(\omega\lambda(x)^2)\right)dx\end{aligned}$$

In order to estimate $\frac{-\lambda(t)\mathcal{F}(\sqrt{\cdot}F_4(t,\cdot\lambda(t)))(\omega\lambda(t)^2)}{\sqrt{\omega}}$ it suffices to multiply the pointwise in ω estimates that we already obtained on $\frac{\mathcal{F}(\sqrt{\cdot}F_4(t,\cdot\lambda(t)))(\omega\lambda(t)^2)}{\omega}$, by $\sqrt{\omega}\lambda(t)$, and then take the $L^2(\rho(\omega\lambda(t)^2)d\omega)$ norm. In fact, we only need to consider the region $\frac{2}{\sqrt{\omega}\lambda(t)} \leq 1$, since, in the other regions, the factor $\sqrt{\omega}\lambda(t)$ that we multiply by, is less than 2. Doing this procedure for each of the terms appearing in (4.24), (4.25), and (4.26), and combining the resulting estimates, we get

$$|\frac{-\lambda(t)\mathcal{F}(\sqrt{\cdot}F_4(t,\cdot\lambda(t)))(\omega\lambda(t)^2)}{\sqrt{\omega}}|$$
$$\leq \frac{C}{\omega^{5/2}\lambda(t)^3t^2\log^{1-2\alpha b}(t)} + \frac{C\lambda(t)^{1/2}|a(\omega\lambda(t)^2)|}{\omega^{3/4}t^2\log^{1-2\alpha b}(t)}, \quad \frac{2}{\sqrt{\omega}\lambda(t)} \leq 1$$

and

$$\left(\int_{\frac{4}{\lambda(t)^2}}^{\infty} \rho(\omega\lambda(t)^2)\frac{d\omega}{\omega^5\lambda(t)^6t^4\log^{2-4\alpha b}(t)}\right)^{1/2} \leq \frac{C}{t^2\log^{1+b-2\alpha b}(t)}$$

$$\left(\int_{\frac{4}{\lambda(t)^2}}^{\infty} \frac{|a(\omega\lambda(t)^2)|^2}{\omega^{3/2}}\rho(\omega\lambda(t)^2)d\omega\right)^{1/2} \cdot \frac{C}{t^2\log^{1+\frac{b}{2}-2\alpha b}(t)} \leq \frac{C}{t^2\log^{1+b-2\alpha b}(t)}$$

Then, by the remarks preceding these estimates, we get

$$||\frac{\lambda(t)\mathcal{F}(\sqrt{\cdot}F_4(t,\cdot\lambda(t)))(\omega\lambda(t)^2)}{\sqrt{\omega}}||_{L^2(\rho(\omega\lambda(t)^2)d\omega)} \leq \frac{C(\log(\log(t)))^2}{t^2\log^{1+b-2\alpha b}(t)}$$

Then, we use the identical procedure as was used to obtain (4.32), and get

$$||\lambda(t)\int_t^{\infty}\frac{\sin((t-x)\sqrt{\omega})}{\omega}\partial_x^2\left(\mathcal{F}(\sqrt{\cdot}F_4(x,\cdot\lambda(x)))(\omega\lambda(x)^2)\right)dx||_{L^2(\rho(\omega\lambda(t)^2)d\omega)}$$
$$\leq \frac{C(\log(\log(t)))^2}{t^3\log^{2b+1-2\alpha b}(t)}$$

Combining these, we get

$$||\sqrt{\omega}\lambda(t)\int_t^{\infty}\frac{\sin((t-x)\sqrt{\omega})}{\sqrt{\omega}}\mathcal{F}(\sqrt{\cdot}F_4(x,\cdot\lambda(x)))(\omega\lambda(x)^2)dx||_{L^2(\rho(\omega\lambda(t)^2)d\omega)}$$
$$\leq \frac{C(\log(\log(t)))^2}{t^2\log^{1+b-2\alpha b}(t)}$$

Next, we consider

$$\begin{aligned}&\sqrt{\omega}\lambda(t)\partial_t\left(\int_t^{\infty}\frac{\sin((t-x)\sqrt{\omega})}{\sqrt{\omega}}\mathcal{F}(\sqrt{\cdot}F_4(x,\cdot\lambda(x)))(\omega\lambda(x)^2)dx\right)\\ &= \frac{-\lambda(t)\partial_t\left(\mathcal{F}(\sqrt{\cdot}F_4(t,\cdot\lambda(t)))(\omega\lambda(t)^2)\right)}{\sqrt{\omega}}\\ &\quad -\int_t^{\infty}\lambda(t)\cdot\frac{\cos((t-x)\sqrt{\omega})}{\sqrt{\omega}}\partial_x^2\left(\mathcal{F}(\sqrt{\cdot}F_4(x,\cdot\lambda(x)))(\omega\lambda(x)^2)\right)dx\end{aligned} \tag{4.33}$$

For $\frac{-\lambda(t)\left(\mathcal{F}(\sqrt{\cdot}\partial_t(F_4(t,\cdot\lambda(t))))(\omega\lambda(t)^2)\right)}{\sqrt{\omega}}$, we again need only multiply the pointwise estimates for $\frac{\mathcal{F}(\sqrt{\cdot}\partial_t(F_4(t,\cdot\lambda(t))))(\omega\lambda(t)^2)}{\omega}$ (which were previously inferred from pointwise estimates for $\frac{\mathcal{F}(\sqrt{\cdot}(F_4(t,\cdot\lambda(t))))(\omega\lambda(t)^2)}{\omega}$ and noting the symbol-type nature of the estimate (3.174)) by $\sqrt{\omega}\lambda(t)$ in the region $\sqrt{\omega}\lambda(t) \geq 2$, and then take the $L^2(\rho(\omega\lambda(t)^2)d\omega)$ norm. This results in

$$||-\sqrt{\omega}\lambda(t)\left(\frac{\mathcal{F}(\sqrt{\cdot}\partial_t(F_4(t,\cdot\lambda(t))))(\omega\lambda(t)^2)}{\omega}\right)||_{L^2(\rho(\omega\lambda(t)^2)d\omega)} \leq \frac{C(\log(\log(t)))^2}{t^3\log^{1+b-2\alpha b}(t)}$$

Note, however, that doing this same procedure for the term

$$-\sqrt{\omega}\lambda(t)\left(\frac{\mathcal{F}(\sqrt{\cdot}F_4(t,\cdot\lambda(t)))'(\omega\lambda(t)^2)\cdot 2\omega\lambda(t)\lambda'(t)}{\omega}\right)$$

would result in an estimate which is not square integrable against the measure $\rho(\omega\lambda(t)^2)d\omega$. Instead, we have to integrate by parts in an appropriate r integral, to gain extra decay in ω. In particular, we first make the decomposition as in (4.27). Then, we can multiply pointwise estimates on the first term in the decomposition (4.27) by $\sqrt{\omega}\lambda(t)$, and proceed as with our previous estimates. This results in a contribution to the overall $L^2(\rho(\omega\lambda(t)^2)d\omega)$ norm of

$$-\sqrt{\omega}\lambda(t)\left(\frac{\mathcal{F}(\sqrt{\cdot}F_4(t,\cdot\lambda(t)))'(\omega\lambda(t)^2)\cdot 2\omega\lambda(t)\lambda'(t)}{\omega}\right)$$

bounded above by:

$$\begin{aligned}&||-2\lambda(t)^2\lambda'(t)\sqrt{\omega}\left(\int_0^{\frac{2}{\sqrt{\omega}\lambda(t)}}\sqrt{r}F_4(t,r\lambda(t))\partial_2\phi(r,\omega\lambda(t)^2)dr\right)||_{L^2(\rho(\omega\lambda(t)^2)d\omega)}\\&\leq\frac{C(\log(\log(t)))^2}{t^3\log^{b+2-2\alpha b}(t)}\end{aligned}$$

For the second term in the decomposition as in (4.27),namely,

$$-2\sqrt{\omega}\lambda(t)^2\lambda'(t)\int_{\frac{2}{\sqrt{\omega}\lambda(t)}}^{\infty}\sqrt{r}F_4(t,r\lambda(t))\partial_2\phi(r,\omega\lambda(t)^2)dr$$

we start with

$$\partial_2\phi(r,\xi) = (\partial_2\phi(r,\xi))_0 + (\partial_2\phi(r,\xi))_1$$

where

$$(\partial_2\phi(r,\xi))_1 = \text{Re}\left(\frac{a(\xi)ir}{\xi^{3/4}}e^{ir\sqrt{\xi}}\sigma(r\sqrt{\xi},r)\right)$$

and

$$|(\partial_2\phi(r,\xi))_0| \leq \frac{C|a(\xi)|}{\xi^{5/4}}, \quad r\sqrt{\xi}\geq 2$$

Then,

$$\begin{aligned}
&-2\lambda(t)^2\lambda'(t)\sqrt{\omega}\int_{\frac{2}{\sqrt{\omega}\lambda(t)}}^{\infty}\sqrt{r}F_4(t,r\lambda(t))(\partial_2\phi(r,\omega\lambda(t)^2))_1 dr\\
&=-2\lambda(t)^2\lambda'(t)\sqrt{\omega}\text{Re}\\
&\qquad\left(\frac{a(\omega\lambda(t)^2)i}{\omega^{3/4}\lambda(t)^{3/2}}\int_{\frac{2}{\sqrt{\omega}\lambda(t)}}^{\infty}r^{3/2}F_4(t,r\lambda(t))e^{ir\sqrt{\omega}\lambda(t)}\sigma(r\sqrt{\omega}\lambda(t),r)dr\right)\\
&=\frac{-2\lambda(t)^{1/2}\lambda'(t)}{\omega^{1/4}}\text{Re}\left(\frac{-a(\omega\lambda(t)^2)2^{3/2}F_4(t,\frac{2}{\sqrt{\omega}})\sigma(2,\frac{2}{\sqrt{\omega}\lambda(t)})e^{2i}}{\omega^{5/4}\lambda(t)^{5/2}}\right)\\
&+\frac{2\lambda(t)^{1/2}\lambda'(t)}{\omega^{1/4}}\text{Re}\\
&\qquad\left(a(\omega\lambda(t)^2)\int_{\frac{2}{\sqrt{\omega}\lambda(t)}}^{\infty}\frac{e^{ir\sqrt{\omega}\lambda(t)}}{\sqrt{\omega}\lambda(t)}\partial_r\left(r^{3/2}F_4(t,r\lambda(t))\sigma(r\sqrt{\omega}\lambda(t),r)\right)dr\right)
\end{aligned}$$

So,

$$\begin{aligned}
&|-2\sqrt{\omega}\lambda(t)^2\lambda'(t)\int_{\frac{2}{\sqrt{\omega}\lambda(t)}}^{\infty}\sqrt{r}F_4(t,r\lambda(t))\partial_2\phi(r,\omega\lambda(t)^2)dr|\\
&\leq\frac{C|\lambda'(t)||a(\omega\lambda(t)^2)|}{\omega^{3/2}\lambda(t)^2}|F_4(t,\frac{2}{\sqrt{\omega}})|\\
&+\frac{C|\lambda'(t)||a(\omega\lambda(t)^2)|}{\lambda(t)^{1/2}\omega^{3/4}}\int_{\frac{2}{\sqrt{\omega}\lambda(t)}}^{\infty}r^{1/2}\left(|F_4(t,r\lambda(t))|+r\lambda(t)|\partial_2F_4(t,r\lambda(t))|\right)dr
\end{aligned}$$

where we used the estimates on σ following from proposition 4.6 of [**12**].

This gives

$$\begin{aligned}
&||-2\sqrt{\omega}\lambda(t)^2\lambda'(t)\int_{\frac{2}{\sqrt{\omega}\lambda(t)}}^{\infty}\sqrt{r}F_4(t,r\lambda(t))\partial_2\phi(r,\omega\lambda(t)^2)dr||_{L^2((\frac{4}{\lambda(t)^2},\infty),\rho(\omega\lambda(t)^2)d\omega)}\\
&\leq\frac{C|\lambda'(t)|}{\lambda(t)^2}\left(\int_0^{\lambda(t)}|F_4(t,y)|^2y^3dy\right)^{1/2}\\
&+\frac{C|\lambda'(t)|}{\lambda(t)^{1/2}}\left(\int_{\frac{4}{\lambda(t)^2}}^{\infty}\frac{1}{\omega^{3/2}}\left(\int_0^{\infty}\frac{r^{3/2}}{t^2\log^{1-2\alpha b}(t)(r^2+1)^2}dr\right)^2d\omega\right)^{1/2}\\
&\leq\frac{C}{t^3\log^{b+2-2\alpha b}(t)}
\end{aligned}$$

Combining these estimates, and appropriately using (4.30), for the region $\sqrt{\omega}\lambda(t)\leq 2$, we get

$$\begin{aligned}
&||-\sqrt{\omega}\lambda(t)\left(\frac{\mathcal{F}(\sqrt{\cdot}F_4(t,\cdot\lambda(t)))'(\omega\lambda(t)^2)\cdot 2\omega\lambda(t)\lambda'(t)}{\omega}\right)||_{L^2(\rho(\omega\lambda(t)^2)d\omega)}\\
&\leq\frac{C(\log(\log(t)))^2}{t^3\log^{b+2-2b\alpha}(t)}
\end{aligned}\tag{4.34}$$

Now, we start to treat the term inside the x integral in (4.33). Here, we expand

$$\begin{aligned}&\partial_x^2\left(\mathcal{F}(\sqrt{\cdot}F_4(x,\cdot\lambda(x)))(\omega\lambda(x)^2)\right)\\&=\mathcal{F}(\sqrt{\cdot}F_4(x,\cdot\lambda(x)))''(\omega\lambda(x)^2)\cdot(2\omega\lambda(x)\lambda'(x))^2\\&\quad+2\mathcal{F}(\sqrt{\cdot}\partial_x(F_4(x,\cdot\lambda(x))))'(\omega\lambda(x)^2)\cdot 2\omega\lambda(x)\lambda'(x)\\&\quad+\mathcal{F}(\sqrt{\cdot}F_4(x,\cdot\lambda(x)))'(\omega\lambda(x)^2)\cdot 2\omega((\lambda'(x))^2+\lambda(x)\lambda''(x))\\&\quad+\mathcal{F}(\sqrt{\cdot}\partial_x^2(F_4(x,\cdot\lambda(x))))(\omega\lambda(x)^2)\end{aligned}$$

We start by considering

$$\frac{\sqrt{\omega}\lambda(t)}{\omega}\cdot\left(2\mathcal{F}(\sqrt{\cdot}\partial_x(F_4(x,\cdot\lambda(x))))'(\omega\lambda(x)^2)\cdot 2\omega\lambda(x)\lambda'(x)\right)$$

Recall that the last term we estimated was

$$-\sqrt{\omega}\lambda(t)\cdot 2\lambda(t)\lambda'(t)\mathcal{F}(\sqrt{\cdot}F_4(t,\cdot\lambda(t)))'(\omega\lambda(t)^2)$$

and we used only (3.187), as well as (3.188), because we needed to integrate by parts in the r variable in one of the terms. We then repeat the same procedure, with the only difference being

$$\partial_1 F_4(x,r\lambda(x))+r\lambda'(x)\partial_2 F_4(x,r\lambda(x))\text{ replacing }F_4(t,r\lambda(t))$$

and

$$\begin{aligned}&\lambda(t)\partial_{12}F_4(x,r\lambda(x))+\lambda'(x)\partial_2F_4(x,r\lambda(x))+r\lambda(x)\lambda'(x)\partial_2^2F_4(x,r\lambda(x))\\&\text{replacing }\lambda(x)\partial_2F_4(x,r\lambda(x))\end{aligned}$$

By noting the symbol-type nature of the estimate (3.174), we get

$$\begin{aligned}&||\sqrt{\omega}\lambda(t)\cdot 4\lambda(x)\lambda'(x)\mathcal{F}(\sqrt{\cdot}\partial_x(F_4(x,\cdot\lambda(x))))'(\omega\lambda(x)^2)||_{L^2(\rho(\omega\lambda(t)^2)d\omega)}\\&\leq C\left(\frac{\lambda(t)}{\lambda(x)}\right)^2||\sqrt{\omega}\lambda(x)\cdot 4\lambda(x)\lambda'(x)\mathcal{F}(\sqrt{\cdot}\partial_x(F_4(x,\cdot\lambda(x))))'(\omega\lambda(x)^2)||_{L^2(\rho(\omega\lambda(x)^2)d\omega)}\\&\leq C\left(\frac{\lambda(t)}{\lambda(x)}\right)^2\frac{(\log(\log(x)))^2}{x^4\log^{b+2-2b\alpha}(x)}\end{aligned}$$

and this gives

$$\begin{aligned}&||-\lambda(t)\int_t^\infty\frac{\cos((t-x)\sqrt{\omega})}{\sqrt{\omega}}\\&\qquad\left(2\mathcal{F}(\sqrt{\cdot}\partial_x(F_4(x,\cdot\lambda(x))))'(\omega\lambda(x)^2)\cdot 2\omega\lambda(x)\lambda'(x)\right)dx||_{L^2(\rho(\omega\lambda(t)^2)d\omega)}\\&\leq\frac{C(\log(\log(t)))^2}{t^3\log^{b+2-2b\alpha}(t)}\end{aligned}$$

Next, we consider

$$\frac{\mathcal{F}(\sqrt{\cdot}F_4(x,\cdot\lambda(x)))'(\omega\lambda(x)^2)\cdot 2\omega((\lambda'(x))^2+\lambda(x)\lambda''(x))\cdot\sqrt{\omega}\lambda(t)}{\omega}$$

We treat this term identically to how (4.34) was treated, noting that the only difference between the two terms is a coefficient which depends on absolute constants and λ. We therefore get

$$\begin{aligned}&||-\lambda(t)\int_t^\infty \frac{\cos((t-x)\sqrt{\omega})}{\sqrt{\omega}}\left(\mathcal{F}(\sqrt{\cdot}F_4(x,\cdot\lambda(x)))'(\omega\lambda(x)^2)\right.\\ &\qquad \left.\cdot 2\omega((\lambda'(x))^2+\lambda(x)\lambda''(x))\right)dx||_{L^2(\rho(\omega\lambda(t)^2)d\omega)}\\ &\leq \int_t^\infty \frac{C}{x}\left(\frac{\lambda(t)}{\lambda(x)}\right)^2 \frac{(\log(\log(x)))^2}{x^3\log^{b+2-2b\alpha}(x)}dx\\ &\leq \frac{C(\log(\log(t)))^2}{t^3\log^{b+2-2b\alpha}(t)}\end{aligned}$$

Next, we study

$$\frac{\sqrt{\omega}\lambda(t)}{\omega}\cdot\mathcal{F}(\partial_x^2\left(\sqrt{\cdot}F_4(x,\cdot\lambda(x))\right))(\omega\lambda(x)^2)$$

by multiplying our previous pointwise in ω estimates on $\frac{\mathcal{F}(\partial_x^2\left(\sqrt{\cdot}F_4(x,\cdot\lambda(x))\right))(\omega\lambda(x)^2)}{\omega}$ by $\left(\frac{\lambda(t)}{\lambda(x)}\right)\sqrt{\omega}\lambda(x)$. We need only check the contributions to

$$\left(\frac{\lambda(t)}{\lambda(x)}\right)\cdot||\frac{\sqrt{\omega}\lambda(x)}{\omega}\cdot\mathcal{F}(\partial_x^2\left(\sqrt{\cdot}F_4(x,\cdot\lambda(x))\right))(\omega\lambda(x)^2)||_{L^2(\rho(\omega\lambda(t)^2)d\omega)}$$

coming from the region $\sqrt{\omega}\lambda(x)\geq 2$, just as for previous terms:

The integrals to check are:

$$\begin{aligned}&\left(\frac{\lambda(t)}{\lambda(x)}\right)\left(\int_{\frac{4}{\lambda(x)^2}}^\infty \frac{\rho(\omega\lambda(t)^2)d\omega}{\omega^4\lambda(x)^4x^8\log^{2b+2N-4}(x)}\right)^{1/2}\\ &\leq C\left(\frac{\lambda(t)}{\lambda(x)}\right)^2\left(\int_{\frac{4}{\lambda(x)^2}}^\infty \frac{\rho(\omega\lambda(x)^2)d\omega}{\omega^4\lambda(x)^4x^8\log^{2b+2N-4}(x)}\right)^{1/2}\\ &\leq C\left(\frac{\lambda(t)}{\lambda(x)}\right)^2\frac{1}{x^4\log^{2b+N-2}(x)}\end{aligned}$$

$$\begin{aligned}&\left(\frac{\lambda(t)}{\lambda(x)}\right)^2\left(\int_{\frac{4}{\lambda(x)^2}}^\infty \frac{\omega\lambda(x)^2d\omega}{\omega^5}\right)^{1/2}\frac{1}{\lambda(x)^3x^4\log^{1-2\alpha b}(x)}\\ &\leq C\left(\frac{\lambda(t)}{\lambda(x)}\right)^2\frac{1}{x^4\log^{b+1-2\alpha b}(x)}\end{aligned}$$

and

$$\left(\frac{\lambda(t)}{\lambda(x)}\right)^2\left(\int_{\frac{4}{\lambda(x)^2}}^\infty \frac{d\omega}{\omega^{3/2}x^8\log^{b+2-4\alpha b}(x)}\right)^{1/2}\leq C\left(\frac{\lambda(t)}{\lambda(x)}\right)^2\frac{1}{x^4\log^{b+1-2\alpha b}(x)}$$

In total, we get

$$\begin{aligned}&||\sqrt{\omega}\lambda(t)\frac{\mathcal{F}(\partial_x^2(\sqrt{\cdot}F_4(x,\cdot\lambda(x))))(\omega\lambda(x)^2)}{\omega}||_{L^2(\rho(\omega\lambda(t)^2)d\omega)}\\ &\leq C\left(\frac{\lambda(t)}{\lambda(x)}\right)^2\frac{(\log(\log(x)))^2}{x^4\log^{1+b-2\alpha b}(x)}\end{aligned}$$

and this gives

$$||\int_t^\infty \cos((t-x)\sqrt{\omega})\mathcal{F}(\sqrt{\cdot}\partial_x^2(F_4(x,\cdot\lambda(x))))(\omega\lambda(x)^2)\frac{\sqrt{\omega}\lambda(t)}{\omega}dx||_{L^2(\rho(\omega\lambda(t)^2)d\omega)}$$
$$\leq \frac{C(\log(\log(t)))^2}{t^3\log^{1+b-2\alpha b}(t)}$$

The last term we need to consider here is

$$\frac{\sqrt{\omega}\lambda(t)}{\omega}\mathcal{F}(\sqrt{\cdot}F_4(x,\cdot\lambda(x)))''(\omega\lambda(x)^2)\cdot(2\omega\lambda(x)\lambda'(x))^2$$

We already estimated

$$\frac{\mathcal{F}(\sqrt{\cdot}F_4(x,\cdot\lambda(x)))''(\omega\lambda(x)^2)\cdot(2\omega\lambda(x)\lambda'(x))^2}{\omega}$$

So, we need only prove new estimates in the region $\sqrt{\omega}\lambda(x)\geq 2$ (by writing $\sqrt{\omega}\lambda(t)=\left(\frac{\lambda(t)}{\lambda(x)}\right)\cdot\sqrt{\omega}\lambda(x)$).

So, we again write

$$\begin{aligned}&\mathcal{F}(\sqrt{\cdot}F_4(x,\cdot\lambda(x)))''(\omega\lambda(x)^2)\\ &=\int_0^{\frac{2}{\sqrt{\omega}\lambda(x)}}\sqrt{r}F_4(x,r\lambda(x))\partial_2^2\phi(r,\omega\lambda(x)^2)dr\\ &+\int_{\frac{2}{\sqrt{\omega}\lambda(x)}}^\infty\sqrt{r}F_4(x,r\lambda(x))\partial_2^2\phi(r,\omega\lambda(x)^2)dr\end{aligned}\tag{4.35}$$

and for the second line of (4.35), we simply multiply our previous pointwise in ω estimate on $\left(\int_0^{\frac{2}{\sqrt{\omega}\lambda(x)}}\sqrt{r}F_4(x,r\lambda(x))\partial_2^2\phi(r,\omega\lambda(x)^2)dr\right)\cdot\frac{(2\omega\lambda(x)\lambda'(x))^2}{\omega}$ by $\frac{\lambda(t)}{\lambda(x)}\sqrt{\omega}\lambda(x)$ in the region $\sqrt{\omega}\lambda(x)\geq 2$, and estimate the $L^2(\rho(\omega\lambda(t)^2)d\omega)$ norm, as before. We have

$$\sqrt{\omega}\lambda(x)\cdot\frac{\omega^2\lambda(x)^2\lambda'(x)^2}{\omega}\cdot|\int_0^{\frac{2}{\sqrt{\omega}\lambda(x)}}\sqrt{r}F_4(x,r\lambda(x))\partial_2^2\phi(r,\omega\lambda(x)^2)dr|$$
$$\leq\frac{C}{x^4\log^{3-2b\alpha}(x)\omega^{5/2}\lambda(x)^3},\quad \sqrt{\omega}\lambda(x)\geq 2$$

and this leads to

$$||\sqrt{\omega}\lambda(t)\left(\frac{\int_0^{\frac{2}{\sqrt{\omega}\lambda(x)}}\sqrt{r}F_4(x,r\lambda(x))\partial_2^2\phi(r,\omega\lambda(x)^2)dr}{\omega}\cdot\omega^2\lambda(x)^2(\lambda'(x))^2\right)||_{L^2(\rho(\omega\lambda(t)^2)d\omega)}$$
$$\leq C\left(\frac{\lambda(t)^2}{\lambda(x)^2}\right)\frac{\log(\log(x)))^2}{x^4\log^{3+b-2b\alpha}(x)}$$

For the third line of (4.35), we can not simply multiply our pointwise in ω estimates on

$$\left(\int_{\frac{2}{\sqrt{\omega}\lambda(x)}}^\infty\sqrt{r}F_4(x,r\lambda(x))\partial_2^2\phi(r,\omega\lambda(x)^2)dr\right)\cdot\frac{(2\omega\lambda(x)\lambda'(x))^2}{\omega}$$

by $\sqrt{\omega}\lambda(t)$, since doing so would result in an estimate that is not square integrable against the measure $\rho(\omega\lambda(t)^2)d\omega$. So, we have to integrate by parts in appropriate

r integrals, just as in a previous situation. To be precise, we consider

$$\int_{\frac{2}{\sqrt{\omega}\lambda(x)}}^{\infty} \sqrt{r}F_4(x, r\lambda(x))\partial_2^2\phi(r, \omega\lambda(x)^2)dr \tag{4.36}$$

and write, for $r\sqrt{\xi} \geq 2$:

(4.37)

$$\begin{aligned}\partial_2^2\phi(r,\xi) = 2\mathrm{Re}\Bigg(\Bigg(&\frac{a''(\xi)}{\xi^{1/4}}\sigma(r\sqrt{\xi},r) + 2a'(\xi)\left(\frac{-\sigma(r\sqrt{\xi},r)}{4\xi^{5/4}} + \frac{r\partial_1\sigma(r\sqrt{\xi},r)}{2\xi^{3/4}}\right)\\
&+ a(\xi)\left(\frac{5\sigma(r\sqrt{\xi},r)}{16\xi^{9/4}} - \frac{r\partial_1\sigma(r\sqrt{\xi},r)}{2\xi^{7/4}} + \frac{ir^2\partial_1\sigma(r\sqrt{\xi},r)}{2\xi^{5/4}} + \frac{r^2\partial_1^2\sigma(r\sqrt{\xi},r)}{4\xi^{5/4}}\right)\Bigg)e^{ir\sqrt{\xi}}\Bigg)\\
&+ 2\mathrm{Re}\left(\left(2\frac{a'(\xi)ir\sigma(r\sqrt{\xi},r)}{2\xi^{3/4}} - \frac{ira(\xi)\sigma(r\sqrt{\xi},r)}{2\xi^{7/4}}\right)e^{ir\sqrt{\xi}}\right)\\
&- 2\mathrm{Re}\left(\frac{a(\xi)r^2}{4\xi^{5/4}}e^{ir\sqrt{\xi}}\sigma(r\sqrt{\xi},r)\right)\end{aligned}$$

For each term on the third line of (4.37), inserted into (4.36), we will integrate by parts once in r. The insertion of the first term on the third line of (4.37) gives

$$\begin{aligned}&2\mathrm{Re}\left(\int_{\frac{2}{\sqrt{\omega}\lambda(x)}}^{\infty} \sqrt{r}F_4(x,r\lambda(x))\cdot\frac{a'(\omega\lambda(x)^2)ir}{\omega^{3/4}\lambda(x)^{3/2}}\sigma(r\sqrt{\omega}\lambda(x),r)e^{ir\sqrt{\omega}\lambda(x)}dr\right)\\
&= 2\mathrm{Re}\Bigg(\frac{ia'(\omega\lambda(x)^2)}{\omega^{3/4}\lambda(x)^{3/2}}\Bigg(\frac{-2^{3/2}F_4(x,\frac{2}{\sqrt{\omega}})}{\omega^{3/4}\lambda(x)^{3/2}i\sqrt{\omega}\lambda(x)}\sigma(2,\frac{2}{\sqrt{\omega}\lambda(x)})e^{2i}\\
&\quad - \int_{\frac{2}{\sqrt{\omega}\lambda(x)}}^{\infty}\frac{e^{ir\sqrt{\omega}\lambda(x)}}{i\sqrt{\omega}\lambda(x)}\partial_r\left(r^{3/2}F_4(x,r\lambda(x))\sigma(r\sqrt{\omega}\lambda(x),r)\right)dr\Bigg)\Bigg)\end{aligned} \tag{4.38}$$

Then we estimate each term separately:

$$\begin{aligned}&|2\mathrm{Re}\left(\frac{ia'(\omega\lambda(x)^2)}{\omega^{3/4}\lambda(x)^{3/2}}\left(\frac{-2^{3/2}F_4(x,\frac{2}{\sqrt{\omega}})}{\omega^{3/4}\lambda(x)^{3/2}i\sqrt{\omega}\lambda(x)}\sigma(2,\frac{2}{\sqrt{\omega}\lambda(x)})e^{2i}\right)\right)|\\
&\leq C\frac{|a'(\omega\lambda(x)^2)|}{\omega^{3/4}\lambda(x)^{3/2}}\frac{|F_4(x,\frac{2}{\sqrt{\omega}})|}{\omega^{3/4}\lambda(x)^{3/2}\sqrt{\omega}\lambda(x)}\end{aligned}$$

The contribution to the $L^2((\frac{4}{\lambda(x)^2},\infty),\rho(\omega\lambda(t)^2)d\omega)$ norm of the integrand of (4.33) due to the above term is thus estimated by

$$\begin{aligned}&\left(\int_{\frac{4}{\lambda(x)^2}}^{\infty}\left(\frac{|a'(\omega\lambda(x)^2)|}{\omega^{3/4}\lambda(x)^{3/2}}\frac{|F_4(x,\frac{2}{\sqrt{\omega}})|}{\omega^{3/4}\lambda(x)^{3/2}\sqrt{\omega}\lambda(x)}\right)^2\omega^4\cdot\left(\frac{\sqrt{\omega}\lambda(t)}{\omega}\right)^2\frac{\rho(\omega\lambda(t)^2)d\omega}{\left(x^2\log^{4b+2}(x)\right)^2}\right)^{1/2}\\
&\leq C\left(\frac{\lambda(t)}{\lambda(x)}\right)^2\left(\int_0^{\lambda(x)}y^6|F_4(x,y)|^2\frac{dy}{y^3}\right)^{1/2}\frac{1}{\lambda(x)^5x^2\log^{4b+2}(x)}\\
&\leq C\left(\frac{\lambda(t)}{\lambda(x)}\right)^2\frac{1}{x^4\log^{b+3-2b\alpha}(x)}\end{aligned}$$

For the next term of (4.38), we use Proposition 4.6 of [**12**] again, to get

$$\begin{aligned}&|2\text{Re}\left(\frac{ia'(\omega\lambda(x)^2)}{\omega^{3/4}\lambda(x)^{3/2}}\left(-\int_{\frac{2}{\sqrt{\omega}\lambda(x)}}^{\infty}\frac{e^{ir\sqrt{\omega}\lambda(x)}}{i\sqrt{\omega}\lambda(x)}\partial_r\left(r^{3/2}F_4(x,r\lambda(x))\sigma(r\sqrt{\omega}\lambda(x),r)\right)dr\right)\right)|\\&\leq\frac{C|a(\omega\lambda(x)^2)|}{\omega^{7/4}\lambda(x)^{7/2}}\frac{1}{\sqrt{\omega}\lambda(x)}\int_{\frac{2}{\sqrt{\omega}\lambda(x)}}^{\infty}\sqrt{r}\left(|F_4(x,r\lambda(x))|+r\lambda(x)|(\partial_2F_4)(x,r\lambda(x))|\right)dr\\&\leq\frac{C|a(\omega\lambda(x)^2)|}{\omega^{9/4}\lambda(x)^{9/2}x^2\log^{1-2b\alpha}(x)}\end{aligned}$$

Then, the contribution of this term to the $L^2((\frac{4}{\lambda(x)^2},\infty),\rho(\omega\lambda(t)^2)d\omega)$ norm of the integrand of (4.33) is estimated by:

$$\begin{aligned}&\left(\int_{\frac{4}{\lambda(x)^2}}^{\infty}\frac{\omega^4}{\omega^{9/2}}\left(\frac{\sqrt{\omega}\lambda(x)}{\omega}\right)^2d\omega\right)^{1/2}\left(\frac{\lambda(t)}{\lambda(x)}\right)^2\frac{1}{x^2\log^{1-2b\alpha}(x)}\cdot\frac{1}{\lambda(x)^{9/2}}\cdot\frac{1}{x^2\log^{4b+2}(x)}\\&\leq C\left(\frac{\lambda(t)}{\lambda(x)}\right)^2\frac{1}{x^4\log^{b+3-2b\alpha}(x)}\end{aligned}$$

By comparing the first and second terms on the third line of (4.37), and recalling the symbol-type estimates on a from Proposition 4.7 of [**12**], we can estimate the second term on the third line of 4.37 by repeating the same exact procedure used to estimate the first term of the third line of (4.37).

Next, we treat the fourth line of (4.37). Here, we integrate by parts twice in the r integral resulting from substitution of the fourth line of (4.37) into (4.36). With $\xi=\omega\lambda(t)^2$, we have

$$\begin{aligned}&\int_{\frac{2}{\sqrt{\xi}}}^{\infty}\sqrt{r}F_4(x,r\lambda(x))\cdot 2\text{Re}\left(\frac{-a(\xi)r^2e^{ir\sqrt{\xi}}\sigma(r\sqrt{\xi},r)}{4\xi^{5/4}}\right)dr\\&=\text{Re}\left(\frac{-a(\xi)}{2\xi^{5/4}}\left(\frac{-2^{5/2}F_4(x,\frac{2}{\sqrt{\xi}}\lambda(x))\sigma(2,\frac{2}{\sqrt{\xi}})e^{2i}}{i\xi^{7/4}}\right.\right.\\&\left.\left.+\frac{1}{i\sqrt{\xi}}\cdot\left(\partial_r\left(r^{5/2}F_4(x,r\lambda(x))\sigma(r\sqrt{\xi},r)\right)\frac{e^{ir\sqrt{\xi}}}{i\sqrt{\xi}}\right)|_{r=\frac{2}{\sqrt{\xi}}}\right)\right)\\&+\text{Re}\left(\frac{-a(\xi)}{2\xi^{5/4}}\cdot\frac{-1}{\xi}\int_{\frac{2}{\sqrt{\xi}}}^{\infty}e^{ir\sqrt{\xi}}\partial_r^2\left(r^{5/2}F_4(x,r\lambda(x))\sigma(r\sqrt{\xi},r)\right)dr\right)\end{aligned}$$

Next, we again note the symbol-type character of the estimate (3.174), to get

(4.39)

$$\begin{aligned}&|\int_{\frac{2}{\sqrt{\xi}}}^{\infty}\sqrt{r}F_4(x,r\lambda(x))2\text{Re}\left(\frac{-a(\xi)r^2e^{ir\sqrt{\xi}}\sigma(r\sqrt{\xi},r)}{4\xi^{5/4}}\right)dr|\cdot 4\omega^2\lambda(x)^2\lambda'(x)^2\frac{\sqrt{\omega}\lambda(x)}{\omega}\\&\leq\frac{C\omega^2\lambda(x)}{\sqrt{\omega}x^2\log^{4b+2}(x)}\left(\frac{|a(\omega\lambda(x)^2)|}{\omega^3\lambda(x)^6}F_{4,est}(x,\frac{2\lambda(x)}{\sqrt{\xi}})\right.\\&\left.+\frac{|a(\omega\lambda(x)^2)|}{\omega^{9/4}\lambda(x)^{9/2}}\int_{\frac{2}{\sqrt{\omega}\lambda(x)}}^{\infty}r^{1/2}F_{4,est}(x,r\lambda(x))dr\right)\end{aligned}$$

where $F_{4,est}$ is the expression which appears on the right-hand side of (3.174). Then, the same procedure used to treat (4.38) also applies to treat (4.39), and we get

$$\begin{aligned}&||\int_{\frac{2}{\sqrt{\xi}}}^{\infty} \sqrt{r}F_4(x,r\lambda(x))2\text{Re}\left(\frac{-a(\xi)r^2e^{ir\sqrt{\xi}}\sigma(r\sqrt{\xi},r)}{4\xi^{5/4}}\right)dr\\&\quad\cdot 4\omega^2\lambda(x)^2\lambda'(x)^2\frac{\sqrt{\omega}\lambda(x)}{\omega}\left(\frac{\lambda(t)}{\lambda(x)}\right)||_{L^2((\frac{4}{\lambda(x)^2},\infty),\rho(\omega\lambda(t)^2)d\omega)}\\&\leq C\left(\frac{\lambda(t)}{\lambda(x)}\right)^2\frac{1}{x^4\log^{b+3-2b\alpha}(x)}\end{aligned}$$

Next, we study the first two lines of (4.37), which are given below:

$$\begin{aligned}(\partial_2^2\phi(r,\xi))_2 &= 2\text{Re}\left(\left(\frac{a''(\xi)}{\xi^{1/4}}\sigma(r\sqrt{\xi},r)+2a'(\xi)\left(\frac{-\sigma(r\sqrt{\xi},r)}{4\xi^{5/4}}+\frac{r\partial_1\sigma(r\sqrt{\xi},r)}{2\xi^{3/4}}\right)\right.\right.\\&\left.\left.+\ a(\xi)\left(\frac{5\sigma(r\sqrt{\xi},r)}{16\xi^{9/4}}-\frac{r\partial_1\sigma(r\sqrt{\xi},r)}{2\xi^{7/4}}+\frac{ir^2\partial_1\sigma(r\sqrt{\xi},r)}{2\xi^{5/4}}+\frac{r^2\partial_1^2\sigma(r\sqrt{\xi},r)}{4\xi^{5/4}}\right)\right)e^{ir\sqrt{\xi}}\right)\end{aligned}$$

We note that

$$|(\partial_2^2\phi(r,\xi))_2|\leq\frac{C|a(\xi)|}{\xi^{9/4}},\quad r\geq\frac{2}{\sqrt{\xi}}$$

So,

$$\begin{aligned}&|\int_{\frac{2}{\sqrt{\omega\lambda(x)}}}^{\infty}\sqrt{r}F_4(x,r\lambda(x))(\partial_2^2\phi(r,\omega\lambda(x)^2))_2dr|\\&\leq\frac{C|a(\omega\lambda(x)^2)|}{\omega^{9/4}\lambda(x)^{9/2}}\int_{\frac{2}{\sqrt{\omega\lambda(x)}}}^{\infty}\sqrt{r}|F_4(x,r\lambda(x))|dr\end{aligned}$$

Finally,

$$\begin{aligned}&\left(\int_{\frac{4}{\lambda(x)^2}}^{\infty}\rho(\omega\lambda(t)^2)\left(\left(\int_{\frac{2}{\sqrt{\omega\lambda(x)}}}^{\infty}\sqrt{r}F_4(x,r\lambda(x))(\partial_2^2\phi(r,\omega\lambda(x)^2))_2dr\right)\cdot 4\omega^2\lambda(x)^2\lambda'(x)^2\frac{\sqrt{\omega}\lambda(t)}{\omega}\right)^2d\omega\right)^{1/2}\\&\leq C\left(\frac{\lambda(t)}{\lambda(x)}\right)^2\frac{1}{x^4\log^{b+3-2b\alpha}(x)}\end{aligned}$$

Combining all of our estimates, we get

$$\begin{aligned}&||\sqrt{\omega}\lambda(t)\left(\frac{\mathcal{F}(\sqrt{\cdot}F_4(x,\cdot\lambda(x)))''(\omega\lambda(x)^2)}{\omega}\omega^2\lambda(x)^2\lambda'(x)^2\right)||_{L^2(\rho(\omega\lambda(t)^2)d\omega)}\\&\leq C\left(\frac{\lambda(t)}{\lambda(x)}\right)^2\frac{\log(\log(x))^2}{x^4\log^{3+b-2b\alpha}(x)}\end{aligned}$$

and

(4.40)

$$\begin{aligned}&||-\int_t^{\infty}\cos((t-x)\sqrt{\omega})\frac{\sqrt{\omega}\lambda(t)}{\omega}\partial_x^2\left(\mathcal{F}(\sqrt{\cdot}F_4(x,\cdot\lambda(x)))(\omega\lambda(x)^2)\right)dx||_{L^2(\rho(\omega\lambda(t)^2)d\omega)}\\&\leq\frac{C(\log(\log(t)))^2}{t^3\log^{1+b-2\alpha b}(t)}\end{aligned}$$

which imply

$$||\sqrt{\omega}\lambda(t)\partial_t\left(\int_t^\infty \frac{\sin((t-x)\sqrt{\omega})}{\sqrt{\omega}}\mathcal{F}(\sqrt{\cdot}F_4(x,\cdot\lambda(x)))(\omega\lambda(x)^2)dx\right)||_{L^2(\rho(\omega\lambda(t)^2)d\omega)}$$
$$\leq \frac{C(\log(\log(t)))^2}{t^3\log^{1+b-2\alpha b}(t)}$$

The last quantity to estimate is

$$\begin{aligned}
&\omega\lambda(t)^2\left(\int_t^\infty \frac{\sin((t-x)\sqrt{\omega})}{\sqrt{\omega}}\mathcal{F}(\sqrt{\cdot}F_4(x,\cdot\lambda(x)))(\omega\lambda(x)^2)dx\right)\\
&= -\lambda(t)^2\mathcal{F}(\sqrt{\cdot}F_4(t,\cdot\lambda(t)))(\omega\lambda(t)^2)\\
&-\sqrt{\omega}\lambda(t)^2\int_t^\infty \frac{\sin((t-x)\sqrt{\omega})}{\omega}\partial_x^2\left(\mathcal{F}(\sqrt{\cdot}F_4(x,\cdot\lambda(x)))(\omega\lambda(x)^2)\right)dx
\end{aligned} \tag{4.41}$$

We can estimate the third line of (4.41), simply by using the estimates which gave (4.40). For the second line of (4.41), we have, with $\xi=\omega\lambda(t)^2$,

$$\begin{aligned}
&\lambda(t)^2\left(\int_0^\infty \rho(\omega\lambda(t)^2)\left(\mathcal{F}(\sqrt{\cdot}F_4(t,\cdot\lambda(t)))(\omega\lambda(t)^2)\right)^2 d\omega\right)^{1/2}\\
&=\lambda(t)^2\left(\int_0^\infty \rho(\xi)\left(\mathcal{F}(\sqrt{\cdot}F_4(t,\cdot\lambda(t)))(\xi)\right)^2\frac{d\xi}{\lambda(t)^2}\right)^{1/2}\\
&=\lambda(t)\left(\int_0^\infty r\left(F_4(t,r\lambda(t))\right)^2dr\right)^{1/2}\\
&\leq C\lambda(t)\left(\int_0^{\frac{\log^N(t)}{\lambda(t)}} \frac{r^3\lambda(t)^2dr}{\lambda(t)^8(r^2+1)^4t^4\log^{6b+2-4b\alpha}(t)}\right.\\
&\qquad\left.+\int_0^{\frac{t}{2\lambda(t)}}\frac{r^3\lambda(t)^2dr}{\lambda(t)^8(r^2+1)^4t^4\log^{10b+4N-4}(t)}\right)^{1/2}\\
&\leq \frac{C}{t^2\log^{b+1-2b\alpha}(t)}+\frac{C}{t^2\log^{3b+2N-2}(t)}\\
&\leq \frac{C}{t^2\log^{b+1-2b\alpha}(t)}
\end{aligned}$$

So,

$$||\omega\lambda(t)^2\int_t^\infty \frac{\sin((t-x)\sqrt{\omega})}{\sqrt{\omega}}\mathcal{F}(\sqrt{\cdot}F_4(x,\cdot\lambda(x)))(\omega\lambda(x)^2)dx||_{L^2(\rho(\omega\lambda(t)^2)d\omega)}$$
$$\leq \frac{C}{t^2\log^{1+b-2b\alpha}(t)}$$

which finishes the proof of the lemma □

To proceed, we quickly translate our estimates (3.3) and (3.4), by noting

$$\begin{aligned}
&||\mathcal{F}(\sqrt{\cdot}\left(F_5+F_6\right)(x,\cdot\lambda(x)))(\omega\lambda(x)^2)||^2_{L^2(\rho(\omega\lambda(x)^2)d\omega)}\\
&=\int_0^\infty \frac{r}{\lambda(x)^4}\left(F_5+F_6\right)^2(x,r)dr
\end{aligned}$$

$$\begin{aligned}&||\sqrt{\omega}\lambda(x)\mathcal{F}(\sqrt{\cdot}\,(F_5+F_6)\,(x,\cdot\lambda(x)))(\omega\lambda(x)^2)||^2_{L^2(\rho(\omega\lambda(x)^2)d\omega)}\\&=\frac{1}{\lambda(x)^2}\int_0^\infty(L((F_5+F_6)\,(x,\cdot\lambda(x))))^2(R)RdR\end{aligned}$$

and

$$|L(f)(r)|\leq C\left(|f'(r)|+\frac{|f(r)|}{r}\right)$$

So, we have

$$||\mathcal{F}(\sqrt{\cdot}\,(F_5+F_6)\,(x,\cdot\lambda(x)))(\omega\lambda(x)^2)||_{L^2(\rho(\omega\lambda(x)^2)d\omega)}\leq\frac{C}{x^4\log^{3b+2N-1}(x)}$$

and

$$||\sqrt{\omega}\lambda(x)\mathcal{F}(\sqrt{\cdot}\,(F_5+F_6)\,(x,\cdot\lambda(x)))(\omega\lambda(x)^2)||_{L^2(\rho(\omega\lambda(x)^2)d\omega)}\leq C\frac{\log^{6+b}(x)}{x^{35/8}}$$

Now, we recall $F(t,r)=F_4(t,r)+F_5(t,r)+F_6(t,r)$, and estimate the following quantities:

(4.42)

$$\begin{aligned}&||\int_t^\infty\frac{\sin((t-x)\sqrt{\omega})}{\sqrt{\omega}}\mathcal{F}(\sqrt{\cdot}F(x,\cdot\lambda(x)))(\omega\lambda(x)^2)dx||_{L^2(\rho(\omega\lambda(t)^2)d\omega)}\\&=||\int_t^\infty\frac{\sin((t-x)\sqrt{\omega})}{\sqrt{\omega}}(\mathcal{F}(\sqrt{\cdot}F_4(x,\cdot\lambda(x)))(\omega\lambda(x)^2)\\&\qquad+\mathcal{F}(\sqrt{\cdot}(F_5+F_6)(x,\cdot\lambda(x)))(\omega\lambda(x)^2))dx||_{L^2(\rho(\omega\lambda(t)^2)d\omega)}\\&\leq||\int_t^\infty\frac{\sin((t-x)\sqrt{\omega})}{\sqrt{\omega}}\mathcal{F}(\sqrt{\cdot}F_4(x,\cdot\lambda(x)))(\omega\lambda(x)^2)dx||_{L^2(\rho(\omega\lambda(t)^2)d\omega)}\\&+||\int_t^\infty\frac{\sin((t-x)\sqrt{\omega})}{\sqrt{\omega}}\mathcal{F}(\sqrt{\cdot}(F_5+F_6)(x,\cdot\lambda(x)))(\omega\lambda(x)^2)dx||_{L^2(\rho(\omega\lambda(t)^2)d\omega)}\\&\leq\frac{C(\log(\log(t)))^2}{t^2\log^{b+1-2\alpha b}(t)}\\&+C\int_t^\infty(x-t)(||\mathcal{F}(\sqrt{\cdot}(F_5+F_6)(x,\cdot\lambda(x)))(\omega\lambda(x)^2)||_{L^2(\rho(\omega\lambda(t)^2)d\omega)})dx\\&\leq\frac{C(\log(\log(t)))^2}{t^2\log^{b+1-2\alpha b}(t)}+C\int_t^\infty(x-t)\left(\frac{\lambda(t)}{\lambda(x)}\right)\frac{dx}{x^4\log^{3b+2N-1}(x)}\\&\leq\frac{C(\log(\log(t)))^2}{t^2\log^{b+1-2\alpha b}(t)}\end{aligned}$$

where we used the fact that,

$$\begin{aligned}&||\mathcal{F}(\sqrt{\cdot}\,(F_5+F_6)\,(x,\cdot\lambda(x)))(\omega\lambda(x)^2)||_{L^2(\rho(\omega\lambda(t)^2)d\omega)}\\&=\left(\int_0^\infty\left(\mathcal{F}(\sqrt{\cdot}\,(F_5+F_6)\,(x,\cdot\lambda(x)))(\omega\lambda(x)^2)\right)^2\frac{\rho(\omega\lambda(t)^2)}{\rho(\omega\lambda(x)^2)}\rho(\omega\lambda(x)^2)d\omega\right)^{1/2}\\&\leq C\frac{\lambda(t)}{\lambda(x)}||\mathcal{F}(\sqrt{\cdot}\,(F_5+F_6)\,(x,\cdot\lambda(x)))(\omega\lambda(x)^2)||_{L^2(\rho(\omega\lambda(x)^2)d\omega)}\end{aligned}$$

where we used (4.23).

Similarly,

$$\begin{aligned}
&||\int_t^\infty \cos((t-x)\sqrt{\omega})\mathcal{F}(\sqrt{\cdot}F(x,\cdot\lambda(x)))(\omega\lambda(x)^2)dx||_{L^2(\rho(\omega\lambda(t)^2)d\omega)}\\
&\leq ||\int_t^\infty \cos((t-x)\sqrt{\omega})\mathcal{F}(\sqrt{\cdot}F_4(x,\cdot\lambda(x)))(\omega\lambda(x)^2)dx||_{L^2(\rho(\omega\lambda(t)^2)d\omega)}\\
&+||\int_t^\infty \cos((t-x)\sqrt{\omega})\mathcal{F}(\sqrt{\cdot}\left(F_5+F_6\right)(x,\cdot\lambda(x)))(\omega\lambda(x)^2)dx||_{L^2(\rho(\omega\lambda(t)^2)d\omega)}\\
&\leq \frac{C(\log(\log(t)))^2}{t^3\log^{b+1-2\alpha b}(t)}\\
&+C\int_t^\infty \left(\frac{\lambda(t)}{\lambda(x)}\right)\left(||\mathcal{F}(\sqrt{\cdot}\left(F_5+F_6\right)(x,\cdot\lambda(x)))(\omega\lambda(x)^2)||_{L^2(\rho(\omega\lambda(x)^2)d\omega)}\right)dx\\
&\leq \frac{C(\log(\log(t)))^2}{t^3\log^{b+1-2\alpha b}(t)}+\frac{C}{t^3\log^{3b+2N-1}(t)}\\
&\leq \frac{C(\log(\log(t)))^2}{t^3\log^{b+1-2\alpha b}(t)}
\end{aligned}$$

$$\begin{aligned}
&||\sqrt{\omega}\lambda(t)\int_t^\infty \frac{\sin((t-x)\sqrt{\omega})}{\sqrt{\omega}}\mathcal{F}(\sqrt{\cdot}F(x,\cdot\lambda(x)))(\omega\lambda(x)^2)dx||_{L^2(\rho(\omega\lambda(t)^2)d\omega)}\\
&\leq ||\sqrt{\omega}\lambda(t)\int_t^\infty \frac{\sin((t-x)\sqrt{\omega})}{\sqrt{\omega}}\mathcal{F}(\sqrt{\cdot}F_4(x,\cdot\lambda(x)))(\omega\lambda(x)^2)dx||_{L^2(\rho(\omega\lambda(t)^2)d\omega)}\\
&+||\sqrt{\omega}\lambda(t)\int_t^\infty \frac{\sin((t-x)\sqrt{\omega})}{\sqrt{\omega}}\mathcal{F}(\sqrt{\cdot}\left(F_5+F_6\right)(x,\cdot\lambda(x)))(\omega\lambda(x)^2)dx||_{L^2(\rho(\omega\lambda(t)^2)d\omega)}\\
&\leq \frac{C(\log(\log(t)))^2}{t^2\log^{1+b-2\alpha b}(t)}\\
&+C\lambda(t)\int_t^\infty \left(\frac{\lambda(t)}{\lambda(x)}\right)\left(||\mathcal{F}(\sqrt{\cdot}\left(F_5+F_6\right)(x,\cdot\lambda(x)))(\omega\lambda(x)^2)||_{L^2(\rho(\omega\lambda(x)^2)d\omega)}\right)dx\\
&\leq \frac{C(\log(\log(t)))^2}{t^2\log^{1+b-2\alpha b}(t)}+\frac{C}{t^3\log^{4b+2N-1}(t)}\\
&\leq \frac{C(\log(\log(t)))^2}{t^2\log^{1+b-2\alpha b}(t)}
\end{aligned}$$

$$\begin{aligned}
&|| \int_t^\infty \cos((t-x)\sqrt{\omega})\sqrt{\omega}\lambda(x)\left(\frac{\lambda(t)}{\lambda(x)}\right)\mathcal{F}(\sqrt{\cdot}F(x,\cdot\lambda(x)))(\omega\lambda(x)^2)dx||_{L^2(\rho(\omega\lambda(t)^2)d\omega)}\\
&\leq || \int_t^\infty \cos((t-x)\sqrt{\omega})\sqrt{\omega}\lambda(x)\left(\frac{\lambda(t)}{\lambda(x)}\right)\mathcal{F}(\sqrt{\cdot}F_4(x,\cdot\lambda(x)))(\omega\lambda(x)^2)dx||_{L^2(\rho(\omega\lambda(t)^2)d\omega)}\\
&+|| \int_t^\infty \cos((t-x)\sqrt{\omega})\sqrt{\omega}\lambda(x)\left(\frac{\lambda(t)}{\lambda(x)}\right)\\
&\qquad\qquad \mathcal{F}(\sqrt{\cdot}\,(F_5+F_6)\,(x,\cdot\lambda(x)))(\omega\lambda(x)^2)dx||_{L^2(\rho(\omega\lambda(t)^2)d\omega)}\\
&\leq \frac{C(\log(\log(t)))^2}{t^3\log^{1+b-2\alpha b}(t)}\\
&+C\int_t^\infty \left(\frac{\lambda(t)}{\lambda(x)}\right)^2\left(||\sqrt{\omega}\lambda(x)\mathcal{F}(\sqrt{\cdot}\,(F_5+F_6)\,(x,\cdot\lambda(x)))(\omega\lambda(x)^2)||_{L^2(\rho(\omega\lambda(x)^2)d\omega)}\right)dx\\
&\leq \frac{C(\log(\log(t)))^2}{t^3\log^{1+b-2\alpha b}(t)}+C\int_t^\infty\left(\frac{\lambda(t)}{\lambda(x)}\right)^2\frac{\log^{b+6}(x)}{x^{35/8}}dx\\
&\leq \frac{C(\log(\log(t)))^2}{t^3\log^{1+b-2\alpha b}(t)}
\end{aligned}$$

and, finally,

$$\begin{aligned}
&|| \int_t^\infty \sin((t-x)\sqrt{\omega})\sqrt{\omega}\lambda(t)^2\mathcal{F}(\sqrt{\cdot}F(x,\cdot\lambda(x)))(\omega\lambda(x)^2)dx||_{L^2(\rho(\omega\lambda(t)^2)d\omega)}\\
&\leq || \int_t^\infty \sin((t-x)\sqrt{\omega})\sqrt{\omega}\lambda(t)^2\mathcal{F}(\sqrt{\cdot}F_4(x,\cdot\lambda(x)))(\omega\lambda(x)^2)dx||_{L^2(\rho(\omega\lambda(t)^2)d\omega)}\\
&+|| \int_t^\infty \sin((t-x)\sqrt{\omega})\sqrt{\omega}\lambda(t)^2\mathcal{F}(\sqrt{\cdot}(F_5+F_6)\,(x,\cdot\lambda(x)))(\omega\lambda(x)^2)dx||_{L^2(\rho(\omega\lambda(t)^2)d\omega)}\\
&\leq \frac{C}{t^2\log^{1+b-2\alpha b}(t)}\\
&+C\int_t^\infty \lambda(t)\left(\frac{\lambda(t)}{\lambda(x)}\right)^2\left(||\lambda(x)\sqrt{\omega}\mathcal{F}(\sqrt{\cdot}(F_5+F_6)\,(x,\cdot\lambda(x)))(\omega\lambda(x)^2)||_{L^2(\rho(\omega\lambda(x)^2)d\omega)}\right)dx\\
&\leq \frac{C}{t^2\log^{1+b-2\alpha b}(t)}+C\int_t^\infty \frac{\lambda(t)^3}{\lambda(x)^2}\frac{\log^{b+6}(x)}{x^{35/8}}dx\\
&\leq \frac{C}{t^2\log^{1+b-2\alpha b}(t)}
\end{aligned}$$

4.0.5. Setup of the final iteration. Let ϵ be given by

$$\epsilon = 2b+\frac{1}{2}(1-2\alpha b)$$

Note that $2b+\frac{1}{2}>\epsilon>2b$. Also, note that (3.5), (3.6), and (3.7) show that

$$||\frac{v_{corr}(x,R\lambda(x))}{R\lambda(x)}||^2_{L^\infty}+||\frac{v_{corr}(x,R\lambda(x))}{R\lambda(x)^2(1+R^2)}||_{L^\infty}\leq \frac{C}{x^2\log^{\epsilon-2b}(x)} \tag{4.43}$$

$$1+||\frac{v_{corr}(x,R\lambda(x))}{R}||_{L^\infty}+||\partial_R(v_{corr}(x,R\lambda(x)))||_{L^\infty}\leq C \tag{4.44}$$

(4.45)
$$
\begin{aligned}
&||\frac{v_{corr}(x,R\lambda(x))\partial_R(v_{corr}(x,R\lambda(x)))}{R\lambda(x)^2}||_{L^\infty_R((0,1))}\\
&+||\frac{v_{corr}(x,R\lambda(x))\partial_R(v_{corr}(x,R\lambda(x)))}{R^2\lambda(x)^2}||_{L^\infty_R((1,\infty))}+||\frac{\partial_R(v_{corr}(x,R\lambda(x)))}{(1+R^2)\lambda(x)^2}||_{L^\infty}\\
&\leq \frac{C}{x^2\log^{\epsilon-2b}(x)}
\end{aligned}
$$

Let $(Z,||\cdot||_Z)$ be the normed vector space defined as follows. Z is the set of (equivalence classes) of measurable functions $y:[T_0,\infty)\times(0,\infty)\to\mathbb{R}$ such that

$$y(t,\omega)t^2\log^{\frac{\epsilon}{2}}(t)\sqrt{\rho(\omega\lambda(t)^2)}\langle\omega\lambda(t)^2\rangle\in C^0_t([T_0,\infty),L^2(d\omega))$$

$$\partial_t y(t,\omega)t^3\log^{\frac{\epsilon}{2}}(t)\langle\sqrt{\omega}\lambda(t)\rangle\sqrt{\rho(\omega\lambda(t)^2)}\in C^0_t([T_0,\infty),L^2(d\omega))$$

and $||y||_Z<\infty$ where

(4.46)
$$
\begin{aligned}
||y||_Z=\sup_{t\geq T_0}\Big(&t^2\log^{\frac{\epsilon}{2}}(t)\left(||y(t)||_{L^2(\rho(\omega\lambda(t)^2)d\omega)}\right.\\
&\left.+\,||\lambda(t)\sqrt{\omega}y(t)||_{L^2(\rho(\omega\lambda(t)^2)d\omega)}+||\omega\lambda(t)^2y(t)||_{L^2(\rho(\omega\lambda(t)^2)d\omega)}\right)\\
&+t^3\log^{\frac{\epsilon}{2}}(t)\left(||\partial_t y(t)||_{L^2(\rho(\omega\lambda(t)^2)d\omega)}+||\lambda(t)\sqrt{\omega}\partial_t y(t)||_{L^2(\rho(\omega\lambda(t)^2)d\omega)}\right)\Big)
\end{aligned}
$$

Define T on Z by

$$
\begin{aligned}
T(y)(t,\omega)=-\int_t^\infty\frac{\sin((t-x)\sqrt{\omega})}{\sqrt{\omega}}&\left(F_2(y)(x,\omega)-\mathcal{F}(\sqrt{\cdot}F(x,\cdot\lambda(x)))(\omega\lambda(x)^2)\right.\\
&\left.-\mathcal{F}(\sqrt{\cdot}F_3(u(y))(x,\cdot\lambda(x)))(\omega\lambda(x)^2)\right)dx
\end{aligned}
$$

Note that a fixed point of T is a solution to (4.3) with 0 Cauchy data at infinity. We will prove the following proposition which implies that T indeed has a fixed point in $\overline{B_1(0)}\subset Z$.

PROPOSITION 4.3. *There exists $T_4>0$ such that, for all $T_0>T_4$, T is a strict contraction on $\overline{B}_1(0)\subset Z$*

We use (4.42), Propositions 4.1 and 4.2, and the equations (4.43), (4.44), and (4.45), to get the following (note that in the following estimates, $C>0$ denotes a constant (which might involve C_ρ) whose value may change from line to line, but which is *independent* of T_0). Also, for ease of notation, we will denote $F_3(u(y))$ by

F_3 until otherwise mentioned.

$$\begin{aligned}
&||T(y)(t)||_{L^2(\rho(\omega\lambda(t)^2)d\omega)}\\
&\leq \frac{C(\log(\log(t)))^2}{t^2\log^{b+1-2\alpha b}(t)}+C\int_t^\infty \frac{\lambda(t)}{\lambda(x)}\frac{1}{x^3\log^{\frac{\epsilon}{2}+1}(x)}dx\\
&+C\int_t^\infty x\left(\frac{\lambda(t)}{\lambda(x)}\right)\left(\frac{1}{x^2\log^{\frac{\epsilon}{2}}(x)}\left(\frac{1}{x^2\log^{\epsilon-2b}(x)}\right)+\frac{1}{\lambda(x)x^4\log^{\epsilon}(x)}\right)dx\\
&\leq \frac{C(\log(\log(t)))^2}{t^2\log^{b+1-2\alpha b}(t)}+\frac{C}{t^2\log^{\frac{\epsilon}{2}+1}(t)}+\frac{C}{t^2\log^{3\frac{\epsilon}{2}-2b}(t)}+\frac{C}{t^2\log^{\epsilon-b}(t)}\\
&\leq \frac{C(\log(\log(t)))^2}{t^2\log^{b+1-2\alpha b}(t)}+C\frac{\log^{-\epsilon+2b}(t)+\log^{-\frac{\epsilon}{2}+b}(t)+\log^{-1}(t)}{t^2\log^{\frac{\epsilon}{2}}(t)}
\end{aligned}$$

Next,

$$\begin{aligned}
\partial_t T(y)(t,\omega)=-\int_t^\infty \cos((t-x)\sqrt{\omega})\,&\big(F_2(y)(x,\omega)-\mathcal{F}(\sqrt{\cdot}F(x,\cdot\lambda(x)))(\omega\lambda(x)^2)\\
&-\mathcal{F}(\sqrt{\cdot}F_3(x,\cdot\lambda(x)))(\omega\lambda(x)^2)\big)\,dx
\end{aligned}$$

and the same procedure as above gives

$$\begin{aligned}
&||\partial_t T(y)(t)||_{L^2(\rho(\omega\lambda(t)^2)d\omega)}\\
&\leq \frac{C(\log(\log(t)))^2}{t^3\log^{b+1-2\alpha b}(t)}+C\int_t^\infty \frac{\lambda(t)}{\lambda(x)}\frac{1}{x^4\log^{\frac{\epsilon}{2}+1}(x)}dx\\
&+C\int_t^\infty \left(\frac{\lambda(t)}{\lambda(x)}\right)\left(\frac{1}{x^2\log^{\frac{\epsilon}{2}}(x)}\left(\frac{1}{x^2\log^{\epsilon-2b}(x)}\right)+\frac{1}{\lambda(x)x^4\log^{\epsilon}(x)}\right)dx\\
&\leq \frac{C(\log(\log(t)))^2}{t^3\log^{b+1-2\alpha b}(t)}+C\frac{\log^{-\epsilon+2b}(t)+\log^{-\frac{\epsilon}{2}+b}(t)+\log^{-1}(t)}{t^3\log^{\frac{\epsilon}{2}}(t)}
\end{aligned}$$

Similarly,

$$\begin{aligned}
\sqrt{\omega}\lambda(t)T(y)(t,\omega)=-\lambda(t)\int_t^\infty \sin((t-x)\sqrt{\omega})\,&(F_2(y)(x,\omega)\\
&-\mathcal{F}(\sqrt{\cdot}F(x,\cdot\lambda(x)))(\omega\lambda(x)^2)\\
&-\mathcal{F}(\sqrt{\cdot}F_3(x,\cdot\lambda(x)))(\omega\lambda(x)^2))\,dx
\end{aligned}$$

and the identical argument as for the previous two terms gives

$$\begin{aligned}
&||\sqrt{\omega}\lambda(t)T(y)(t,\omega)||_{L^2(\rho(\omega\lambda(t)^2)d\omega)}\\
&\leq \frac{C(\log(\log(t)))^2}{t^2\log^{1+b-2\alpha b}(t)}+C\lambda(t)\frac{\log^{-\epsilon+2b}(t)+\log^{-\frac{\epsilon}{2}+b}(t)+\log^{-1}(t)}{t^3\log^{\frac{\epsilon}{2}}(t)}\\
&\leq \frac{C(\log(\log(t)))^2}{t^2\log^{1+b-2\alpha b}(t)}+C\frac{\log^{-\epsilon+2b}(t)+\log^{-\frac{\epsilon}{2}+b}(t)+\log^{-1}(t)}{t^3\log^{\frac{\epsilon}{2}+b}(t)}
\end{aligned}$$

The next term is

$$\begin{aligned}
&\lambda(t)\sqrt{\omega}\partial_t T(y)(t,\omega)\\
&=-\int_t^\infty \cos((t-x)\sqrt{\omega})\left(\frac{\lambda(t)}{\lambda(x)}\right)\sqrt{\omega}\lambda(x)\,(F_2(y)(x,\omega)\\
&\qquad -\mathcal{F}(\sqrt{\cdot}F(x,\cdot\lambda(x)))(\omega\lambda(x)^2)-\mathcal{F}(\sqrt{\cdot}F_3(x,\cdot\lambda(x)))(\omega\lambda(x)^2))\,dx
\end{aligned}$$

and we get

$$||\lambda(t)\sqrt{\omega}\partial_t T(y)(t,\omega)||_{L^2(\rho(\omega\lambda(t)^2)d\omega)}$$
$$\leq \frac{C(\log(\log(t)))^2}{t^3\log^{1+b-2\alpha b}(t)}+C\int_t^\infty\left(\frac{\lambda(t)}{\lambda(x)}\right)^2\frac{C}{x^4\log^{\frac{\epsilon}{2}+1}(x)}dx$$
$$+C\int_t^\infty\left(\frac{\lambda(t)}{\lambda(x)}\right)^2\frac{1}{x^4\log^{\frac{3\epsilon}{2}-2b}(x)}dx+C\int_t^\infty\left(\frac{\lambda(t)}{\lambda(x)}\right)^2\log^b(x)\frac{1}{x^4\log^{\epsilon}(x)}dx$$
$$\leq \frac{C(\log(\log(t)))^2}{t^3\log^{1+b-2\alpha b}(t)}+C\frac{\log^{b-\frac{\epsilon}{2}}(t)+\log^{-1}(t)+\log^{-\epsilon+2b}(t)}{t^3\log^{\frac{\epsilon}{2}}(t)}$$

$$\omega\lambda(t)^2T(y)(t,\omega)$$
$$=-\int_t^\infty \sin((t-x)\sqrt{\omega})\frac{\lambda(t)^2}{\lambda(x)}\lambda(x)\sqrt{\omega}\left(F_2(y)(x,\omega)-\mathcal{F}(\sqrt{\cdot}F(x,\cdot\lambda(x)))(\omega\lambda(x)^2)\right.$$
$$\left.-\mathcal{F}(\sqrt{\cdot}F_3(x,\cdot\lambda(x)))(\omega\lambda(x)^2)\right)dx$$

and the same procedure as above gives

$$||\omega\lambda(t)^2T(y)(t,\omega)||_{L^2(\rho(\omega\lambda(t)^2)d\omega)}$$
$$\leq\frac{C}{t^2\log^{1+b-2\alpha b}(t)}+C\int_t^\infty\left(\frac{\lambda(t)^2}{\lambda(x)}\right)\left(\frac{\lambda(t)}{\lambda(x)}\right)\frac{C}{x^3\log^{\frac{\epsilon}{2}}(x)}\frac{1}{x\log(x)}dx$$
$$+C\int_t^\infty\left(\frac{\lambda(t)^2}{\lambda(x)}\right)\left(\frac{\lambda(t)}{\lambda(x)}\right)\frac{C}{x^2\log^{\frac{\epsilon}{2}}(x)}\frac{1}{x^2\log^{\epsilon-2b}(x)}dx$$
$$+C\int_t^\infty\left(\frac{\lambda(t)^2}{\lambda(x)}\right)\left(\frac{\lambda(t)}{\lambda(x)}\right)\frac{\log^b(x)}{x^4\log^{\epsilon}(x)}dx$$
$$\leq\frac{C}{t^2\log^{1+b-2\alpha b}(t)}+C\frac{\log^{-b-1}(t)+\log^{-b-\epsilon+2b}(t)+\log^{-\frac{\epsilon}{2}}(t)}{\log^{\frac{\epsilon}{2}}(t)t^3}$$

Moreover, by, for example the Dominated convergence theorem,

$$T(y)(t,\omega)t^2\log^{\frac{\epsilon}{2}}(t)\sqrt{\rho(\omega\lambda(t)^2)}\langle\omega\lambda(t)^2\rangle\in C^0_t([T_0,\infty),L^2(d\omega))$$

and

$$\partial_tT(y)(t,\omega)t^3\log^{\frac{\epsilon}{2}}(t)\langle\sqrt{\omega}\lambda(t)\rangle\sqrt{\rho(\omega\lambda(t)^2)}\in C^0_t([T_0,\infty),L^2(d\omega))$$

So, if T_0 is large enough, then, $T(y)\in\overline{B}_{\frac{1}{2}}(0)\subset Z$ if $y\in\overline{B}_1(0)\subset Z$. We will now show that T is a strict contraction on $\overline{B}_1(0)\subset Z$. Let $y_1,y_2\in Z$ satisfy

$$||y_1||_Z,||y_2||_Z\leq 1$$

$$T(y_1)-T(y_2)=-\int_t^\infty\frac{\sin((t-x)\sqrt{\omega})}{\sqrt{\omega}}\left(F_2(y_1)-F_2(y_2)\right.$$
$$\left.-\left(\mathcal{F}(\sqrt{\cdot}(F_3(u(y_1))-F_3(u(y_2)))(x,\cdot\lambda(x))))(\omega\lambda(x)^2)\right)\right)dx \tag{4.47}$$

First, note that F_2 is linear in y, so

$$F_2(y_1)-F_2(y_2)=F_2(y_1-y_2)$$

Next, we treat F_3 starting with the L_1 terms. We will denote by u_i the function associated to y_i via (4.20). We will also use $\overline{v}_i,w_i$ to denote the functions associated to u_i in the same way $\overline{v}$ and w were used in the above discussion

Recall that

$$(L_1(u_1)-L_1(u_2))(t,r)$$
$$=\left(\frac{\sin(2u_1(t,r))-\sin(2u_2(t,r))}{2r^2}\right)$$
$$\left(\cos(2Q_1(\frac{r}{\lambda(t)}))\left(\cos(2v_{corr}(t,r))-1\right)-\sin(2Q_1(\frac{r}{\lambda(t)}))\sin(2v_{corr}(t,r))\right)$$

Since

$$|\sin(2u_1)-\sin(2u_2)|\leq 2|u_1-u_2|$$

we get (after estimating in terms of u_i and then translating to y_i in exactly the same manner as done above)

$$||\mathcal{F}(\sqrt{\cdot}(L_1(u_1)-L_1(u_2))(t,\cdot\lambda(t)))(\omega\lambda(t)^2)||_{L^2(\rho(\omega\lambda(t)^2)d\omega)}$$
$$\leq C||y_1(t)-y_2(t)||_{L^2(\rho(\omega\lambda(t)^2)d\omega)}\left(||\frac{v_{corr}(t,R\lambda(t))}{R\lambda(t)}||_{L^\infty}^2+||\frac{v_{corr}(t,R\lambda(t))}{R(1+R^2)\lambda(t)^2}||_{L^\infty}\right)$$

Next, we estimate

$$|\partial_R\left((L_1(u_1)-L_1(u_2))(t,R\lambda(t))\right)|$$
$$\leq\frac{C\left(|\overline{v}_1-\overline{v}_2|\cdot|\partial_R\overline{v}_1|+|\partial_R(\overline{v}_1-\overline{v}_2)|\right)}{R^2\lambda(t)^2}\left(|v_{corr}(t,R\lambda(t))|^2+\frac{R}{(1+R^2)}|v_{corr}(t,R\lambda(t))|\right)$$
$$+\frac{C|\overline{v}_1-\overline{v}_2|}{R^3\lambda(t)^2}\left(|v_{corr}(t,R\lambda(t))|^2+\frac{R|v_{corr}(t,R\lambda(t))|}{(1+R^2)}\right)$$
$$+\frac{C|\overline{v}_1-\overline{v}_2|}{R^2\lambda(t)^2}\left(|v_{corr}(t,R\lambda(t))\partial_R(v_{corr}(t,R\lambda(t)))|+\frac{R|\partial_R(v_{corr}(t,R\lambda(t)))|}{(1+R^2)}\right)$$

Then, we get

$$||\partial_R((L_1(u_1)-L_1(u_2))(t,R\lambda(t)))||_{L^2(RdR)}$$
$$\leq C\left(||\overline{v}_1-\overline{v}_2||_{L^2(RdR)}+||L(\overline{v}_1-\overline{v}_2)||_{L^2(RdR)}\right)$$
$$\cdot\left(\left(1+||\partial_R(\overline{v}_1(t,R\lambda(t)))||_{L^2(RdR)}\right)\left(||\frac{v_{corr}(t,R\lambda(t))}{R\lambda(t)}||_{L^\infty}^2+||\frac{v_{corr}(t,R\lambda(t))}{R\lambda(t)^2(1+R^2)}||_{L^\infty}\right)\right.$$
$$+||\frac{v_{corr}(t,R\lambda(t))\partial_R(v_{corr}(t,R\lambda(t)))}{R\lambda(t)^2}||_{L^\infty(R\leq 1)}$$
$$\left.+||\frac{v_{corr}(t,R\lambda(t))\partial_R(v_{corr}(t,R\lambda(t)))}{R^2\lambda(t)^2}||_{L^\infty(R\geq 1)}+||\frac{\partial_R(v_{corr}(t,R\lambda(t)))}{\lambda(t)^2(1+R^2)}||_{L^\infty}\right)$$

Similarly,

$$|\frac{L_1(u_1)(t,R\lambda(t))-L_1(u_2)(t,R\lambda(t))}{R}|$$
$$\leq\frac{C|\overline{v}_1-\overline{v}_2|(t,R)}{R^3\lambda(t)^2}\left(|v_{corr}(t,R\lambda(t))|^2+\frac{R|v_{corr}(t,R\lambda(t))|}{(R^2+1)}\right)$$

We combine these to estimate $||L\left((L_1(u_1)-L_1(u_2))(t,R\lambda(t))\right)||_{L^2(RdR)}$, and then, as in the previous estimates, translate the right-hand side in terms of y_i, and use the estimates on our ansatz to get

$$||\lambda(t)\sqrt{\omega}\mathcal{F}(\sqrt{\cdot}\left((L_1(u_1)-L_1(u_2))(t,\cdot\lambda(t))\right))(\omega\lambda(t)^2)||_{L^2(\rho(\omega\lambda(t)^2)d\omega)}$$
$$\leq C\frac{||y_1-y_2||_{L^2(\rho(\omega\lambda(t)^2)d\omega)}+||\sqrt{\omega}\lambda(t)(y_1-y_2)||_{L^2(\rho(\omega\lambda(t)^2)d\omega)}}{t^2\log^{\epsilon-2b}(t)}$$
$$\leq C\frac{||y_1-y_2||_Z}{t^4\log^{\frac{3\epsilon}{2}-2b}(t)}$$

Now, we treat the nonlinear terms. First, note that, if

$$n_1(x) = \sin(2x) - 2x$$

then, by the mean value theorem,

$$\begin{aligned}|n_1(x_1) - n_1(x_2)| &\leq |x_1 - x_2|\max_{\theta\in[0,1]}|n_1'(\theta x_1 + (1-\theta)x_2)| \\ &\leq |x_1 - x_2|\max_{\theta\in[0,1]}|4\sin^2(\theta x_1 + (1-\theta)x_2)| \\ &\leq C|x_1 - x_2|\left(|x_1|^2 + |x_2|^2\right)\end{aligned}$$

Similarly,

$$|\left(\cos(2x_1) - 1\right) - \left(\cos(2x_2) - 1\right)| \leq C\left(|x_1| + |x_2|\right)|x_1 - x_2|$$

So,

$$\begin{aligned}|(N(u_1) - N(u_2))(t, R\lambda(t))| &\leq C\frac{|\overline{v}_1 - \overline{v}_2|\left(\overline{v}_1^2 + \overline{v}_2^2\right)}{R^2\lambda(t)^2} \\ &+ C\frac{|\overline{v}_1 - \overline{v}_2|\left(|\overline{v}_1| + |\overline{v}_2|\right)}{R^2\lambda(t)^2}\left(|Q_1(R)| + |v_{corr}(t, R\lambda(t))|\right)\end{aligned}$$

If $R < 1$, we use the estimates (4.14) to get

$$\frac{|\overline{v}_1(t,R)|}{\sqrt{R}\lambda(t)} \leq C\frac{||\overline{v}_1(t)||_{\dot{H}^1_e}}{\lambda(t)} + C\frac{||L^*L\overline{v}_1(t)||_{L^2(RdR)}}{\lambda(t)}$$

and if $R > 1$, then, we use (4.17) to get

$$\frac{|\overline{v}_1(t,R)|}{\sqrt{R}\lambda(t)} \leq \frac{C}{\lambda(t)}||\overline{v}_1(t)||_{\dot{H}^1_e}$$

In total, we obtain

$$\begin{aligned}&||N(u_1) - N(u_2)(t, \cdot\lambda(t))||_{L^2(RdR)} \\ &\leq C||\overline{v}_1 - \overline{v}_2||_{\dot{H}^1_e}\left(||\langle\omega\lambda(t)^2\rangle y_1||^2_{L^2(\rho(\omega\lambda(t)^2)d\omega)} + ||\langle\omega\lambda(t)^2\rangle y_2||^2_{L^2(\rho(\omega\lambda(t)^2)d\omega)}\right) \\ &+ C||\overline{v}_1 - \overline{v}_2||_{\dot{H}^1_e}\left(\frac{||\langle\sqrt{\omega}\lambda(t)\rangle y_1||_{L^2(\rho(\omega\lambda(t)^2)d\omega)} + ||\langle\sqrt{\omega}\lambda(t)\rangle y_2||_{L^2(\rho(\omega\lambda(t)^2)d\omega)}}{\lambda(t)}\right) \\ &\quad\left(1 + ||\frac{v_{corr}(t, R\lambda(t))}{R}||_{L^\infty}\right)\end{aligned}$$

which gives

$$||\mathcal{F}(\sqrt{\cdot}(N(u_1) - N(u_2))(t, \cdot\lambda(t)))(\omega\lambda(t)^2)||_{L^2(\rho(\omega\lambda(t)^2)d\omega)} \leq \frac{C}{\lambda(t)}\frac{||y_1 - y_2||_Z}{t^4\log^\epsilon(t)}$$

where we used the estimates on the ansatz, and the fact that

$$||y_i||_Z \leq 1, \quad i = 1, 2$$

Next, we will estimate $\partial_R ((N(u_1) - N(u_2))(t, R\lambda(t)))$, treating the following expression one line at a time.

$$\begin{aligned}
&\partial_R ((N(u_1) - N(u_2))(t, R\lambda(t))) \\
&= \frac{\cos(2Q_1(R))}{2R^2\lambda(t)^2} (2(\cos(2\overline{v}_1(t,R)) - 1)\partial_R\overline{v}_1 - 2(\cos(2\overline{v}_2) - 1)\partial_R\overline{v}_2) \\
&+ \partial_R\left(\frac{\cos(2Q_1(R))}{2R^2\lambda(t)^2}\right)(n_1(\overline{v}_1) - n_2(\overline{v}_2)) \\
&- 2\left(\frac{\sin(2Q_1(R) + 2v_{corr}(t, R\lambda(t)))}{2R^2\lambda(t)^2}\right)(\sin(2\overline{v}_1)\partial_R\overline{v}_1 - \sin(2\overline{v}_2)\partial_R\overline{v}_2) \\
&+ (\cos(2\overline{v}_1) - \cos(2\overline{v}_2))\partial_R\left(\frac{\sin(2Q_1(R) + 2v_{corr}(t, R\lambda(t)))}{2R^2\lambda(t)^2}\right)
\end{aligned} \tag{4.48}$$

For $R \leq 1$, we estimate the first line of the right-hand side of (4.48) by

$$\begin{aligned}
&|\frac{\cos(2Q_1(R))}{2R^2\lambda(t)^2} (2(\cos(2\overline{v}_1(t,R)) - 1)\partial_R\overline{v}_1 - 2(\cos(2\overline{v}_2) - 1)\partial_R\overline{v}_2)| \\
&\leq \frac{C}{\lambda(t)^2}||\frac{\overline{v}_1}{\sqrt{R}}||_{L^\infty}^2 |\frac{L(\overline{v}_1 - \overline{v}_2)}{R}| + \frac{C}{\lambda(t)^2}\frac{\overline{v}_1(t,R)^2}{R^2}\frac{|\overline{v}_1 - \overline{v}_2|}{R} \\
&+ \frac{C}{\lambda(t)^2}||\frac{\overline{v}_1 - \overline{v}_2}{\sqrt{R}}||_{L^\infty}||\frac{|\overline{v}_1| + |\overline{v}_2|}{\sqrt{R}}||_{L^\infty}\frac{|L\overline{v}_2|}{R} + \frac{C}{\lambda(t)^2}|\frac{\overline{v}_1 - \overline{v}_2}{R}|\left(\frac{|\overline{v}_1| + |\overline{v}_2|}{R}\right)\frac{|\overline{v}_2|}{R}
\end{aligned}$$

where we used the fact that

$$\begin{aligned}
&2(\cos(2\overline{v}_1(t,R)) - 1)\partial_R\overline{v}_1(t,R) - 2(\cos(2\overline{v}_2(t,R)) - 1)\partial_R\overline{v}_2(t,R) \\
&= 2(\cos(2\overline{v}_1(t,R)) - 1)(\partial_R(\overline{v}_1 - \overline{v}_2)) + 2\partial_R\overline{v}_2(\cos(2\overline{v}_1) - \cos(2\overline{v}_2))
\end{aligned}$$

to get

$$\begin{aligned}
&|2(\cos(2\overline{v}_1(t,R)) - 1)\partial_R\overline{v}_1(t,R) - 2(\cos(2\overline{v}_2(t,R)) - 1)\partial_R\overline{v}_2(t,R)| \\
&\leq C\overline{v}_1(t,R)^2|\partial_R(\overline{v}_1 - \overline{v}_2)| + C|\overline{v}_1 - \overline{v}_2||\partial_R\overline{v}_2|(|\overline{v}_1| + |\overline{v}_2|)
\end{aligned}$$

The second line of the right-hand side of (4.48) is estimated by

$$|\partial_R\left(\frac{\cos(2Q_1(R))}{2R^2\lambda(t)^2}\right)(n_1(\overline{v}_1) - n_2(\overline{v}_2))| \leq \frac{C}{\lambda(t)^2}\left(\frac{\overline{v}_1^2}{R^2} + \frac{\overline{v}_2^2}{R^2}\right)\frac{|\overline{v}_1 - \overline{v}_2|}{R}$$

The third line of the right-hand side of (4.48) is estimated by

$$\begin{aligned}
&|-2\left(\frac{\sin(2Q_1(R) + 2v_{corr}(t, R\lambda(t)))}{2R^2\lambda(t)^2}\right)(\sin(2\overline{v}_1)\partial_R\overline{v}_1 - \sin(2\overline{v}_2)\partial_R\overline{v}_2)| \\
&\leq \frac{C}{\lambda(t)^2}\left(1 + ||\frac{v_{corr}(t, R\lambda(t))}{R}||_{L^\infty}\right) \\
&\quad\left(||\overline{v}_1||_{\dot{H}^1_e}\frac{|L(\overline{v}_1 - \overline{v}_2)|}{R} + ||\overline{v}_1 - \overline{v}_2||_{\dot{H}^1_e}|\frac{L\overline{v}_2}{R}| + \left(\frac{|\overline{v}_1| + |\overline{v}_2|}{R}\right)\frac{|\overline{v}_1 - \overline{v}_2|}{R}\right)
\end{aligned}$$

where we used

$$\sin(2\overline{v}_1)\partial_R\overline{v}_1 - \sin(2\overline{v}_2)\partial_R\overline{v}_2 = \sin(2\overline{v}_1)(\partial_R(\overline{v}_1 - \overline{v}_2)) + (\sin(2\overline{v}_1) - \sin(2\overline{v}_2))\partial_R\overline{v}_2$$

The final line of (4.48) is estimated by

$$(\cos(2\overline{v}_1) - \cos(2\overline{v}_2))\,\partial_R\left(\frac{\sin(2Q_1(R) + 2v_{corr}(t, R\lambda(t)))}{2R^2\lambda(t)^2}\right)$$
$$\leq C\left(\frac{|\overline{v}_1| + |\overline{v}_2|}{R}\right)\frac{|\overline{v}_1 - \overline{v}_2|}{R\lambda(t)^2}\left(1+||\partial_R(v_{corr}(t, R\lambda(t)))||_{L^\infty}+||\frac{v_{corr}(t, R\lambda(t))}{R}||_{L^\infty}\right)$$

Now, we will estimate the contribution of each of these terms to

$$||\partial_R\left((N(u_1) - N(u_2))(t, R\lambda(t))\right)||_{L^2((0,1),RdR)}$$

We use (4.14) to control $||\frac{\overline{v}_1}{\sqrt{R}}||_{L^\infty}$, and get

$$\int_0^1 \frac{C}{\lambda(t)^4}||\frac{\overline{v}_1}{\sqrt{R}}||_{L^\infty}^4\left(\frac{L(\overline{v}_1 - \overline{v}_2)}{R}\right)^2 RdR$$
$$\leq C\lambda(t)^2||\langle\omega\lambda(t)^2\rangle y_1||_{L^2(\rho(\omega\lambda(t)^2)d\omega)}^4||\langle\omega\lambda(t)^2\rangle(y_1 - y_2)||_{L^2(\rho(\omega\lambda(t)^2)d\omega)}^2$$

Next, we get

$$\int_0^1 \frac{C}{\lambda(t)^4}\frac{(\overline{v}_1(t,R))^4}{R^4}\frac{|\overline{v}_1 - \overline{v}_2|^2}{R}dR$$
$$\leq \frac{C}{\lambda(t)^4}\int_0^1 \left(\left(||\overline{v}_1||_{\dot{H}^1_e}^4 + ||L^*L\overline{v}_1||_{L^2(RdR)}^4\right)\left(\log^2(\frac{1}{R}) + 1\right)^2\right)\left(||\overline{v}_1 - \overline{v}_2||_{\dot{H}^1_e}^2 + ||L^*L(\overline{v}_1 - \overline{v}_2)||_{L^2(RdR)}^2\right)$$
$$\cdot R\left(\log^2\left(\frac{1}{R}\right) + 1\right)dR$$
$$\leq C\lambda(t)^2||\langle\omega\lambda(t)^2\rangle y_1||_{L^2(\rho(\omega\lambda(t)^2)d\omega)}^4||\langle\omega\lambda(t)^2\rangle(y_1 - y_2)||_{L^2(\rho(\omega\lambda(t)^2)d\omega)}^2$$

The next term to consider is

$$\int_0^1 \frac{C}{\lambda(t)^4}||\frac{\overline{v}_1 - \overline{v}_2}{\sqrt{R}}||_{L^\infty}^2||\frac{|\overline{v}_1| + |\overline{v}_2|}{\sqrt{R}}||_{L^\infty}^2\frac{(L\overline{v}_2)^2}{R^2}RdR$$
$$\leq \frac{C}{\lambda(t)^4}\left(||\overline{v}_1 - \overline{v}_2||_{\dot{H}^1_e}^2 + ||L^*L(\overline{v}_1 - \overline{v}_2)||_{L^2(RdR)}^2\right)$$
$$\cdot\left(||\overline{v}_1||_{\dot{H}^1_e}^2 + ||L^*L\overline{v}_1||_{L^2(RdR)}^2 + ||\overline{v}_2||_{\dot{H}^1_e}^2 + ||L^*L\overline{v}_2||_{L^2(RdR)}^2\right)\int_0^1 \frac{(L\overline{v}_2(t,R))^2}{R^2}RdR$$
$$\leq C\lambda(t)^2||\langle\omega\lambda(t)^2\rangle(y_1 - y_2)||_{L^2(\rho(\omega\lambda(t)^2)d\omega)}^2$$
$$\cdot\left(||\langle\omega\lambda(t)^2\rangle y_1||_{L^2(\rho(\omega\lambda(t)^2)d\omega)}^2 + ||\langle\omega\lambda(t)^2\rangle y_2||_{L^2(\rho(\omega\lambda(t)^2)d\omega)}^2\right)$$
$$\cdot||\langle\omega\lambda(t)^2\rangle y_2||_{L^2(\rho(\omega\lambda(t)^2)d\omega)}^2$$

Next, we have

$$\frac{C}{\lambda(t)^4}\int_0^1\left(|\frac{\overline{v}_1 - \overline{v}_2}{R}|\left(\frac{|\overline{v}_1| + |\overline{v}_2|}{R}\right)\frac{|\overline{v}_2|}{R}\right)^2 RdR$$
$$\leq \frac{C}{\lambda(t)^4}\left(||\overline{v}_1 - \overline{v}_2||_{\dot{H}^1_e}^2 + ||L^*L(\overline{v}_1 - \overline{v}_2)||_{L^2(RdR)}^2\right)$$
$$\cdot\left(||\overline{v}_1||_{\dot{H}^1_e}^4 + ||L^*L\overline{v}_1||_{L^2(RdR)}^4 + ||\overline{v}_2||_{\dot{H}^1_e}^4 + ||L^*L\overline{v}_2||_{L^2(RdR)}^4\right)\int_0^1\left(\log^2(\frac{1}{R}) + 1\right)^3 RdR$$
$$\leq C\lambda(t)^2||\langle\omega\lambda(t)^2\rangle(y_1 - y_2)||_{L^2(\rho(\omega\lambda(t)^2)d\omega)}^2$$
$$\cdot\left(||\langle\omega\lambda(t)^2\rangle y_1||_{L^2(\rho(\omega\lambda(t)^2)d\omega)}^4 + ||\langle\omega\lambda(t)^2\rangle y_2||_{L^2(\rho(\omega\lambda(t)^2)d\omega)}^4\right)$$

For the second line of the right-hand side of (4.48),

$$\begin{aligned}
&\frac{C}{\lambda(t)^4}\int_0^1 \left(\frac{\overline{v}_1^4}{R^4}+\frac{\overline{v}_2^4}{R^4}\right)\frac{|\overline{v}_1-\overline{v}_2|^2}{R^2} RdR\\
&\leq \frac{C}{\lambda(t)^4}\left(||\overline{v}_1||^4_{\dot{H}^1_e}+||L^*L\overline{v}_1||^4_{L^2(RdR)}+||\overline{v}_2||^4_{\dot{H}^1_e}+||L^*L\overline{v}_2||^4_{L^2(RdR)}\right)\\
&\cdot\left(||\overline{v}_1-\overline{v}_2||^2_{\dot{H}^1_e}+||L^*L(\overline{v}_1-\overline{v}_2)||^2_{L^2(RdR)}\right)\int_0^1\left(\log^2(\frac{1}{R})+1\right)^3 RdR\\
&\leq C\lambda(t)^2||\langle\omega\lambda(t)^2\rangle(y_1-y_2)||^2_{L^2(\rho(\omega\lambda(t)^2)d\omega)}\\
&\cdot\left(||\langle\omega\lambda(t)^2\rangle y_1||^4_{L^2(\rho(\omega\lambda(t)^2)d\omega)}+||\langle\omega\lambda(t)^2\rangle y_2||^4_{L^2(\rho(\omega\lambda(t)^2)d\omega)}\right)
\end{aligned}$$

Next, we treat the third line of the right-hand side of (4.48),

$$\begin{aligned}
&\int_0^1\frac{1}{\lambda(t)^4}\left(1+||\frac{v_{corr}(t,R\lambda(t))}{R}||^2_{L^\infty}\right)\\
&\cdot\left(||\overline{v}_1||^2_{\dot{H}^1_e}\frac{(L(\overline{v}_1-\overline{v}_2))^2}{R^2}+||\overline{v}_1-\overline{v}_2||^2_{\dot{H}^1_e}\frac{(L\overline{v}_2)^2}{R^2}+\frac{(\overline{v}_1)^2+(\overline{v}_2)^2}{R^2}\frac{|\overline{v}_1-\overline{v}_2|^2}{R^2}\right)RdR\\
&\leq C\left(1+||\frac{v_{corr}(t,R\lambda(t))}{R}||^2_{L^\infty}\right)||\langle\omega\lambda(t)^2\rangle(y_1-y_2)||^2_{L^2(\rho(\omega\lambda(t)^2)d\omega)}\\
&\cdot\left(||\langle\omega\lambda(t)^2\rangle y_2||^2_{L^2(\rho(\omega\lambda(t)^2)d\omega)}+||\langle\omega\lambda(t)^2\rangle y_1||^2_{L^2(\rho(\omega\lambda(t)^2)d\omega)}\right)
\end{aligned}$$

The final line of (4.48) is treated as follows.

$$\begin{aligned}
&\int_0^1\left(\frac{|\overline{v}_1|+|\overline{v}_2|}{R}\frac{|\overline{v}_1-\overline{v}_2|}{R\lambda(t)^2}\left(1+||\partial_R(v_{corr}(t,R\lambda(t)))||_{L^\infty}+||\frac{v_{corr}(t,R\lambda(t))}{R}||_{L^\infty}\right)\right)^2 RdR\\
&\leq\frac{C}{\lambda(t)^4}\left(1+||\partial_R(v_{corr}(t,R\lambda(t)))||^2_{L^\infty}+||\frac{v_{corr}(t,R\lambda(t))}{R}||^2_{L^\infty}\right)\\
&\cdot\left(||\overline{v}_1||^2_{\dot{H}^1_e}+||L^*L\overline{v}_1||^2_{L^2(RdR)}+||\overline{v}_2||^2_{\dot{H}^1_e}+||L^*L\overline{v}_2||^2_{L^2(RdR)}\right)\\
&\cdot\left(||\overline{v}_1-\overline{v}_2||^2_{\dot{H}^1_e}+||L^*L(\overline{v}_1-\overline{v}_2)||^2_{L^2(RdR)}\right)\int_0^1\left(\log^2(\frac{1}{R})+1\right)^2 RdR\\
&\leq C\left(1+||\partial_R(v_{corr}(t,R\lambda(t)))||^2_{L^\infty}+||\frac{v_{corr}(t,R\lambda(t))}{R}||^2_{L^\infty}\right)\\
&\cdot||\langle\omega\lambda(t)^2\rangle(y_1-y_2)||^2_{L^2(\rho(\omega\lambda(t)^2)d\omega)}\\
&\cdot\left(||\langle\omega\lambda(t)^2\rangle y_1||^2_{L^2(\rho(\omega\lambda(t)^2)d\omega)}+||\langle\omega\lambda(t)^2\rangle y_2||^2_{L^2(\rho(\omega\lambda(t)^2)d\omega)}\right)
\end{aligned}$$

We combine the above estimates to get

$$\begin{aligned}
&\int_0^1 \left(\partial_R\left((N(u_1)-N(u_2))(t,R\lambda(t))\right)\right)^2 RdR\\
&\leq C\lambda(t)^2\left(||\langle\omega\lambda(t)^2\rangle y_1||^4_{L^2(\rho(\omega\lambda(t)^2)d\omega)}\right.\\
&\qquad\left.+||\langle\omega\lambda(t)^2\rangle y_2||^4_{L^2(\rho(\omega\lambda(t)^2)d\omega)}\right)||\langle\omega\lambda(t)^2\rangle(y_1-y_2)||^2_{L^2(\rho(\omega\lambda(t)^2)d\omega)}\\
&+C\left(1+||\partial_R(v_{corr}(t,R\lambda(t)))||^2_{L^\infty}+||\frac{v_{corr}(t,R\lambda(t))}{R}||^2_{L^\infty}\right)\\
&\cdot||\langle\omega\lambda(t)^2\rangle(y_1-y_2)||^2_{L^2(\rho(\omega\lambda(t)^2)d\omega)}\\
&\cdot\left(||\langle\omega\lambda(t)^2\rangle y_2||^2_{L^2(\rho(\omega\lambda(t)^2)d\omega)}+||\langle\omega\lambda(t)^2\rangle y_1||^2_{L^2(\rho(\omega\lambda(t)^2)d\omega)}\right)
\end{aligned}$$

When $R\geq 1$, we can estimate (4.48) by

$$\begin{aligned}
&|\partial_R\left((N(u_1)-N(u_2))(t,R\lambda(t))\right)|\\
&\leq\frac{C}{\lambda(t)^2}||\overline{v}_1||^2_{\dot{H}^1_e}|L(\overline{v}_1-\overline{v}_2)|+\frac{C}{\lambda(t)^2}||\overline{v}_1||^2_{\dot{H}^1_e}|\overline{v}_1-\overline{v}_2|\\
&+\frac{C}{\lambda(t)^2}||\overline{v}_1-\overline{v}_2||_{\dot{H}^1_e}\left(||\overline{v}_1||_{\dot{H}^1_e}+||\overline{v}_2||_{\dot{H}^1_e}\right)(|L\overline{v}_2|+|\overline{v}_2|)\\
&+\frac{C}{\lambda(t)^2}\left(||\overline{v}_1||^2_{\dot{H}^1_e}+||\overline{v}_2||^2_{\dot{H}^1_e}\right)|\overline{v}_1-\overline{v}_2|\\
&+\frac{C}{\lambda(t)^2}\left(||\overline{v}_1||_{\dot{H}^1_e}+||\overline{v}_2||_{\dot{H}^1_e}\right)|\overline{v}_1-\overline{v}_2|\\
&\cdot\left(1+||\partial_R(v_{corr}(t,R\lambda(t)))||_{L^\infty}+||\frac{v_{corr}(t,R\lambda(t))}{R}||_{L^\infty}\right)\\
&+\frac{C}{\lambda(t)^2}\left(1+||\frac{v_{corr}(t,R\lambda(t))}{R}||_{L^\infty}\right)\\
&\cdot\left(||\overline{v}_1||_{\dot{H}^1_e}|L(\overline{v}_1-\overline{v}_2)|+||\overline{v}_1-\overline{v}_2||_{\dot{H}^1_e}|L\overline{v}_2|+\left(||\overline{v}_1||_{\dot{H}^1_e}+||\overline{v}_2||_{\dot{H}^1_e}\right)|\overline{v}_1-\overline{v}_2|\right)
\end{aligned}$$

Taking the L^2 norm over $R\geq 1$, and combining with our previous estimates, we get

$$\begin{aligned}
&||\partial_R\left((N(u_1)-N(u_2))(t,R\lambda(t))\right)||_{L^2(RdR)}\\
&\leq C\lambda(t)||\langle\omega\lambda(t)^2\rangle(y_1-y_2)||_{L^2(\rho(\omega\lambda(t)^2)d\omega)}\\
&\cdot\left(||\langle\omega\lambda(t)^2\rangle y_1||^2_{L^2(\rho(\omega\lambda(t)^2)d\omega)}+||\langle\omega\lambda(t)^2\rangle y_2||^2_{L^2(\rho(\omega\lambda(t)^2)d\omega)}\right)\\
&+C\left(1+||\frac{v_{corr}(t,R\lambda(t))}{R}||_{L^\infty}+||\partial_R(v_{corr}(t,R\lambda(t)))||_{L^\infty}\right)\\
&\cdot\left(||\langle\omega\lambda(t)^2\rangle y_2||_{L^2(\rho(\omega\lambda(t)^2)d\omega)}+||\langle\omega\lambda(t)^2\rangle y_1||_{L^2(\rho(\omega\lambda(t)^2)d\omega)}\right)\\
&\cdot||\langle\omega\lambda(t)^2\rangle(y_1-y_2)||_{L^2(\rho(\omega\lambda(t)^2)d\omega)}
\end{aligned}$$

It only remains to consider the following.

(4.49)

$$|\frac{N(u_1)(t,R\lambda(t))-N(u_2)(t,R\lambda(t))}{R}|$$
$$\leq C\frac{|\overline{v}_1-\overline{v}_2|\left((\overline{v}_1)^2+(\overline{v}_2)^2\right)}{R^3\lambda(t)^2}+\frac{C\left(|\overline{v}_1|+|\overline{v}_2|\right)|\overline{v}_1-\overline{v}_2|}{R^3\lambda(t)^2}\left(|Q_1(R)|+|v_{corr}(t,R\lambda(t))|\right)$$

So,

$$\int_0^1 \frac{\left(N(u_1)(t,R\lambda(t))-N(u_2)(t,R\lambda(t))\right)^2}{R^2}RdR$$
$$\leq \frac{C}{\lambda(t)^4}\left(||\overline{v}_1-\overline{v}_2||_{\dot{H}^1_e}^2+||L^*L(\overline{v}_1-\overline{v}_2)||_{L^2(RdR)}^2\right)$$
$$\cdot\left(||\overline{v}_1||_{\dot{H}^1_e}^4+||L^*L\overline{v}_1||_{L^2(RdR)}^4+||\overline{v}_2||_{\dot{H}^1_e}^4+||L^*L\overline{v}_2||_{L^2(RdR)}^4\right)\int_0^1\left(\log^2(\frac{1}{R})+1\right)^3RdR$$
$$+\frac{C}{\lambda(t)^4}\left(1+||\frac{v_{corr}(t,R\lambda(t))}{R}||_{L^\infty}^2\right)\left(||\overline{v}_1||_{\dot{H}^1_e}^2+||L^*L\overline{v}_1||_{L^2(RdR)}^2\right.$$
$$\left.+||\overline{v}_2||_{\dot{H}^1_e}^2+||L^*L\overline{v}_2||_{L^2(RdR)}^2\right)$$
$$\cdot\left(||\overline{v}_1-\overline{v}_2||_{\dot{H}^1_e}^2+||L^*L(\overline{v}_1-\overline{v}_2)||_{L^2(RdR)}^2\right)\int_0^1\left(\log^2(\frac{1}{R})+1\right)^2RdR$$
$$\leq C\lambda(t)^2||\langle\omega\lambda(t)^2\rangle(y_1-y_2)||_{L^2(\rho(\omega\lambda(t)^2)d\omega)}^2$$
$$\cdot\left(||\langle\omega\lambda(t)^2\rangle y_1||_{L^2(\rho(\omega\lambda(t)^2)d\omega)}^4+||\langle\omega\lambda(t)^2\rangle y_2||_{L^2(\rho(\omega\lambda(t)^2)d\omega)}^4\right)$$
$$+C\left(1+||\frac{v_{corr}(t,R\lambda(t))}{R}||_{L^\infty}^2\right)$$
$$\cdot||\langle\omega\lambda(t)^2\rangle(y_1-y_2)||_{L^2(\rho(\omega\lambda(t)^2)d\omega)}^2$$
$$\cdot\left(||\langle\omega\lambda(t)^2\rangle y_1||_{L^2(\rho(\omega\lambda(t)^2)d\omega)}^2+||\langle\omega\lambda(t)^2\rangle y_2||_{L^2(\rho(\omega\lambda(t)^2)d\omega)}^2\right)$$

We return to (4.49), and study

$$\left(\int_1^\infty \frac{\left((N(u_1)-N(u_2))(t,R\lambda(t))\right)^2}{R^2}RdR\right)^{1/2}$$
$$\leq\frac{C}{\lambda(t)^2}||\overline{v}_1-\overline{v}_2||_{L^2(RdR)}\left(||\overline{v}_1||_{\dot{H}^1_e}^2+||\overline{v}_2||_{\dot{H}^1_e}^2\right)$$
$$+\frac{C}{\lambda(t)^2}\left(||\overline{v}_1||_{\dot{H}^1_e}+||\overline{v}_2||_{\dot{H}^1_e}\right)||\overline{v}_1-\overline{v}_2||_{L^2(RdR)}\left(1+||\frac{v_{corr}(t,R\lambda(t))}{R}||_{L^\infty}\right)$$

We now combine the above estimates, translating between norms of $\overline{v_i}$ and norms of y_i as previously, and use the fact that $||y_i||_Z\leq 1$, as well as the estimates of the ansatz to get

$$||\lambda(t)\sqrt{\omega}\mathcal{F}(\sqrt{\cdot}(N(u_1)-N(u_2))(t,\cdot\lambda(t)))(\omega\lambda(t)^2)||_{L^2(\rho(\omega\lambda(t)^2)d\omega)}$$
$$\leq C\frac{||y_1-y_2||_Z}{t^4\log^{\epsilon-b}(t)}$$

We now return to (4.47) to get

$$\begin{aligned}
&||(T(y_1)-T(y_2))(t)||_{L^2(\rho(\omega\lambda(t)^2)d\omega)}\\
&\leq C\int_t^\infty x\left(\frac{\lambda(t)}{\lambda(x)}\right)\left(\frac{||y_1-y_2||_Z}{x^4\log^{1+\frac{\epsilon}{2}}(x)}\right)dx+C\int_t^\infty x\left(\frac{\lambda(t)}{\lambda(x)}\right)\frac{||y_1-y_2||_Z}{x^4\log^{3\frac{\epsilon}{2}-2b}(x)}dx\\
&+C\int_t^\infty x\left(\frac{\lambda(t)}{\lambda(x)}\right)\frac{||y_1-y_2||_Z}{x^4\log^{\epsilon-b}(x)}dx\\
&\leq C||y_1-y_2||_Z\frac{\log^{-1}(t)+\log^{-\epsilon+2b}(t)+\log^{-\frac{\epsilon}{2}+b}(t)}{t^2\log^{\frac{\epsilon}{2}}(t)}
\end{aligned}$$

$$\begin{aligned}
&\partial_t(T(y_1)-T(y_2))\\
&=-\int_t^\infty\cos((t-x)\sqrt{\omega})F_2(y_1-y_2)dx\\
&+\int_t^\infty\cos((t-x)\sqrt{\omega})\mathcal{F}(\sqrt{\cdot}(F_3(y_1)-F_3(y_2))(x,\cdot\lambda(x)))(\omega\lambda(x)^2)dx
\end{aligned}$$

So, the same argument as in the previous estimate gives

$$\begin{aligned}
&||\partial_t(T(y_1)-T(y_2))||_{L^2(\rho(\omega\lambda(t)^2)d\omega)}\\
&\leq C||y_1-y_2||_Z\frac{\log^{-1}(t)+\log^{-\epsilon+2b}(t)+\log^{-\frac{\epsilon}{2}+b}(t)}{t^3\log^{\frac{\epsilon}{2}}(t)}
\end{aligned}$$

Similarly,

$$\begin{aligned}
&\sqrt{\omega}\lambda(t)(T(y_1)-T(y_2))\\
&=-\int_t^\infty\lambda(t)\sin((t-x)\sqrt{\omega})F_2(y_1-y_2)dx\\
&+\lambda(t)\int_t^\infty\sin((t-x)\sqrt{\omega})\mathcal{F}(\sqrt{\cdot}(F_3(y_1)-F_3(y_2))(x,\cdot\lambda(x)))(\omega\lambda(x)^2)dx
\end{aligned}$$

and the identical argument gives us

$$\begin{aligned}
&||\sqrt{\omega}\lambda(t)(T(y_1)-T(y_2))(t)||_{L^2(\rho(\omega\lambda(t)^2)d\omega)}\\
&\leq C||y_1-y_2||_Z\frac{\log^{-1}(t)+\log^{-\epsilon+2b}(t)+\log^{-\frac{\epsilon}{2}+b}(t)}{t^3\log^{\frac{\epsilon}{2}+b}(t)}
\end{aligned}$$

Next, we have

$$\begin{aligned}
&\sqrt{\omega}\lambda(t)\partial_t(T(y_1)-T(y_2))\\
&=-\int_t^\infty\left(\frac{\lambda(t)}{\lambda(x)}\right)\cos((t-x)\sqrt{\omega})\lambda(x)\sqrt{\omega}F_2(y_1-y_2)dx\\
&+\int_t^\infty\left(\frac{\lambda(t)}{\lambda(x)}\right)\cos((t-x)\sqrt{\omega})\lambda(x)\sqrt{\omega}\mathcal{F}(\sqrt{\cdot}(F_3(y_1)-F_3(y_2))(x,\cdot\lambda(x)))(\omega\lambda(x)^2)dx
\end{aligned}$$

So,

$$\begin{aligned}&||\sqrt{\omega}\lambda(t)\partial_t(T(y_1)-T(y_2))(t)||_{L^2(\rho(\omega\lambda(t)^2)d\omega)}\\&\leq C\int_t^\infty\left(\frac{\lambda(t)}{\lambda(x)}\right)^2\frac{||y_1-y_2||_Z}{x^4\log^{1+\frac{\epsilon}{2}}(x)}dx+C\int_t^\infty\left(\frac{\lambda(t)}{\lambda(x)}\right)^2\frac{||y_1-y_2||_Z}{x^4\log^{\frac{3\epsilon}{2}-2b}(x)}dx\\&+C\int_t^\infty\left(\frac{\lambda(t)}{\lambda(x)}\right)^2\frac{||y_1-y_2||_Z}{x^4\log^{\epsilon-b}(x)}dx\\&\leq C||y_1-y_2||_Z\frac{\log^{-1}(t)+\log^{-\epsilon+2b}(t)+\log^{-\frac{\epsilon}{2}+b}(t)}{t^3\log^{\frac{\epsilon}{2}}(t)}\end{aligned}$$

The identical procedure shows that

$$\begin{aligned}&||\omega\lambda(t)^2(T(y_1)-T(y_2))(t)||_{L^2(\rho(\omega\lambda(t)^2)d\omega)}\\&\leq C||y_1-y_2||_Z\frac{\log^{-1}(t)+\log^{-\epsilon+2b}(t)+\log^{-\frac{\epsilon}{2}+b}(t)}{t^3\log^{\frac{\epsilon}{2}+b}(t)}\end{aligned}$$

Thus, T is a strict contraction on $\overline{B}_1(0)\subset Z$, for T_0 large enough; so, T has a fixed point, say y_0, in $\overline{B}_1(0)\subset Z$ $\square$

CHAPTER 5

The energy of the solution, and its decomposition as in Theorem 1.1

Let us define

$$v_6(t,r) := \begin{cases} \sqrt{\frac{\lambda(t)}{r}} \left(\mathcal{F}^{-1}(y_0(t, \frac{\cdot}{\lambda(t)^2}))\right)(\frac{r}{\lambda(t)}), & r > 0 \\ 0, \quad r = 0 \end{cases}$$

Note that $v_6(t,\cdot) \in C^0([0,\infty))$, by the same argument as in Lemma (4.1). Inspecting the derivation of (4.3), we see that we have a solution to (1.1):

$$u(t,r) = Q_{\frac{1}{\lambda(t)}}(r) + \sum_{k=1}^{6} v_k(t,r) \tag{5.1}$$

Here, we study the energy, (1.2), of our solution, and prove that it has a decomposition as in the main theorem statement. First, we note that $\partial_t Q_{\frac{1}{\lambda(t)}} \notin L^2((0,\infty), rdr)$, because $\phi_0 \notin L^2((0,\infty), rdr)$, so we have to first capture a delicate cancellation between the large r behavior of $\partial_t Q_{\frac{1}{\lambda(t)}}$ and $\partial_t v_1$ before we can even show that the solution has finite energy. We consider the region $r \geq t$, and use the representation formula for v_1, (3.10), which implies

(5.2)

$$\begin{aligned}
\partial_t v_1(t,r) &= \int_t^\infty \frac{\lambda'''(s)}{r} \int_0^{s-t} \frac{\rho}{\sqrt{(s-t)^2-\rho^2}} \left(1+\frac{r^2-1-\rho^2}{\sqrt{(1+r^2+\rho^2)^2-4r^2\rho^2}}\right) d\rho ds \\
&= \int_t^{t+\frac{r}{6}} \frac{\lambda'''(s)}{r} \int_0^{s-t} \frac{\rho}{\sqrt{(s-t)^2-\rho^2}} \left(1+\frac{r^2-1-\rho^2}{\sqrt{(1+r^2+\rho^2)^2-4r^2\rho^2}}\right) d\rho ds \\
&+ \int_{t+\frac{r}{6}}^\infty \frac{\lambda'''(s)}{r} \int_0^{s-t} \frac{\rho}{(s-t)} \left(1+\frac{r^2-1-\rho^2}{\sqrt{(1+r^2+\rho^2)^2-4r^2\rho^2}}\right) d\rho ds \\
&+ \int_{t+\frac{r}{6}}^\infty \frac{\lambda'''(s)}{r} \int_0^{s-t} \rho \left(\frac{1}{\sqrt{(s-t)^2-\rho^2}} - \frac{1}{(s-t)}\right)\left(1+\frac{r^2-1-\rho^2}{\sqrt{(1+r^2+\rho^2)^2-4r^2\rho^2}}\right) d\rho ds
\end{aligned}$$

For the second line of (5.2), we have

$$\begin{aligned}
&\int_t^{t+\frac{r}{6}} \frac{\lambda'''(s)}{r} \int_0^{s-t} \frac{\rho}{\sqrt{(s-t)^2-\rho^2}} \left(1+\frac{r^2-1-\rho^2}{\sqrt{(1+r^2+\rho^2)^2-4r^2\rho^2}}\right) d\rho ds \\
&= \int_t^{t+\frac{r}{6}} \frac{\lambda'''(s)}{r} \int_0^{s-t} \frac{\rho}{\sqrt{(s-t)^2-\rho^2}} \left(1+1+O\left(\frac{r^2}{(r^2-1-\rho^2)^2}\right)\right) d\rho ds \qquad (5.3) \\
&= 2\int_t^{t+\frac{r}{6}} \frac{\lambda'''(s)}{r}(s-t) ds + E_{\partial_t v_1}(t,r)
\end{aligned}$$

where

$$|E_{\partial_t v_1}(t,r)| \leq \frac{C}{r^3}\int_t^{t+\frac{r}{6}} |\lambda'''(s)|(s-t)ds \leq \frac{C}{r^3 t \log^{b+1}(t)}, \quad r \geq t$$

For the third line of (5.2), we have

$$\begin{aligned}&|\int_{t+\frac{r}{6}}^{\infty} \frac{\lambda'''(s)}{r}\int_0^{s-t} \frac{\rho}{(s-t)}\left(1+\frac{r^2-1-\rho^2}{\sqrt{(1+r^2+\rho^2)^2-4r^2\rho^2}}\right) d\rho ds|\\ &\leq C\int_{t+\frac{r}{6}}^{\infty} \frac{|\lambda'''(s)|}{r(s-t)}\int_0^{\infty} \rho\left(1+\frac{r^2-1-\rho^2}{\sqrt{(1+r^2+\rho^2)^2-4r^2\rho^2}}\right) d\rho ds\\ &\leq \frac{C}{r^2\log^{b+1}(r)}, \quad r\geq t\end{aligned}$$

Finally, the fourth line of (5.2) is treated as follows:

$$\begin{aligned}&|\int_{t+\frac{r}{6}}^{\infty} \frac{\lambda'''(s)}{r}\int_0^{s-t} \rho\left(\frac{1}{\sqrt{(s-t)^2-\rho^2}}-\frac{1}{(s-t)}\right)\left(1+\frac{r^2-1-\rho^2}{\sqrt{(1+r^2+\rho^2)^2-4r^2\rho^2}}\right) d\rho ds|\\ &\leq \frac{C}{r^3\log^{b+1}(r)}\cdot\frac{1}{r}\int_{t+\frac{r}{6}}^{\infty}\int_0^{s-t} \rho\left(\frac{1}{\sqrt{(s-t)^2-\rho^2}}-\frac{1}{(s-t)}\right)\left(1+\frac{r^2-1-\rho^2}{\sqrt{(1+r^2+\rho^2)^2-4r^2\rho^2}}\right) d\rho ds\\ &\leq \frac{C}{r^4\log^{b+1}(r)}\int_0^{\infty} \rho\left(1+\frac{r^2-1-\rho^2}{\sqrt{(1+r^2+\rho^2)^2-4r^2\rho^2}}\right)\int_{\rho+t}^{\infty}\left(\frac{1}{\sqrt{(s-t)^2-\rho^2}}-\frac{1}{(s-t)}\right) ds d\rho\\ &\leq \frac{C}{r^2\log^{b+1}(r)}, \quad r\geq t\end{aligned}$$

Finally, we further treat the first term on the last line of (5.3)

$$\begin{aligned}2\int_t^{t+\frac{r}{6}} \frac{\lambda'''(s)}{r}(s-t)ds &= \frac{2}{r}\left(\frac{r}{6}\lambda''(t+\frac{r}{6})-\lambda'(t+\frac{r}{6})+\lambda'(t)\right)\\ &= \frac{2\lambda'(t)}{r} + E_{\partial_t v_1,1}(t,r)\end{aligned}$$

where

$$|E_{\partial_t v_1,1}(t,r)| \leq \frac{C}{r^2\log^{b+1}(r)}, \quad r\geq t$$

Then, we note that

$$\partial_t Q_{\frac{1}{\lambda(t)}}(r) = \frac{-2r\lambda'(t)}{r^2+\lambda(t)^2}$$

So,

$$|\partial_t\left(Q_{\frac{1}{\lambda(t)}}(r)+v_1(t,r)\right)| \leq \frac{C}{r^2\log^{b+1}(r)}, \quad r\geq t$$

Using the estimates on $\partial_t v_1$, we then get

$$||\partial_t\left(Q_{\frac{1}{\lambda(t)}}(r)+v_1(t,r)\right)||_{L^2(rdr)}^2 \leq \frac{C}{t^2\log^{2b}(t)}$$

Then, we recall

$$E(u,v) = \pi\left(||v||_{L^2(rdr)}^2 + ||u||_{\dot{H}^1_e}^2\right)$$

and note that energy estimates for the equations solved by v_k for $k=3,4,5$ give

$$E(v_3(t),\partial_t v_3(t)) \leq C\left(\int_t^{\infty} ||F_{0,1}(s)||_{L^2(rdr)}ds\right)^2 \leq \frac{C\log(\log(t))}{t^2\log^{2b+2}(t)}$$

$$E(v_4(t), \partial_t v_4(t)) \leq C \left(\int_t^\infty ||v_{4,c}(s)||_{L^2(rdr)} ds \right)^2 \leq \frac{C}{t^2 \log^{4N+4b}(t)}$$

$$E(v_5(t), \partial_t v_5(t)) \leq C \left(\int_t^\infty ||N_2(f_{v_5})(s)||_{L^2(rdr)} ds \right)^2 \leq \frac{C \log^6(t)}{t^{7/2}}$$

Next, we consider v_2 for $b \neq 1$. By Plancherel, we have

$$E(v_2(t), \partial_t v_2(t)) = E(v_2(0), \partial_t v_2(0)) = \frac{16b^2}{\pi(b-1)^2} \int_0^\infty \frac{(\chi_{\leq \frac{1}{4}}(\xi))^2}{\log^{2b-2}(\frac{1}{\xi})} d\xi$$

$$\frac{16b^2}{\pi(b-1)^2} \int_0^{\frac{1}{8}} \frac{d\xi}{\log^{2b-2}(\frac{1}{\xi})} \leq E(v_2, \partial_t v_2) \leq \frac{16b^2}{\pi(b-1)^2} \int_0^{\frac{1}{4}} \frac{d\xi}{\log^{2b-2}(\frac{1}{\xi})}$$

where we used properties of $\chi_{\leq \frac{1}{4}}$. This gives, for $b \neq 1$,

$$\frac{16b^2}{\pi(b-1)^2} \Gamma(3-2b, \log(8)) \leq E(v_2, \partial_t v_2) \leq \frac{16b^2}{\pi(b-1)^2} \Gamma(3-2b, \log(4))$$

By inspection of $v_{2,0}$ for $b=1$, we see that $E(v_2, \partial_t v_2) < \infty$ for $b=1$.

Using our estimates on v_1 and $\partial_r v_1$, we get

$$||v_1||^2_{\dot{H}^1_e} \leq \frac{C}{t^2 \log^{2b}(t)}$$

Finally, we treat the v_6 term in (5.1). Theorem 5.1 of [**12**] (the transferrence identity) shows that

$$\begin{aligned} ||\partial_t v_6||_{L^2(rdr)} &\leq C|| \frac{\lambda'(t)}{\sqrt{r\lambda(t)}} \mathcal{F}^{-1}(y_0(t, \frac{\cdot}{\lambda(t)^2}))(\frac{r}{\lambda(t)})||_{L^2(rdr)} \\ &+ C|| \sqrt{\frac{\lambda(t)}{r}} \mathcal{F}^{-1}(\partial_1 y_0(t, \frac{\cdot}{\lambda(t)^2}))(\frac{r}{\lambda(t)})||_{L^2(rdr)} \\ &+ C|| \frac{\lambda'(t)}{\lambda(t)} \sqrt{\frac{\lambda(t)}{r}} \mathcal{F}^{-1}(\mathcal{K}(y_0(t, \frac{\cdot}{\lambda(t)^2})))(\frac{r}{\lambda(t)})||_{L^2(rdr)} \end{aligned} \tag{5.4}$$

The first line of (5.4) is handled by re-scaling, and applying the L^2 isometry property of $\mathcal{F}(\cdot)$:

$$\begin{aligned} &\int_0^\infty \frac{(\lambda'(t))^2}{r\lambda(t)} \left(\mathcal{F}^{-1}(y_0(t, \frac{\cdot}{\lambda(t)^2}))(\frac{r}{\lambda(t)}) \right)^2 rdr \\ &= (\lambda'(t))^2 \lambda(t)^2 \int_0^\infty |y_0(t,\omega)|^2 \rho(\omega\lambda(t)^2) d\omega \\ &\leq C \frac{\lambda'(t)^2 \lambda(t)^2}{t^4 \log^\epsilon(t)} \end{aligned}$$

The second and third lines of (5.4) are treated by again re-scaling and using the L^2 isometry property of $\mathcal{F}(\cdot)$, as well as the same estimates on $\mathcal{K}$ that we used while estimating F_2 to get

$$\begin{aligned}&\int_0^\infty \frac{\lambda(t)}{r}|\mathcal{F}^{-1}(\partial_1 y_0(t,\frac{\cdot}{\lambda(t)^2}))|^2(\frac{r}{\lambda(t)})rdr\\&+\frac{\lambda'(t)^2}{\lambda(t)}\int_0^\infty \frac{1}{r}|\mathcal{F}^{-1}(\mathcal{K}(y_0(t,\frac{\cdot}{\lambda(t)})))|^2(\frac{r}{\lambda(t)})rdr\\&\leq C\lambda(t)^4||\partial_t y_0(t)||^2_{L^2(\rho(\omega\lambda(t)^2)d\omega)}+C\lambda'(t)^2\lambda(t)^2||y_0(t)||^2_{L^2(\rho(\omega\lambda(t)^2)d\omega)}\\&\leq \frac{C}{t^6\log^{4b+\epsilon}(t)}\end{aligned}$$

This gives

$$||\partial_t v_6||^2_{L^2(rdr)}\leq \frac{C}{t^6\log^{4b+\epsilon}(t)}$$

From the definition of v_6, we have (using the same argument used when estimating F_2, F_3)

$$||v_6(t)||_{\dot{H}^1_e}=||v_6(t,\cdot\lambda(t))||_{\dot{H}^1_e}\leq C\left(||v_6(t,\cdot\lambda(t))||_{L^2(RdR)}+||L(v_6(t,\cdot\lambda(t)))||_{L^2(RdR)}\right)$$

and

$$\begin{aligned}||v_6(t,\cdot\lambda(t))||_{L^2(RdR)}&=\lambda(t)||y_0(t)||_{L^2(\rho(\omega\lambda(t)^2)d\omega)}\\||L(v_6(t,\cdot\lambda(t)))||_{L^2(RdR)}&=\lambda(t)||\sqrt{\omega}\lambda(t)y_0(t)||_{L^2(\rho(\omega\lambda(t)^2)d\omega)}\end{aligned}$$

which gives

$$||v_6(t)||_{\dot{H}^1_e}\leq \frac{C}{t^2\log^{b+\frac{\epsilon}{2}}(t)}$$

Finally, we have

$$E_{\mathrm{WM}}(u,\partial_t u)=\pi\left(||\partial_t u||^2_{L^2(rdr)}+||\frac{\sin(u)}{r}||^2_{L^2(rdr)}+||\partial_r u||^2_{L^2(rdr)}\right)$$

Then, we use

$$\begin{aligned}&\int_0^\infty \frac{\sin^2(u-Q_{\frac{1}{\lambda(t)}}+Q_{\frac{1}{\lambda(t)}})}{r^2}rdr\\&\quad\leq C\left(\int_0^\infty \frac{\sin^2(Q_{\frac{1}{\lambda(t)}})}{r^2}rdr+\int_0^\infty\frac{|u-Q_{\frac{1}{\lambda(t)}}|^2}{r^2}rdr\right)\end{aligned}$$

to get

$$\begin{aligned}E_{\mathrm{WM}}(u,\partial_t u)\leq C\Bigg(&||\partial_t\left(Q_{\frac{1}{\lambda(t)}}+v_1\right)||^2_{L^2(rdr)}+\sum_{k=2}^6||\partial_t v_k||^2_{L^2(rdr)}\\&+||\partial_r Q_{\frac{1}{\lambda(t)}}||^2_{L^2(rdr)}+||\frac{\sin(Q_{\frac{1}{\lambda(t)}})}{r}||^2_{L^2(rdr)}+\sum_{k=1}^6||v_k||^2_{\dot{H}^1_e}\Bigg)\end{aligned}$$

By combining our above estimates, and recalling

$$||\partial_r Q_{\frac{1}{\lambda(t)}}||^2_{L^2(rdr)}+||\frac{\sin(Q_{\frac{1}{\lambda(t)}})}{r}||^2_{L^2(rdr)}=4$$

we get

$$E_{\mathrm{WM}}(u,\partial_t u)<\infty$$

and

$$||\partial_t(u-v_2)||^2_{L^2(rdr)}+||u-Q_{\frac{1}{\lambda(t)}}-v_2||^2_{\dot{H}^1_e}\leq\frac{C}{t^2\log^{2b}(t)}$$

Finally, we note that the remark after the main theorem concerning the regularity of v_6 follows from the definition of the space Z, the continuity of dilation on L^2, and lemma 9.1 of [**12**]. This completes the proof of the main theorem.

APPENDIX A

Proof of Theorem 1.2

In this appendix, we will summarize the extra arguments needed to prove Theorem 1.2. To prove Theorem 1.2, we use a slightly different starting point than for the proof of Theorem 1.1. In particular, fix $b > 0$, $\lambda_{0,0,b} \in \Lambda$, and let $T_0 > C_1$, where $C_1 > 100$ is some sufficiently large constant depending on $\lambda_{0,0,b}$. Then, we define the X norm as in the proof of Theorem 1.1, and write $\lambda = \lambda_{0,0,b} + e_1$, where $e_1 \in \overline{B_1}(0) \subset X$. For all such λ, we then define v_1 exactly as in the main body of the paper. The main difference is a modification of v_2: Let $\psi \in C^\infty([0,\infty))$ satisfy

$$\psi(t) = \begin{cases} 1, & t \geq 200 \\ 0, & t \leq 100 \end{cases} \qquad 0 \leq \psi(t) \leq 1$$

and define

$$F(t) = \left(4\int_t^\infty \frac{\lambda''_{0,0,b}(s)ds}{1+s-t}\right)\psi(t)$$

Then, v_2 is exactly as in the main body of the paper, except with $v_{2,0}$, the initial velocity of v_2, given below

$$\widehat{v_{2,0}}(\xi) = -\frac{1}{\xi\pi}\int_0^\infty F(t)\sin(t\xi)dt$$

The point of this definition is that, by the sine transform inversion, $\widehat{v_{2,0}}$ solves

$$-2\int_0^\infty \sin(t\xi)\xi\widehat{v_{2,0}}(\xi)d\xi = F(t), \quad t \geq 0$$

v_3 is defined exactly as in the main body of the paper. Several estimates on v_k in the main body of the paper used the fact that $\lambda'(x) < 0$, which we no longer have in the setting of this appendix. Unless specified later on, analogs of all of these estimates are still true in the setting of this appendix, and can be proven by instead using that, for some $C_2, C_3 > 0$,

$$\frac{C_2}{\log^b(t)} \leq \lambda(t) \leq \frac{C_3}{\log^b(t)} \tag{A.1}$$

and $t \to \frac{1}{\log^b(t)}$ is decreasing. The definitions of v_4, v_5, and the equation resulting from u_{ansatz} are the same as previously. The inner product of the v_1 linear error term with the re-scaled ϕ_0 is unchanged. On the other hand, for v_2, we have

$$\begin{aligned}&\int_0^\infty \left(\frac{\cos(2Q_1(R))-1}{R^2\lambda(t)^2}\right) v_2(t,R\lambda(t))\phi_0(R)RdR\\&=-2\int_0^\infty \frac{\sin(t\xi)}{\lambda(t)}\xi\widehat{v_{2,0}}(\xi)d\xi-2\int_0^\infty \sin(t\xi)\xi^2\left(K_1(\xi\lambda(t))-\frac{1}{\xi\lambda(t)}\right)\widehat{v_{2,0}}(\xi)d\xi\\&=\frac{4}{\lambda(t)}\int_t^\infty \frac{\lambda''_{0,0,b}(s)ds}{1+s-t}+E_{v_2,ip}(t,\lambda(t)),\quad t\geq T_0\end{aligned}$$

where

$$E_{v_2,ip}(t,\lambda(t))=-2\int_0^\infty \sin(t\xi)\xi^2\left(K_1(\xi\lambda(t))-\frac{1}{\xi\lambda(t)}\right)\widehat{v_{2,0}}(\xi)d\xi$$

and we used the fact that $\psi(t)=1, \quad t\geq 100$. In order to prove the pointwise estimates on v_2, we require some estimates on $\partial_\xi\left(\xi\widehat{v_{2,0}}\right)$. For these, we use

$$\xi\widehat{v_{2,0}}(\xi)=\frac{-1}{\pi}\int_0^\infty F(\frac{\sigma}{\xi})\sin(\sigma)\frac{d\sigma}{\xi}$$

and can then differentiate under the integral sign, using the symbol type estimates on $\lambda_{0,0,b}$. The most crucial point to mention regarding the pointwise behavior of v_2 is the analogue of (3.49), since it is this which allows us to prove the crucial near origin estimates on F_4. In the setting of this appendix, we prove the analogue of (3.49) by noting, for $r\leq\frac{t}{2}$,

(A.2)

$$\begin{aligned}v_2(t,r)&=\int_0^\infty J_1(r\xi)\sin(t\xi)\widehat{v_{2,0}}(\xi)d\xi\\&=\frac{r}{\pi}\int_0^\pi\frac{\sin^2(\theta)}{2}\int_0^\infty \xi\left(\sin(\xi(t+r\cos(\theta)))+\sin(\xi(t-r\cos(\theta)))\right)\widehat{v_{2,0}}(\xi)d\xi d\theta\\&=\frac{-r}{4\pi}\int_0^\pi \sin^2(\theta)\left(F(t+r\cos(\theta))+F(t-r\cos(\theta))\right)d\theta\end{aligned}$$

and then using

$$F(t\pm r\cos(\theta))=F(t)+\text{Err}(t,r),\quad |\text{Err}(t,r)|\leq\frac{Cr}{t^3\log^b(t)},\quad r\leq\frac{t}{2},\quad t\geq T_0$$

The arguments for derivatives of v_2 are done similarly. The linear error term associated to v_3 is studied similarly to the main body of the paper. The modulation equation for λ then has the same form as previously, except with the modified definition of $E_{v_2,ip}$ given above, and the replacement

$$\frac{4b}{\lambda(t)t^2\log^b(t)}\to\frac{4}{\lambda(t)}\int_t^\infty\frac{\lambda''_{0,0,b}(s)ds}{1+s-t}$$

Recalling that $\lambda=\lambda_{0,0,b}+e_1$, we then substitute $e_1(t)=\lambda_{0,1}(t)+e(t)$, where

$$\lambda_{0,1}(t)=\int_t^\infty\int_{t_1}^\infty\frac{\lambda''_{0,0,b}(t_2)\log(\lambda_{0,0,b}(t_2)))}{\log(t_2)}dt_2dt_1$$

and proceed exactly as previously. The crucial kernel estimate (3.69) was previously proven using an argument which used $\lambda'_{0,0}(x)+\lambda'_{0,1}(x)\leq 0$, which is no longer true in this context. On the other hand, (3.69) is still true in the context of this appendix, and is proven with the same calculation of $\partial_{st}k(s,t)$ as before. Instead of using $\lambda'_{0,0}(t)+\lambda'_{0,1}(t)\leq 0$, we simply use

$$-\lambda_0'(-t)(1-\alpha)\lambda_0(-t)^{-\alpha}+1\geq\frac{7}{8},\quad t\leq -T_0$$

which follows from the lower bound on T_0 imposed at the beginning of the entire argument of this appendix, and where $\lambda_0(t)=\lambda_{0,0,b}(t)+\lambda_{0,1}(t)$. Similarly, using the fact that λ_0 is comparable to $\frac{1}{\log^b(t)}$, which is a decreasing function, we carry out the same procedure as previously, to obtain the important resolvent kernel estimate (3.72). In this context, the analog of (3.72) has some absolute constant C, rather than precisely 2, appearing on the right-hand side. Recalling the comment just above (A.1), the rest of the estimates on v_k required to construct $\lambda(t)$ and estimate $\lambda^{(k)}(t),\quad k\leq 4$ follow from an argument similar to that used previously.

Recall that the modulation equation in the setting of this appendix is of the same form as the one in the main body of the paper. In addition, the definition of v_1 is the same in both settings. Finally, (A.2) is true. Combined, these imply that we can prove the crucial pointwise estimates on F_4, and its derivatives with the same procedure as before. Also, all other subsections involved in the construction of the ansatz section can be established similarly. Finally, we can then complete the argument exactly as before, with $E(v_e,\partial_t\left(Q_{\frac{1}{\lambda(t)}}+v_e\right))$ having the same decay in t as previously.

Bibliography

[1] I. Bejenaru, J. Krieger, and D. Tataru, *A codimension-two stable manifold of near soliton equivariant wave maps*, Anal. PDE **6** (2013), no. 4, 829–857, DOI 10.2140/apde.2013.6.829. MR3092731

[2] R. Côte, C. E. Kenig, A. Lawrie, and W. Schlag, *Characterization of large energy solutions of the equivariant wave map problem: I*, Amer. J. Math. **137** (2015), no. 1, 139–207, DOI 10.1353/ajm.2015.0002. MR3318089

[3] R. Côte, C. E. Kenig, A. Lawrie, and W. Schlag, *Characterization of large energy solutions of the equivariant wave map problem: II*, Amer. J. Math. **137** (2015), no. 1, 209–250, DOI 10.1353/ajm.2015.0003. MR3318090

[4] R. Donninger and J. Krieger, *Nonscattering solutions and blowup at infinity for the critical wave equation*, Math. Ann. **357** (2013), no. 1, 89–163, DOI 10.1007/s00208-013-0898-1. MR3084344

[5] C. Gao and J. Krieger, *Optimal polynomial blow up range for critical wave maps*, Commun. Pure Appl. Anal. **14** (2015), no. 5, 1705–1741, DOI 10.3934/cpaa.2015.14.1705. MR3359541

[6] I. S. Gradshteyn and I. M. Ryzhik, *Table of integrals, series, and products*, 5th ed., Academic Press, Inc., San Diego, CA, 1996. CD-ROM version 1.0 for PC, MAC, and UNIX computers. MR1398882

[7] G. Gripenberg, S.-O. Londen, and O. Staffans, *Volterra integral and functional equations*, Encyclopedia of Mathematics and its Applications, vol. 34, Cambridge University Press, Cambridge, 1990, DOI 10.1017/CBO9780511662805. MR1050319

[8] J. Jendrej, *Construction of two-bubble solutions for energy-critical wave equations*, Amer. J. Math. **141** (2019), no. 1, 55–118, DOI 10.1353/ajm.2019.0002. MR3904767

[9] J. Jendrej and A. Lawrie, *Two-bubble dynamics for threshold solutions to the wave maps equation*, Invent. Math. **213** (2018), no. 3, 1249–1325, DOI 10.1007/s00222-018-0804-2. MR3842064

[10] H. Koch, D. Tataru, and M. Vişan, *Dispersive equations and nonlinear waves*, Oberwolfach Seminars, vol. 45, Birkhäuser/Springer, Basel, 2014. Generalized Korteweg-de Vries, nonlinear Schrödinger, wave and Schrödinger maps. MR3618884

[11] J. Krieger and S. Miao, *On the stability of blowup solutions for the critical corotational wave-map problem*, Duke Math. J. **169** (2020), no. 3, 435–532, DOI 10.1215/00127094-2019-0053. MR4065147

[12] J. Krieger, W. Schlag, and D. Tataru, *Renormalization and blow up for charge one equivariant critical wave maps*, Invent. Math. **171** (2008), no. 3, 543–615, DOI 10.1007/s00222-007-0089-3. MR2372807

[13] J. Krieger, W. Schlag, and D. Tataru, *Renormalization and blow up for the critical Yang-Mills problem*, Adv. Math. **221** (2009), no. 5, 1445–1521, DOI 10.1016/j.aim.2009.02.017. MR2522426

[14] J. Krieger, W. Schlag, and D. Tataru, *Slow blow-up solutions for the $H^1(\mathbb{R}^3)$ critical focusing semilinear wave equation*, Duke Math. J. **147** (2009), no. 1, 1–53, DOI 10.1215/00127094-2009-005. MR2494455

[15] A. Lawrie and S.-J. Oh, *A refined threshold theorem for $(1+2)$-dimensional wave maps into surfaces*, Comm. Math. Phys. **342** (2016), no. 3, 989–999, DOI 10.1007/s00220-015-2513-7. MR3465437

[16] P. Raphaël and I. Rodnianski, *Stable blow up dynamics for the critical co-rotational wave maps and equivariant Yang-Mills problems*, Publ. Math. Inst. Hautes Études Sci. **115** (2012), 1–122, DOI 10.1007/s10240-011-0037-z. MR2929728

[17] I. Rodnianski and J. Sterbenz, *On the formation of singularities in the critical* O(3) σ*-model*, Ann. of Math. (2) **172** (2010), no. 1, 187–242, DOI 10.4007/annals.2010.172.187. MR2680419

[18] C. Rodriguez, *Threshold dynamics for corotational wave maps*, Anal. PDE **14** (2021), no. 7, 2123–2161, DOI 10.2140/apde.2021.14.2123. MR4353567

[19] J. Shatah and A. S. Tahvildar-Zadeh, *On the Cauchy problem for equivariant wave maps*, Comm. Pure Appl. Math. **47** (1994), no. 5, 719–754, DOI 10.1002/cpa.3160470507. MR1278351

[20] J. Sterbenz and D. Tataru, *Energy dispersed large data wave maps in* $2+1$ *dimensions*, Comm. Math. Phys. **298** (2010), no. 1, 139–230, DOI 10.1007/s00220-010-1061-4. MR2657817

[21] J. Sterbenz and D. Tataru, *Regularity of wave-maps in dimension* $2+1$, Comm. Math. Phys. **298** (2010), no. 1, 231–264, DOI 10.1007/s00220-010-1062-3. MR2657818

Editorial Information

To be published in the *Memoirs*, a paper must be correct, new, nontrivial, and significant. Further, it must be well written and of interest to a substantial number of mathematicians. Piecemeal results, such as an inconclusive step toward an unproved major theorem or a minor variation on a known result, are in general not acceptable for publication.

Papers appearing in *Memoirs* are generally at least 80 and not more than 200 published pages in length. Papers less than 80 or more than 200 published pages require the approval of the Managing Editor of the Transactions/Memoirs Editorial Board. Published pages are the same size as those generated in the style files provided for $\mathcal{A}\mathcal{M}\mathcal{S}$-LaTeX.

Information on the backlog for this journal can be found on the AMS website starting from `http://www.ams.org/memo`.

A Consent to Publish is required before we can begin processing your paper. After a paper is accepted for publication, the Providence office will send a Consent to Publish and Copyright Agreement to all authors of the paper. By submitting a paper to the *Memoirs*, authors certify that the results have not been submitted to nor are they under consideration for publication by another journal, conference proceedings, or similar publication.

Information for Authors

Memoirs is an author-prepared publication. Once formatted for print and on-line publication, articles will be published as is with the addition of AMS-prepared frontmatter and backmatter. Articles are not copyedited; however, confirmation copy will be sent to the authors.

Initial submission. The AMS uses Centralized Manuscript Processing for initial submissions. Authors should submit a PDF file using the Initial Manuscript Submission form found at `www.ams.org/submission/memo`, or send one copy of the manuscript to the following address: Centralized Manuscript Processing, MEMOIRS OF THE AMS, 201 Charles Street, Providence, RI 02904-2294 USA. If a paper copy is being forwarded to the AMS, indicate that it is for *Memoirs* and include the name of the corresponding author, contact information such as email address or mailing address, and the name of an appropriate Editor to review the paper (see the list of Editors below).

The paper must contain a *descriptive title* and an *abstract* that summarizes the article in language suitable for workers in the general field (algebra, analysis, etc.). The *descriptive title* should be short, but informative; useless or vague phrases such as "some remarks about" or "concerning" should be avoided. The *abstract* should be at least one complete sentence, and at most 300 words. Included with the footnotes to the paper should be the 2020 *Mathematics Subject Classification* representing the primary and secondary subjects of the article. The classifications are accessible from `www.ams.org/msc/`. The Mathematics Subject Classification footnote may be followed by a list of *key words and phrases* describing the subject matter of the article and taken from it. Journal abbreviations used in bibliographies are listed in the latest *Mathematical Reviews* annual index. The series abbreviations are also accessible from `www.ams.org/msnhtml/serials.pdf`. To help in preparing and verifying references, the AMS offers MR Lookup, a Reference Tool for Linking, at `www.ams.org/mrlookup/`.

Electronically prepared manuscripts. The AMS encourages electronically prepared manuscripts, with a strong preference for $\mathcal{A}\mathcal{M}\mathcal{S}$-LaTeX. To this end, the Society has prepared $\mathcal{A}\mathcal{M}\mathcal{S}$-LaTeX author packages for each AMS publication. Author packages include instructions for preparing electronic manuscripts, samples, and a style file that generates the particular design specifications of that publication series.

Authors may retrieve an author package for *Memoirs of the AMS* from `www.ams.org/journals/memo/memoauthorpac.html`. The *AMS Author Handbook* is available in PDF format from the author package link. The author package can also be obtained free of charge by sending email to `tech-support@ams.org` or from the Publication Division,

American Mathematical Society, 201 Charles St., Providence, RI 02904-2294, USA. When requesting an author package, please specify the publication in which your paper will appear. Please be sure to include your complete mailing address.

After acceptance. The source files for the final version of the electronic manuscript should be sent to the Providence office immediately after the paper has been accepted for publication. The author should also submit a PDF of the final version of the paper to the editor, who will forward a copy to the Providence office.

Accepted electronically prepared files can be submitted via the web at `www.ams.org/submit-book-journal/`, sent via FTP, or sent on CD to the Electronic Prepress Department, American Mathematical Society, 201 Charles Street, Providence, RI 02904-2294 USA. TEX source files and graphic files can be transferred over the Internet by FTP to the Internet node `ftp.ams.org` (130.44.1.100). When sending a manuscript electronically via CD, please be sure to include a message indicating that the paper is for the *Memoirs*.

Electronic graphics. Comprehensive instructions on preparing graphics are available at `www.ams.org/authors/journals.html`. A few of the major requirements are given here.

Submit files for graphics as EPS (Encapsulated PostScript) files. This includes graphics originated via a graphics application as well as scanned photographs or other computer-generated images. If this is not possible, TIFF files are acceptable as long as they can be opened in Adobe Photoshop or Illustrator.

Authors using graphics packages for the creation of electronic art should also avoid the use of any lines thinner than 0.5 points in width. Many graphics packages allow the user to specify a "hairline" for a very thin line. Hairlines often look acceptable when proofed on a typical laser printer. However, when produced on a high-resolution laser imagesetter, hairlines become nearly invisible and will be lost entirely in the final printing process.

Screens should be set to values between 15% and 85%. Screens which fall outside of this range are too light or too dark to print correctly. Variations of screens within a graphic should be no less than 10%.

Any graphics created in color will be rendered in grayscale for the printed version unless color printing is authorized by the Managing Editor and the Publisher. In general, color graphics will appear in color in the online version.

Inquiries. Any inquiries concerning a paper that has been accepted for publication should be sent to `memo-query@ams.org` or directly to the Electronic Prepress Department, American Mathematical Society, 201 Charles St., Providence, RI 02904-2294 USA.

Editors

This journal is designed particularly for long research papers, normally at least 80 pages in length, and groups of cognate papers in pure and applied mathematics. Papers intended for publication in the *Memoirs* should be addressed to one of the following editors. The AMS uses Centralized Manuscript Processing for initial submissions to AMS journals. Authors should follow instructions listed on the Initial Submission page found at www.ams.org/memo/memosubmit.html.

SELECTED PUBLISHED TITLES IN THIS SERIES

1404 **Luchezar Stoyanov,** Spectral Properties of Ruelle Transfer Operators for Regular Gibbs Measures and Decay of Correlations for Contact Anosov Flows, 2023

1403 **Akinari Hoshi, Ming-chang Kang, and Aiichi Yamasaki,** Multiplicative Invariant Fields of Dimension ≤ 6, 2023

1402 **Roberto Frigerio and Marco Moraschini,** Gromov's Theory of Multicomplexes with Applications to Bounded Cohomology and Simplicial Volume, 2023

1401 **Jang Soo Kim, Kyu-Hwan Lee, and Se-jin Oh,** Weight Multiplicities and Young Tableaux Through Affine Crystals, 2023

1400 **Tarek M. Elgindi and In-Jee Jeong,** On Singular Vortex Patches, I: Well-Posedness Issues, 2023

1399 **Fan Gao, Freydoon Shahidi, and Dani Szpruch,** Local Coefficients and Gamma Factors for Principal Series of Covering Groups, 2023

1398 **David A. Croydon, Tsuyoshi Kato, Makiko Sasada, and Satoshi Tsujimoto,** Dynamics of the Box-Ball System with Random Initial Conditions via Pitman's Transformation, 2023

1397 **Anand P. Patel and Ashvin A. Swaminathan,** Inflectionary Invariants for Isolated Complete Intersection Curve Singularities, 2023

1396 **Margaret Bilu,** Motivic Euler Products and Motivic Height Zeta Functions, 2023

1395 **Sebastian Casalaina-Martin, Samuel Grushevsky, Klaus Hulek, and Radu Laza,** Cohomology of the Moduli Space of Cubic Threefolds and Its Smooth Models, 2023

1394 **Valentin Blomer, Étienne Fouvry, Emmanuel Kowalski, Philippe Michel, Djordje Milićević, and Will Sawin,** The Second Moment Theory of Families of L-Functions–The Case of Twisted Hecke L-Functions, 2023

1393 **Óscar Domínguez and Sergey Tikhonov,** Function Spaces of Logarithmic Smoothness: Embeddings and Characterizations, 2023

1392 **André Gil Henriques, David Penneys, and James Tener,** Planar Algebras in Braided Tensor Categories, 2023

1391 **Tiago J. Fonseca,** Higher Ramanujan Equations and Periods of Abelian Varieties, 2023

1390 **Tomasz Downarowicz and Guohua Zhang,** Symbolic Extensions of Amenable Group Actions and the Comparison Property, 2023

1389 **João C. A. Barata, Christian D. Jäkel, and Jens Mund,** The $\mathcal{P}(\varphi)_2$ Model on de Sitter Space, 2023

1388 **Guozhen Lu, Jiawei Shen, and Lu Zhang,** Multi-Parameter Hardy Spaces Theory and Endpoint Estimates for Multi-Parameter Singular Integrals, 2023

1387 **François Ledrappier and Lin Shu,** The Regularity of the Linear Drift in Negatively Curved Spaces, 2023

1386 **Zhaobing Fan, Chun-Ju Lai, Yiqiang Li, Li Luo, and Weiqiang Wang,** Affine Hecke Algebras and Quantum Symmetric Pairs, 2023

1385 **D. Kaledin,** Mackey Profunctors, 2022

1384 **Matt Bainbridge, John Smillie, and Barak Weiss,** Horocycle Dynamics: New Invariants and Eigenform Loci in the Stratum $\mathcal{H}(1,1)$, 2022

1383 **Chris Kottke and Michael Singer,** Partial Compactification of Monopoles and Metric Asymptotics, 2022

1382 **Jenny Fuselier, Ling Long, Ravi Ramakrishna, Holly Swisher, and Fang-Ting Tu,** Hypergeometric Functions Over Finite Fields, 2022

1381 **Henri Berestycki and Grégoire Nadin,** Asymptotic Spreading for General Heterogeneous Fisher-KPP Type Equations, 2022

For a complete list of titles in this series, visit the
AMS Bookstore at **www.ams.org/bookstore/memoseries/**.